How to Count

Robert A. Beeler

How to Count

An Introduction to Combinatorics and Its Applications

Robert A. Beeler
Department of Mathematics and Statistics
East Tennessee State University
Johnson City
Tennessee
USA

ISBN 978-3-319-35508-5 ISBN 978-3-319-13844-2 (eBook)
DOI 10.1007/978-3-319-13844-2

Springer Cham Heidelberg New York Dordrecht London

Softcover reprint of the hardcover 1st edition 2015

Printed on acid-free paper

Springer is part of Springer Science+Business Media (www.springer.com)

Preface

The goal of this book is to provide a reasonably self-contained introduction to combinatorics. For this reason, this book assumes no knowledge of combinatorics. It does however assume that the reader has been introduced to elementary proof techniques and mathematical reasoning. These modest prerequisites are typically developed at the late sophomore or early junior level. Students wishing to improve their skills in such areas are referred to *Mathematical Proofs: A Transition to Advanced Mathematics* by Chartrand et al. [14].

This text is aimed at the junior or senior undergraduate level. There is a strong emphasis on computation, problem solving, and proof technique. In particular, there is a particular emphasis on combinatorial proofs for reasons discussed in Sect. 1.6. In addition, this book is written as a "problem based" approach to combinatorics. In each section, specific problems are introduced. Students are then guided in finding the solution to not only the original problem, but a number of variations. Hence, there are a number of examples throughout each section. Often these examples require the student to not only apply the new material, but to implement information developed in previous sections. For this reason, students are generally expected to have a working mastery of the key concepts developed in previous sections before proceeding. In particular, the basic Principle of Inclusion and Exclusion and the Multiplication Principle are used repeatedly.

Intuitive descriptions of abstract concepts (such as generating functions) are provided. In addition, supplementary reading on several topics are suggested throughout the text. Hence, this text lends itself not only to a traditional combinatorics course, but also to honors classes or undergraduate research.

There are a number of exercises provided at the end of each section. These exercises range from simple computations (in other words, evaluate a formula for a given set of values) to more advanced proofs. Most of the exercises are modeled after examples in the book allowing the student to refer through the text for insight. However, other exercises require deeper problem solving skills. In particular, many of the exercises make use of the key ideas of the Principle of Inclusion and Exclusion and the Multiplication Principle. This helps to reinforce these skills.

The first seven chapters form the core of a typical one semester course in combinatorics. Of these chapters, Sects. 2.6, 2.7, 3.2, and 3.7 are not required for the remainder of the first seven chapters. Instructors wishing to provide a more theoretical introduction may wish to include Chap. 8 on Pólya theory. In which case, Sect. 2.7 should be covered before introducing this material. Instructors wishing to provide a more applied introduction may wish to sprinkle material on probability from Chap. 9 throughout their course. Instructors may also wish to use the material on combinatorial designs (Chap. 10) to provide more applications. Instructors wishing to provide an introduction to graph theory (for instance, in a course on discrete mathematics) may wish to incorporate material from Chap. 11 as well.

The author welcomes any constructive suggestions on the improvement of future versions of this text.

East Tennessee State University, 2015. Robert A. Beeler, Ph.D.,

Acknowledgments

I would first like to thank my family, D. Beeler, L. Beeler, J. Beeler, and P. Keck for their love and support throughout my life. I would like to thank my colleagues R. Gardner, A. Godbole, T. Haynes, M. Helfgott, D. Knisley, R. Price, and E. Seier for encouraging me to finish this manuscript. Finally, I wish to acknowledge some of the excellent math teachers in my career. In particular, I would like to thank N. Calkin, J. Dydak, R. Jamison, G. Matthews, R. Sharp, C. Wagner, D. Vinson, and J. Xiong.

Contents

List of Figures

List of Tables

Chapter 1
Preliminaries

1.1 What is Combinatorics?

Put simply, *combinatorics* is the mathematics of counting. It may seem odd to devote an entire book to counting, especially since we all learned to count as children. More precisely, combinatorics is the mathematics of *combinations*. This being the case, combinatorics has numerous applications to experimental design, probability theory, game theory, and computer science.

Some questions that arise in combinatorics include:

(i) If you have five books and want to place three on a shelf, in how many ways can this be done?

(ii) If you have n books and want to place k on a shelf, in how many ways can this be done?

(iii) How many words of length n can be constructed from the alphabet {a,b} such that no word has two adjacent a's?

(iv) In a five-card poker hand, how many ways are there to get three of a kind?

(v) How many ways can n married couples be seated around a circular dinner table (with $2n$ seats) such that sexes must alternate?

(vi) How many ways can n married couples be seated around a circular dinner table (with $2n$ seats) such that sexes must alternate and no one can sit next to their own spouse?

(vii) If n people check their hats at the theater and the claim tickets are lost, in how many ways can the hats be distributed in such a way that no one receives their own hat?

(viii) How many ways are there to make change for a dollar?

(ix) Give general formulas for $\frac{d^n}{dx^n} f(x)g(x)$ and $\frac{d^n}{dx^n} f(g(x))$.

As we see from the above list, combinatorial questions often require little mathematical vocabulary to state. Further more, an examination of these questions can often begin by simply listing out all of the possibilities. In fact, while you are learning combinatorics, you *should* begin these problems by listing out all of the possibilities.

As an example, we will solve problem (i). An obvious question occurs: Do we care about the *order* of the three books, or simply which books are placed on the

R. A. Beeler, *How to Count*, DOI 10.1007/978-3-319-13844-2_1

shelf? A less obvious question is: Are the books all different? For this example, we will assume that the books are distinct and that we do care about which order the books are placed on the shelf. For simplicity, we will denote the books {A, B, C, D, E}. Since order is important, the combination ABC will denote the placement of Book A on the left of the shelf, Book B in the middle, and Book C on the right.

Begin by listing all combinations that begin with *A*:

ABC ABD ABE ACB ACD ACE

ADB ADC ADE AEB AEC AED

Notice that this list is organized in alphabetical order. This makes it very easy to check that we have listed all of the possibilities. The remaining combinations (in alphabetical order) are:

BAC BAD BAE BCA BCD BCE
BDA BDC BDE BEA BEC BED
CAB CAD CAE CBA CBD CBE
CDA CDB CDE CEA CEB CED
DAB DAC DAE DBA DBC DBE
DCA DCB DCE DEA DEB DEC
EAB EAC EAD EBA EBC EBD
ECA ECB ECD EDA EDB EDC

The advantage of this approach is that it is intuitive. In fact, elementary school students (given enough time) could solve this problem by brute force enumeration. However, this does have a major disadvantage. While it works well enough for small numbers, it would be unreasonable to use this approach for larger values. For instance, if you have 10 books and want to place five on the shelf, then there are over 30,000 possible ways that the books could be placed on the shelf. To make matter worse, the general problem involving n books and placing k on the shelf would be completely untractable with this method. In this book, we will find solutions to the problems listed above as well as numerous others.

Exercise 1.1.1 Suppose you have four books, denoted A, B, C, and D. List all possible ways to place two on a shelf.

Exercise 1.1.2 Suppose that Alice, Bob, Chad, Diane, and Edward are eligible to be officers in their club. The three offices are president, vice-president, and secretary. List all possible ways in which the officers can be selected.

1.2 Induction and Contradiction

In this section, we give two commonly used methods of mathematical proof, namely proofs by induction and proofs by contradiction. These methods will be used sporadically throughout this book.

Inductive proofs are only valid for propositions which deal with whole numbers. In a proof by induction, we first show that the proposition holds for some k. This is called the *basis step*. We then assume that the proposition holds for some $n \geq k$. This is called the *inductive hypothesis*. In general, we can assume that the claim holds for all $\ell \leq n$ (this is often referred to as *strong induction*). We then show, assuming the inductive hypothesis, that the proposition holds for $n+1$. By the *Principle of Mathematical Induction*, the proposition hold for all $n \geq k$.

Proposition 1.2.1 *The sum of the first n positive integers is* $\frac{n(n+1)}{2}$.

Proof (Basis Step) If $n = 1$, then the sum of the first n integers is $1 = \frac{1(2)}{2}$.
(Inductive Hypothesis) Assume that for some n,

$$1 + \cdots + n = \frac{n(n+1)}{2}.$$

By the inductive hypothesis,

$$\begin{aligned} 1 + \cdots + n + (n+1) &= \frac{n(n+1)}{2} + (n+1) \\ &= \frac{n(n+1)}{2} + \frac{2(n+1)}{2} \\ &= \frac{n+1}{2}(n+2) = \frac{(n+1)(n+2)}{2}. \end{aligned}$$

By the Principle of Mathematical Induction, the proposition holds for all n. ■

In previous proposition, we labeled the basis step and inductive hypothesis for emphasis. In the future, we will avoid this convention.

Proposition 1.2.2 *Suppose that the sequence* $\{F_n\}$ *satisfies* $F_n = F_{n-1} + F_{n-2}$ *with* $F_0 = 0$ *and* $F_1 = 1$. *It follows that*

$$F_n = \frac{1}{\sqrt{5}}\left[\left(\frac{1+\sqrt{5}}{2}\right)^n - \left(\frac{1-\sqrt{5}}{2}\right)^n\right].$$

Proof Note that

$$F_0 = 0 = \frac{1}{\sqrt{5}}\left[\left(\frac{1+\sqrt{5}}{2}\right)^0 - \left(\frac{1-\sqrt{5}}{2}\right)^0\right]$$

and

$$F_1 = 1 = \frac{1}{\sqrt{5}}\left[\left(\frac{1+\sqrt{5}}{2}\right)^1 - \left(\frac{1-\sqrt{5}}{2}\right)^1\right]$$

$$= \frac{1}{\sqrt{5}}\sqrt{5} = 1.$$

Thus the result holds for $n = 0$ and $n = 1$. Suppose that for some n, the result holds for F_n and F_{n-1}.

We need only confirm that the result holds for F_{n+1}. Note that $F_{n+1} = F_n + F_{n-1}$. By inductive hypothesis, we have

$$F_{n+1} = F_n + F_{n-1}$$

$$= \frac{1}{\sqrt{5}}\left[\left(\frac{1+\sqrt{5}}{2}\right)^n - \left(\frac{1-\sqrt{5}}{2}\right)^n\right] + \frac{1}{\sqrt{5}}\left[\left(\frac{1+\sqrt{5}}{2}\right)^{n-1} - \left(\frac{1-\sqrt{5}}{2}\right)^{n-1}\right]$$

$$= \frac{1}{\sqrt{5}}\left\{\left[\left(\frac{1+\sqrt{5}}{2}\right)^n + \left(\frac{1+\sqrt{5}}{2}\right)^{n-1}\right] - \left[\left(\frac{1-\sqrt{5}}{2}\right)^n + \left(\frac{1-\sqrt{5}}{2}\right)^{n-1}\right]\right\}$$

$$= \frac{1}{\sqrt{5}}\left\{\left(\frac{1+\sqrt{5}}{2}\right)^{n-1}\left[\frac{1+\sqrt{5}}{2} + 1\right] - \left(\frac{1-\sqrt{5}}{2}\right)^{n-1}\left[\frac{1-\sqrt{5}}{2} + 1\right]\right\}$$

$$= \frac{1}{\sqrt{5}}\left[\left(\frac{1+\sqrt{5}}{2}\right)^{n-1}\left(\frac{3+\sqrt{5}}{2}\right) - \left(\frac{1-\sqrt{5}}{2}\right)^{n-1}\left(\frac{3-\sqrt{5}}{2}\right)\right]$$

$$= \frac{1}{\sqrt{5}}\left[\left(\frac{1+\sqrt{5}}{2}\right)^{n-1}\left(\frac{1+\sqrt{5}}{2}\right)^2 - \left(\frac{1-\sqrt{5}}{2}\right)^{n-1}\left(\frac{1-\sqrt{5}}{2}\right)^2\right]$$

$$= \frac{1}{\sqrt{5}}\left[\left(\frac{1+\sqrt{5}}{2}\right)^{n+1} - \left(\frac{1-\sqrt{5}}{2}\right)^{n+1}\right].$$

Therefore, the result holds by the Principle of Mathematical Induction. ■

In a *proof by contradiction* we begin by assuming that the proposition is *false*. We then show that this assumption leads to a falsehood. Two of the best examples of a proof by contradiction are also the oldest.

Proposition 1.2.3 *There is no rational number whose square equals 2.*

Proof Assume to the contrary that $2 = \frac{p^2}{q^2}$ where p and q are integers such that $q \neq 0$. If p and q share a common factor, then we can simply cancel it out. Hence, we can assume without loss of generality that p and q share no common factor. So $2q^2 = p^2$. This implies that p is even, say $p = 2k$. Thus, $2q^2 = 4k^2$ implies $q^2 = 2k^2$. This implies that q must also be even, contrary to p and q sharing no common factor. ■

Recall that a *prime number* is an integer greater than one that is only divisible by one and itself.

Proposition 1.2.4 *There are an infinite number of primes.*

Proof Suppose to the contrary that there are only finitely many primes, say $p_1, p_2, ..., p_n$. Define $m = (p_1...p_n) + 1$. Note that m is larger than any prime. Hence it must be divisible by some prime, p_i. Further note that the product $p_1...p_n$ is divisible by p_i. Thus $m - (p_1...p_n)$ is divisible by p_i. However, $m - (p_1...p_n) = 1$ and one is not divisible by p_i, a contradiction. ■

There are some theorems that require both induction and contradiction to prove. Perhaps the most famous (and useful) theorem requiring both is given below.

Theorem 1.2.5 *(The Fundamental Theorem of Arithmetic) Every integer* $n \geq 2$ *can be written uniquely as a product of prime powers. That is,*

$$n = p_1^{m_1}...p_k^{m_k},$$

where the p_i *are prime numbers,* $p_i < p_{i+1}$ *for all* i*, and* $m_i \geq 1$ *for all* i*. Further, this representation is unique.*

We leave the proof of this theorem as an exercise for the reader.

Exercise 1.2.6 Prove that:

$$1^2 + \cdots + n^2 = \frac{n(n+1)(2n+1)}{6}.$$

Exercise 1.2.7 Suppose that the sequence $\{R_n\}$ satisfies $R_n = 5R_{n-1} - 6R_{n-2}$ with $R_0 = 0$ and $R_1 = 1$. Prove that $R_n = 3^n - 2^n$.

Exercise 1.2.8 Prove that there is no rational number whose square equals 3.

Exercise 1.2.9 Recall that $log_a(b) = c$ is equivalent to $a^c = b$. Prove that $log_2(3)$ cannot be expressed as the ratio of integers.

Exercise 1.2.10 Prove the Fundamental Theorem of Arithmetic.

1.3 Sets

A *set* is a collection of distinct objects. The objects in the set are often referred to as the *elements* of the set. For instance, if $A = \{2, 3, 5, 7, 11\}$, then the elements of A are $2, 3, 5, 7$, and 11. If A is a set and x is an element of A, then we denote this situation by $x \in A$. A set with exactly k elements is called a *k-set* or a *k-element set*. Conversely, if x is not an element of A, then we denote this by $x \notin A$. Sets are defined by the elements that they contain. The *empty set*, denoted $\emptyset$, is the set containing no elements. It is not necessary to restrict ourselves to sets of numbers. During the course of this book, we will consider sets of arrangements, sets of words, etcetera.

A *multiset* is a collection of not necessarily distinct objects. The advantage of multisets over traditional sets is that it allows us to keep track of how many times each object is being used. For instance, suppose that we hand out two gumdrops to Alice, three gumdrops to Bob, and one gumdrop to Chad. The set of people that have received gumdrops is {Alice, Bob, Chad}. However, if we wish to know not only who has received gumdrops but how many gumdrops they have received, then a multiset that includes each person once for each gumdrop they receive would be most applicable. In this case, the appropriate multiset is {Alice, Alice, Bob, Bob, Bob, Chad}.

Note that in a set, the order of the elements is not important. So, $\{2, 3, 5, 7, 11\}$ and $\{5, 11, 2, 7, 3\}$ are the same set. However, when we list sets of numbers, we will list the elements in increasing order by convention.

Let A and B be sets. If, for all $x \in B$, we have that $x \in A$, then we say that B is a *subset* of A. This is denoted $B \subseteq A$. If $B \subseteq A$ and there exists $x \in A$ such that $x \notin B$, then we say that B is a *proper subset* of A and denote this by $B \subset A$. If $B \subseteq A$ and $A \subseteq B$, then the two sets are *equal*. This is denoted $A = B$.

Remark 1.3.1 The empty set, $\emptyset$, is a subset of every set. Every set is a subset of itself.

For example, let $A = \{2, 3, 5, 7, 11\}$. Note that $\emptyset$, $\{3\}$, $\{2, 3\}$, $\{2, 5, 11\}$, and $\{2, 3, 5, 7, 11\}$ are all subsets of A. Further, $A \subset \{1, ..., 12\}$.

Often it is convenient to simply describe a set rather than list all of its elements. For instance, $A = \{2, 3, 5, 7, 11\}$ is the set of the first five prime numbers. In other cases, a symbolic representation may be more appropriate. For example, $B = \{1, 3, 5, 7, 9, 11\}$ can be represented by $B = \{2x + 1 : x = 0, 1, ..., 5\}$. This is especially important when dealing with large sets.

Of particular interest to combinatorialists is the *cardinality* of a set. The cardinality of a set A is the number of elements in A. The cardinality of the set A is denoted $|A|$.

Example 1.3.2 Let $A = \{2, 3, 5, 7, 11\}$, $B = \{2, 3, 11\}$, $C = \{1, ..., 12\}$, and $D = \{3, 5, 12, 18\}$. Find the cardinality of each set. Which sets are subsets of the others?

Solution Note that $|A| = 5$, $|B| = 3$, $|C| = 12$, and $|D| = 4$. Note that $B \subset A \subset C$ and that $|B| < |A| < |C|$. We will generalize this result below. □

When discussing sets, it is often convenient to describe a *universal set* which contains all relevant sets. We will denote this universal set by U.

A useful tool in exploring the relationship between sets is a *Venn diagram*. At the most basic level, a Venn diagram consists of a box representing the universal set U and circles representing the different sets involved (see Fig. 1.1). More complicated relationships are illustrated by shading the areas involved.

We now define several operations on sets. Note that while we define these operations in terms of two sets, these definitions can easily be generalized to any arbitrary number of sets.

Definition 1.3.3 Let A and B be sets.

(i) The *union* of A and B, denoted $A \cup B$, is the set of all elements in either A or B. In other words, $A \cup B = \{x : x \in A \text{ or } x \in B\}$. See Fig. 1.2.
(ii) The *intersection* of A and B, denoted $A \cap B$, is the set of all elements that are in both A and B. In other words, $A \cap B = \{x : x \in A \text{ and } x \in B\}$. See Fig. 1.3.

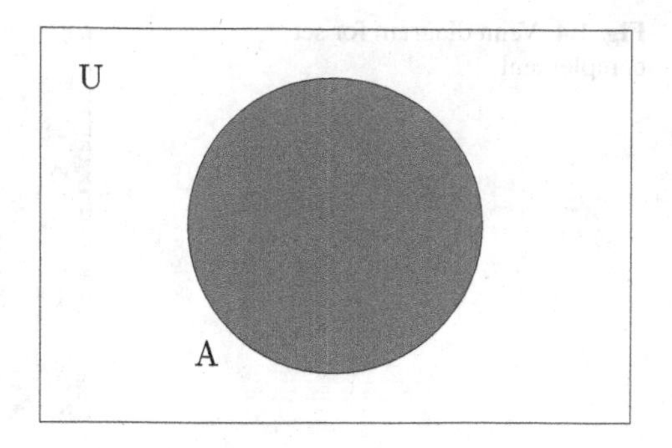

Fig. 1.1 A basic Venn diagram

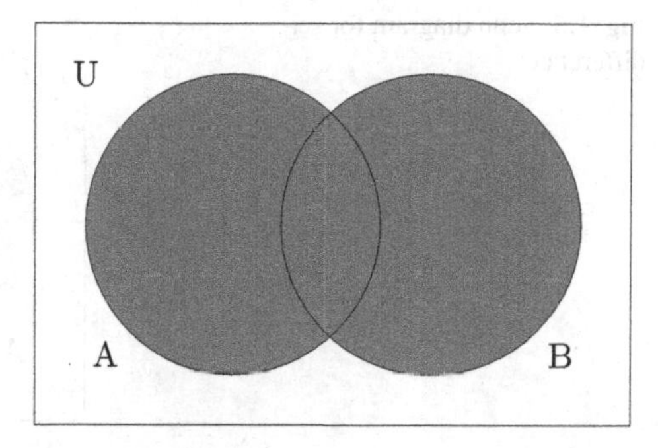

Fig. 1.2 Venn diagram for set union

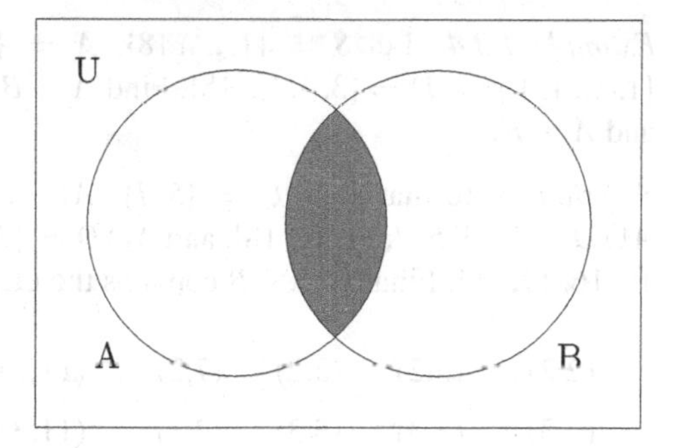

Fig. 1.3 Venn diagram for set intersection

(iii) The *complement* of A, denoted A^c or $\overline{A}$, is the set of all elements not in A. In other words, $x \in A^c$ if and only if $x \in U$ but $x \notin A$. See Fig. 1.4.

(iv) The *set difference*, denoted $A - B$, is the set of all elements of A that are not in B. In other words, $A - B = A \cap B^c = \{x : x \in A \text{ and } x \in B^c\}$. See Fig. 1.5.

(v) The *Cartesian product* of A and B, denoted $A \times B$ is the set of all ordered pairs (a, b) where $a \in A$ and $b \in B$.

A and B are said to be *disjoint* if they share no common elements, in other words, $A \cap B = \emptyset$.

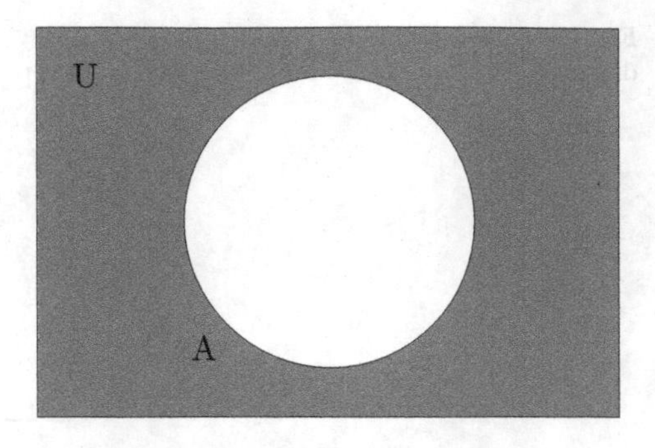

Fig. 1.4 Venn diagram for set complement

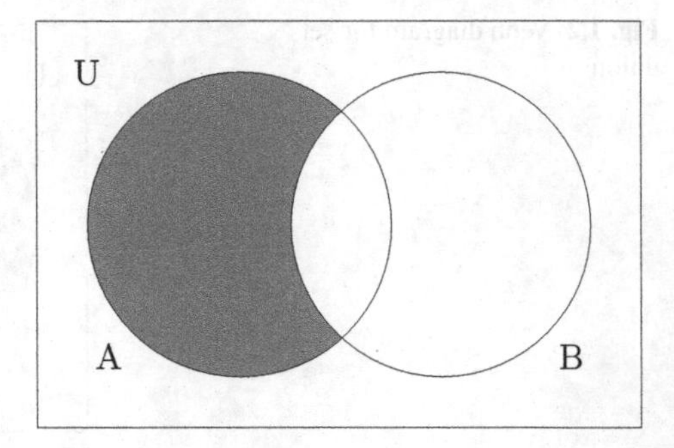

Fig. 1.5 Venn diagram for set difference

Example 1.3.4 Let $S = \{1, ..., 18\}$, $A = \{2, 3, 5, 7, 11\}$, $B = \{2, 3, 11\}$, $C = \{1, ..., 12\}$, and $D = \{3, 5, 12, 18\}$. Find $A - B$, $A - D$, $D - A$, $A \cup D$, $A \cap D$, A^c, and $A \times B$.

Solution Note that $A - B = \{5, 7\}$, $A - D = \{2, 7, 11\}$, $D - A = \{12, 18\}$, $A \cup D = \{2, 3, 5, 7, 11, 12, 18\}$, and $A \cap D = \{3, 5\}$. $A^c = \{1, 4, 6, 8, 9, 10, 12, 13, 14, 15, 16, 17, 18\}$. Finally, $A \times B$ contains the elements:

$$\begin{array}{ccccc} (2,2) & (3,2) & (5,2) & (7,2) & (11,2) \\ (2,3) & (3,3) & (5,3) & (7,3) & (11,3) \\ (2,11) & (3,11) & (5,11) & (7,11) & (11,11). \end{array}$$

□

In the above example, we had that $|A \times B| = |A||B|$. This is true in general because we can think of $|A \times B|$ as the area of a rectangle with sides length $|A|$ and $|B|$, respectively. This will be further generalized later.

A well-known and useful proposition follows.

Proposition 1.3.5 *(DeMorgan's Law) Let A and B be sets.*

(i) $(A \cup B)^c = A^c \cap B^c$.
(ii) $(A \cap B)^c = A^c \cup B^c$.

Proof To show that two sets are equal, we must show that they are subsets of each other.

(i) Let $x \in (A \cup B)^c$. Thus $x \notin A \cup B$. From this it follows that x is in neither A nor B. In other words $x \notin A$ and $x \notin B$. Hence, $x \in A^c$ and $x \in B^c$. Ergo, $x \in A^c \cap B^c$. So, by definition, $(A \cup B)^c \subseteq A^c \cap B^c$.

Conversely, let $x \in A^c \cap B^c$. Hence, $x \in A^c$ and $x \in B^c$. Thus, $x \notin A$ and $x \notin B$. From this it follows that $x \notin A \cup B$. Ergo, $x \in (A \cup B)^c$. By definition, $A^c \cap B^c \subseteq (A \cup B)^c$. Thus $A^c \cap B^c = (A \cup B)^c$.

(ii) Left as an exercise to the reader. ∎

Of particular interest is how to compute the cardinality of $A \cup B$ and $A - B$ based on the cardinality of A and B.

Proposition 1.3.6 *(The Addition Principle) If A and B are disjoint sets, then $|A \cup B| = |A| + |B|$.*

Proof Let $x \in A \cup B$. Since A and B are disjoint, it follows that $x \in A$ or $x \in B$, but not both. Hence, for every element counted by $|A \cup B|$, it is counted exactly once by either $|A|$ or $|B|$ (but not both). From this it follows that $|A \cup B| = |A| + |B|$. ∎

Unfortunately, two sets will often share common elements. Thus it is important to also consider the case when A and B are not disjoint.

Proposition 1.3.7 *(The Subtraction Principle) Let A and B be sets such that $B \subseteq A$. It follows that $|A - B| = |A| - |B|$.*

Proof By definition, B and $A - B$ are disjoint sets. By the Addition Principle, $|B \cup (A - B)| = |B| + |A - B|$. Since $B \subseteq A$, we have that $B \cup (A - B) = A$ (see Exercise 1.3.15). Thus $|A| = |B| + |A - B|$. From this it follows that $|A - B| = |A| - |B|$. ∎

This will allow us to create a more generalized Addition Principle.

Theorem 1.3.8 *(Principle of Inclusion and Exclusion) For any sets A and B we have*

$$|A \cup B| = |A| + |B| - |A \cap B|.$$

Proof Note that $A \cup B = (A - (A \cap B)) \cup B$ and that $A \cap B \subseteq A$ (see Exercise 1.3.16). Since $A - (A \cap B)$ and B are disjoint sets, it follows that $|(A - (A \cap B)) \cup B| = |A - (A \cap B)| + |B|$ by the Addition Principle. Because $A \cap B \subseteq A$, we have $|A - (A \cap B)| = |A| - |A \cap B|$ by the Subtraction Principle. From this it follows that:

$$\begin{aligned} |A \cup B| &= |(A - (A \cap B)) \cup B| \\ &= |A - (A \cap B)| + |B| = |A| - |A \cap B| + |B| \\ &= |A| + |B| - |A \cap B|. \end{aligned}$$

∎

This can be more intuitively explained using the Venn diagram in Fig. 1.6. The elements in A are shaded with vertical lines while the elements in B are shaded with horizontal lines. Hence the elements in $A \cap B$ are shaded with a "cross-hatch" (the mixture of vertical and horizontal lines). Thus the elements of $A \cap B$ are counted twice. They are counted once when the vertical elements are counted in $|A|$ and once

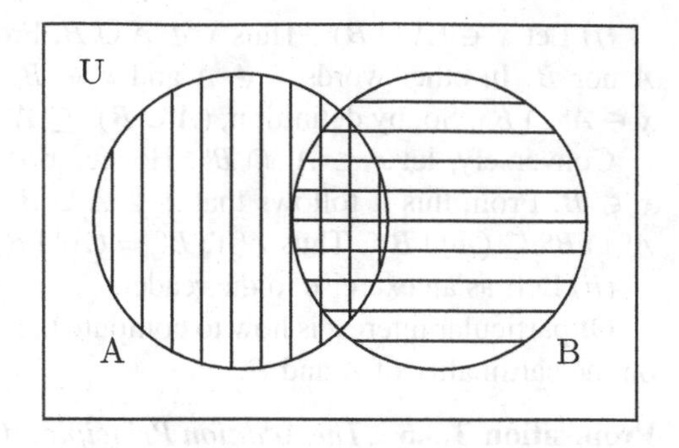

Fig. 1.6 Venn diagram illustrating the Principle of Inclusion and Exclusion

when the horizontal elements are counted in $|B|$. Thus we must subtract off $|A \cap B|$ to compensate for this. Hence $|A \cup B| = |A| + |B| - |A \cap B|$.

Example 1.3.9 Each child at a party has either punch or cake, possibly both. Sixteen children have cake. Nine children have punch. There are five children that have both punch and cake. How many children were at the party?

Solution Let A be the set of children who had cake at the party. Let B be the set of children who had punch at the party. Since each child had either punch or cake, the number of children at the party is given by $|A \cup B|$. So by the Principle of Inclusion and Exclusion, we have

$$|A \cup B| = |A| + |B| - |A \cap B| = 16 + 9 - 5 = 20.$$

□

We can also compute the cardinality of the Cartesian product.

Proposition 1.3.10 *Let $A_1,...,A_n$ be sets. The cardinality of $A_1 \times \cdots \times A_n$ is given by $|A_1|...|A_n|$.*

Proof We proceed by induction on n. If $n = 1$, then the claim is obvious. Suppose for some n, $|A_1 \times \cdots \times A_n| = |A_1|...|A_n|$. Consider the case with $n+1$ sets. Note that:

$$A_1 \times \cdots \times A_n \times A_{n+1} = (A_1 \times \cdots \times A_n) \times A_{n+1}.$$

By inductive hypothesis, $|A_1 \times ... \times A_n| = |A_1|...|A_n|$. Thus there are $|A_1|...|A_n|$ choices for the first n entries in the Cartesian product. There are $|A_{n+1}|$ choices for the final entry in the Cartesian product. Since this last entry is chosen independently from the others, it follows that $|A_1 \times ... \times A_n \times A_{n+1}| = |A_1|...|A_n||A_{n+1}|$. Thus the proposition holds by the Principle of Mathematical Induction. ■

Given a set A, we may be interested in all possible subsets of A. The *power set* of A, denoted $P(A)$, is the collection of all subsets of A. For instance, if $A = \{0, 1, 2\}$, then $P(A) = \{\emptyset, \{0\}, \{1\}, \{2\}, \{0, 1\}, \{0, 2\}, \{1, 2\}, \{0, 1, 2\}\}$.

We end this section by listing several sets that will be used frequently throughout this book:

(i) $[n] = \{1, ..., n\}$;
(ii) $\mathbb{Z} = \{..., -3, -2, -1, 0, 1, 2, 3, ...\}$, in other words, the set of *integers*;
(iii) $\mathbb{Z}^+ = \{1, 2, ...\}$, in other words, the set of *positive integers*;
(iv) $\mathbb{N} = \{0, 1, ...\}$, in other words, the set of *non-negative integers*;
(v) $\mathbb{Q} = \{p/q : p, q \in \mathbb{Z}, q \neq 0\}$, in other words, the set of *rational numbers*;
(vi) $\mathbb{R}$, the set of *real numbers*;
(vii) $\mathbb{C} = \{a + bi : a, b \in \mathbb{R}\}$, where i is the imaginary unit $\sqrt{-1}$. In other words, $\mathbb{C}$ is the set of *complex numbers*.

In combinatorics, our primary interest will be in $[n]$, $\mathbb{Z}$, $\mathbb{Z}^+$, and $\mathbb{N}$. However, we will revisit $\mathbb{Q}$, $\mathbb{R}$, and $\mathbb{C}$ in the solutions of several problems later in this text. Often, these sets are used in an intermediate step.

Exercise 1.3.11 Find the cardinality of the following sets: $A = \{1, 4, 9, 16, 25\}$, $B = \{1, 9, 25\}$, $C = \{1, ..., 25\}$, and $D = \{2, 4, 9, 30\}$.

Exercise 1.3.12 Let $S = [30]$, $A = \{1, 4, 9, 16, 25\}$ and $D = \{2, 4, 9, 30\}$. Find the following: $A \cup D$, $A \cap D$, $A - D$, $(A \cup D)^c$, and $A \times D$.

Exercise 1.3.13 At a barbecue, each guest has either a hamburger or a hot dog, possibly both. Twenty-five guests have a hamburger. Eighteen guests have a hot dog. Ten guests have both a hamburger and a hot dog. How many guests were at the barbecue?

Exercise 1.3.14 Prove the second part of DeMorgan's Law, in other words, prove $(A \cap B)^c = A^c \cup B^c$.

Exercise 1.3.15 Prove that if $B \subseteq A$, then $B \cup (A - B) = A$.

Exercise 1.3.16 Prove that $A \cup B = (A - (A \cap B)) \cup B$ and $A \cap B \subseteq A$ for all sets A and B.

Exercise 1.3.17 Prove that if $B \subseteq A$, then $|A| \geq |B|$.

1.4 Functions

In this section, we consider functions on sets. A *function* f is a mapping from a set A to a set B. If f is a function mapping A to B, then we denote this by $f : A \rightarrow B$. Let $f(A) = \{f(x) : x \in A\}$. We call $f(A)$ the *image* of A under f. Note that $f(A) \subseteq B$. We say that A is the *domain* of f. In the case where $f(A) \subseteq A$, we say that f is a function on A.

Example 1.4.1 Let $A = \{-1, 0, 1, 2\}$ and $f(x) = x^2$. Find the image of A under f.

Solution Note that $f(-1) = f(1) = 1$, $f(0) = 0$, and $f(2) = 4$. So the image of A under f is $f(A) = \{0, 1, 4\}$. □

A particularly useful function is the *floor function* or the *greatest integer function*. The domain of the floor function is the set of real numbers. Given $x \in \mathbb{R}$, the floor of x, denoted $\lfloor x \rfloor$, is the largest integer less than or equal to x. Essentially, the floor function "rounds down" to the nearest integer.

Example 1.4.2 Find the floor of each of the following real numbers:

(i) $x = 67$;
(ii) $y = 22/7$;
(iii) $z = -13/9$.

Solution

(i) Since 67 is an integer, it is the floor function of itself. Hence, $\lfloor 67 \rfloor = 67$.
(ii) Note that 22/7 is $3 + (1/7)$. Thus, the greatest integer less than or equal to 22/7 is 3. Thus, $\lfloor 22/7 \rfloor = 3$.
(iii) Here, $-13/9$ is approximately -1.44. It follows that the greatest integer less than or equal to $-13/9$ is -2. Ergo, $\lfloor -13/9 \rfloor = -2$. □

There is an analogous function called the *ceiling function* or the *least integer function*. The ceiling function of x is denoted $\lceil x \rceil$. This returns the smallest integer greater than or equal to x.

In many cases, we cannot give a succinct formula for a mapping f. For instance suppose that $A = \{1, 2, 3, 4\}$ with $f(1) = 3$, $f(2) = 0$, $f(3) = 7$, and $f(4) = 2$. While there are polynomials that will give the required mapping, we prefer the more intuitive notation:

$$\begin{pmatrix} 1 & 2 & 3 & 4 \\ 3 & 0 & 7 & 2 \end{pmatrix}.$$

Here, the top line of the array lists the elements of A. The second line lists their corresponding values under f.

If for every $x, y \in A$, $f(x) = f(y)$ implies $x = y$, then we say that f is an *injective* function. Injective functions are also called *one-to-one* functions. In high-school algebra, injective functions pass a *horizontal line test*. In other words, if a horizontal line is passed through the graph of the function, then it will pass through at most one point on the graph.

A *surjective* function is a mapping from A to B with the property that for every $b \in B$, there exists $a \in A$ such that $f(a) = b$. A surjective function from A to B is also called *onto* B. A function from A to B that is both injective and surjective is called a *bijection* from A to B.

Example 1.4.3 Consider the following functions which have domain $[5] = \{1, 2, 3, 4, 5\}$. Which are injections? Which are onto the set [5]? Which are bijections from [5] to itself?

(i)

$$\begin{pmatrix} 1 & 2 & 3 & 4 & 5 \\ 6 & 4 & 2 & 3 & 1 \end{pmatrix};$$

(ii)

$$\begin{pmatrix} 1 & 2 & 3 & 4 & 5 \\ 3 & 4 & 1 & 2 & 2 \end{pmatrix};$$

(iii)

$$\begin{pmatrix} 1 & 2 & 3 & 4 & 5 \\ 3 & 5 & 1 & 2 & 4 \end{pmatrix};$$

(iv)

$$\begin{pmatrix} 1 & 2 & 3 & 4 & 5 \\ 3 & 3 & 5 & 1 & 2 \end{pmatrix}.$$

Solution

(i) This function is injective. Its image is $\{1, 2, 3, 4, 6\}$. Since no element maps to 5, it is not onto [5].
(ii) This function is onto the set [4]. It is not onto [5] as no element maps to 5. It is not injective as both 4 and 5 are mapped to 2.
(iii) This is a bijection from [5] to itself.
(iv) This function is not injective as two elements are mapped to 3. It is not onto the set [5] as 4 does not appear in the image. □

We can use properties of functions to derive bounds on the cardinality of sets.

Theorem 1.4.4 *Let A and B be sets. Let f be a function from A to B.*

(i) If f is injective, then $|A| \leq |B|$.
(ii) If f is surjective, then $|A| \geq |B|$.
(iii) If f is a bijection, then $|A| = |B|$.

Proof Suppose that $|A| = n$, with $A = \{a_1, ..., a_n\}$. Note that $f(A) = \{f(a_1), ..., f(a_n)\}$. By definition, since f is a function from A to B, it follows that $f(A) \subseteq B$.

(i) Since f is injective, it follows that for $i \neq j$, we have that $f(a_i) \neq f(a_j)$. Hence there are n distinct elements in $f(A)$. Since $f(A) \subseteq B$, it follows from Exercise 1.3.17 that $n = |f(A)| \leq |B|$.
(ii) Since f is a surjective function from A to B, it follows that for every $b \in B$ there is a (not necessarily unique) $a_i \in A$ such that $f(a_i) = b$. Therefore, $B \subseteq f(A)$. It follows from Exercise 1.3.17 that $|B| \leq |f(A)| = n$.
(iii) Follows immediately from (i), (ii), and the above definitions.

■

Theorem 1.4.4 has broader implications than what is immediately evident. Suppose that we know $|A|$ and wish to know $|B|$. If we can find a bijection from A to B, then we have counted the set B. Further, if $|A| = |B|$ then there should exist a bijection relating the elements of A to B. However, this bijection is not always easy to find, as we will see later.

Exercise 1.4.5 Let $A = \{1, 2, 3, 4\}$ and $f(x) = x^3 - 1$. Find the image of A under f.

Exercise 1.4.6 Let $A = \{1, 2, 3, 4\}$ and $f(x) = x^3 - 1$. Write f in array notation.

Exercise 1.4.7 Let $A = [4]$ and $B = [3]$. Which of the following functions are injective? Which of the following functions are onto B? Which are bijections from A to itself?

$$\begin{pmatrix} 1 & 2 & 3 & 4 \\ 2 & 5 & 1 & 4 \end{pmatrix} \quad \begin{pmatrix} 1 & 2 & 3 & 4 \\ 2 & 3 & 3 & 1 \end{pmatrix}$$

$$\begin{pmatrix} 1 & 2 & 3 & 4 \\ 2 & 3 & 1 & 4 \end{pmatrix} \quad \begin{pmatrix} 1 & 2 & 3 & 4 \\ 5 & 3 & 3 & 1 \end{pmatrix}$$

Exercise 1.4.8 Show that if σ and τ are bijections on a set S, then the composition $\sigma \circ \tau$ is also a bijection on S.

1.5 The Pigeonhole Principle

One of the simplest and most intuitive principles in mathematics is also one of the most useful. Put simply, the *Pigeonhole Principle* states that if you have more pigeons than holes to place them in, then at least one hole must contain more than one pigeon (see Fig. 1.7). We now give a more formal statement of this idea.

Theorem 1.5.1 *(The Pigeonhole Principle) Let $n, k \in \mathbb{N}$ such that $n > k$. If there are n objects and k containers, then at least one container must contain more than one object.*

Proof Let A be the set of n objects and B be the set of k containers. Let f be a function that maps objects to containers. In other words, $f : A \to B$. If no container contains two or more objects, then it follows that f is injective. Thus by Theorem 1.4.4 we have that $n = |A| \le |B| = k$. However, this is contrary to the assumption that $n > k$. ■

The Pigeonhole Principle implies a very simple, but unexpected result. Namely, it implies that there are two people living in New York City that have the same number of hairs on their head. On average, humans have about 150,000 hairs on their head. Therefore, we can comfortably assume that there is no one with more than two million hairs on their head. Further there are over 8 million people currently living in New

Fig. 1.7 Illustration of the Pigeonhole Principle

York City. We let the number of hairs on a person's head to be the "pigeonholes" and each person in New York to be a "pigeon." Since $k \leq 2000000$ and $n \geq 8000001$, it follows that there are two people living in New York City that have the same number of hairs on their head by the Pigeonhole Principle. We now give additional examples using the Pigeonhole Principle.

Example 1.5.2 Suppose that we have unmatched pairs of white, blue, black, grey, and tan socks in a drawer. What is the minimum number of socks that must be removed from the drawer before we are guaranteed of finding a matching pair?

Solution The "pigeonholes" in this case are the five colors. The socks removed from the drawer are the "pigeons" which must be sorted according to color. If we have six socks (the pigeons), then the Pigeonhole Principle guarantees that at least two must be the same color. □

Example 1.5.3 Suppose that a manufacturer makes at least one bookcase every day and makes no more than $\frac{7}{6}$ bookcases per day over a period of 24 days. Show that there is a period of consecutive days in which the manufacturer completes *exactly* 19 bookcases.

Solution Let a_i be the number of bookcases completed at the end of the ith day. Since the manufacturer makes at least one bookcase each day and makes no more than $\frac{7}{6}$ bookcases per day, we have that:

$$1 \leq a_1 < a_2 < \cdots < a_{23} < a_{24} \leq \frac{7}{6}(24) = 28.$$

This implies that:

$$a_1 + 19 < a_2 + 19 < \cdots < a_{23} + 19 < a_{24} + 19 \leq 47.$$

Now we consider the numbers

$$a_1, ..., a_{24}, a_1 + 19, ..., a_{24} + 19.$$

These 48 numbers will be our "pigeons." Each of these numbers is between 1 and 47 (these numbers are our "pigeonholes"). Thus, by the Pigeonhole Principle, two of these numbers must be equal. Note that the numbers a_1,...,a_{24} are all distinct. Similarly, the numbers $a_1 + 19$,...,$a_{24} + 19$ are all distinct. Finally, $a_i \neq a_i + 19$. Thus there exists $i \neq j$ such that $a_i = a_j + 19$. Hence, between day i and day j, the manufacturer completes exactly 19 bookcases. □

Proposition 1.5.4 *Let $A = \{a_1, ..., a_n\}$ be an ordered sequence of positive integers. There exists a consecutive sum (in other words, $s_{i,j} = a_i + \cdots + a_j$) that is divisible by n.*

Proof Note that when dividing by n, the possible remainders are $0, 1, ..., n - 1$. Consider the consecutive sums $s_{1,j} = a_1 + \cdots + a_j$ (these will be the "pigeons"). If one of the $s_{1,j}$ is divisible by n, then we are done. So without loss of generality, we may assume that none of $s_{1,j}$ are divisible by n. Thus when dividing $s_{1,j}$ by n the possible remainders are $1, ..., n - 1$ (these are the pigeonholes). As $n > n - 1$, there must be $i > j$ such that $s_{1,i}$ and $s_{1,j}$ leave the same remainder when divided by n by the Pigeonhole Principle. In other words, $s_{1,i} = q_i n + r$ and $s_{1,j} = q_j n + r$. From this it follows that $s_{1,i} - s_{1,j} = a_{j+1} + \cdots + a_i$ is a consecutive sum. Moreover,

$$s_{1,i} - s_{1,j} = q_i n + r - (q_j n + r) = n(q_i - q_j).$$

So this consecutive sum is divisible by n. ■

Theorem 1.5.5 *(The Generalized Pigeonhole Principle) If m pigeons are placed in k pigeonholes, then at least one pigeonhole will contain more than $\lfloor \frac{m-1}{k} \rfloor$ pigeons.*

Proof If $m \leq k$, then $\lfloor \frac{m-1}{k} \rfloor = 0$ and the claim is obvious. Suppose that $nk + 1 \leq m \leq (n + 1)k$. Note that $\lfloor \frac{m-1}{k} \rfloor = n$. Distribute $(n - 1)k$ pigeons by placing $n - 1$ pigeons into each of the k holes. Thus we have at least $nk + 1 - (nk - k) = k + 1$ pigeons to place in k holes. Thus by the Pigeonhole Principle, at least two of the remaining pigeons must be placed in one of the holes. Hence there is a hole that contains at least $n + 1$ pigeons. ■

Note that the Generalized Pigeonhole Principle implies that there are at least four people living in New York City that have the same number of hairs on their head.

Table 1.1 An example that shows $R(3,3) \geq 6$

Guest	Acquaintances
A	B, E
B	A, C
C	B, D
D	C, E
E	A, D

The astute reader should note that any version of the Pigeonhole Principle only tells us the *existence* of a pigeonhole having multiple pigeons. It gives us no indication of how to find this particular hole or that this hole would be unique.

Proposition 1.5.6 *In any group of six people there are three people that mutually know each other or three people that mutually do not know each other.*

Proof Take any individual in the group, call them A. The remaining five people (the pigeons) can be placed into one of two categories (the holes). Namely, they either know A or they do not. Thus by the generalized Pigeonhole Principle, there are at least three people that know A or at least three people that do not know A. Without loss of generality, suppose that A_1, A_2, and A_3 all know A. If none of A_1, A_2, and A_3 know each other, then we have a set of three people who do not know each other. If at least two of them know each other, say A_1 and A_2, then A, A_1, and A_2 form a set of three mutually acquainted people. ■

Proposition 1.5.6 is a special case of what is known as *Ramsey theory*. In general, the *Ramsey number* $R(r,s)$ is the smallest integer n such that in any group with n people, at least r people mutually know each other or at least s people mutually do not know each other. By above, we know that in any group of six people, at least three people mutually know each other or three people mutually do not know each other. Does the same hold true for a group of five people? In a group of five people, say {A, B, C, D, E}, we can have a series of acquaintances such as given in Table 1.1. So, $R(3,3) = 6$ by Table 1.1 and Proposition 1.5.6. Even for relatively small values of r and s, the number $R(r,s)$ is not known (for instance $R(5,5)$ is not known). Ramsey theory is a very broad area of mathematical research with many variations. The interested reader is referred to any of the excellent texts (such as the book by Graham, Rothschild, and Spencer [23]) on the subject for more information.

Exercise 1.5.7 Suppose that Alice does at least one math problem per day and solves no more than $\frac{9}{8}$ problems per day over a period of 32 days. Show that there exists a period of consecutive days when she completes *exactly* 27 problems.

Exercise 1.5.8 Suppose that Bob paints at least one picture per day and paints no more than $\frac{13}{12}$ pictures per day over a period of 36 days. Show that there exists a period of consecutive days when he completes *exactly* 31 pictures.

Exercise 1.5.9 Show that any sequence of $n^2 + 1$ distinct integers contains an increasing subsequence of $n + 1$ terms or a decreasing sequence of $n + 1$ terms.

Exercise 1.5.10 Show that in every group of n people, where $n \geq 2$, there are at least two people that know the same number of people within the group. We assume that A knows himself and that A knows B if and only if B knows A.

Exercise 1.5.11 Suppose there is a set of n men and $n + 1$ women to be seated around a circular table. Show that if $n \geq 2$, there is at least one person that will be seated between two women.

1.6 The Method of Combinatorial Proof

In this section, we solve another problem that was introduced in the Sect. 1.1. In particular, we solve the problem of how many words of length n can be constructed from the alphabet {a,b} in such a way that there are no two adjacent a's.

The solution of this problem will give us two things. First, it will give us a combinatorial interpretation for a famous mathematical sequence. Second, it will allow us to introduce the method of *combinatorial proof*.

A combinatorial proof exhibits a specific bijection between the set of interest and two or more smaller sets. Usually, the smaller sets will be easier to count than the set of interest. Ideally, they will be sets that we have already counted.

There are several advantages to this method of proof. The problem with an inductive proof (see Sect. 1.2) is that you must know that the result is true when you begin to prove it. In a combinatorial proof, a result is discovered and proved simultaneously.

Many of the identities in this book can be proven using mathematical induction or algebraic manipulation. Unfortunately, these methods can be very messy and do not give any indication as to *why* the result is true. In a combinatorial proof, we avoid messy algebraic manipulations and give a clear indication as to why the result is true. For this reason, combinatorial proofs are often easier to remember in the long term.

Finally, many combinatorial proofs "condition" the set by considering some restriction on the elements of the set. This being the case, it is not unusual to discover new, or unexpected, combinatorial identities by conditioning a set in clever ways.

With this in mind, we determine the number of words of length n from the alphabet {a,b} that have no adjacent a's. First, note that when a combinatorialist discusses "words," they generally do not require the "words" to be in any human dictionary. This being the case, we consider a *word* to be any string of letters. With this in mind, we begin this problem by listing all the acceptable words for small values of n.

When $n = 0$, there is one acceptable word, namely, the word with no letters. The concept of an empty word may seem a bit counterintuitive, however, it is a concept that we will revisit often in the early chapters of this book.

When $n = 1$, there are two acceptable words, namely a and b. When $n = 2$, there are three acceptable words, namely ab, ba, and bb. When $n = 3$, there are five acceptable words:

aba abb

bab bba bbb.

When $n = 4$, there are eight acceptable words:

abab abba abbb

baba babb bbab bbba bbbb.

You may have already recognized this sequence of values as the *Fibonacci sequence*. The Fibonacci sequence, denoted $\{F_n\}$, is the sequence defined by $F_1 = 1$, $F_2 = 1$, $F_n = F_{n-1} + F_{n-2}$ for all $n \geq 2$. The sequence we generated above is also the Fibonacci sequence, with a different indexing. While we only needed to list a handful of examples to determine the underlying pattern, in practice it may be necessary to list dozens, even hundreds, of examples before realizing the pattern.

You may have also noticed that we have organized our list of words in a particular way. Notice that we have listed all the words beginning with 'a,' followed by all the words that begin with 'b.' As any acceptable word will begin with either 'a' or 'b' (and no word can begin with both 'a' and 'b'), this list is exhaustive. Hence, we will be conditioning the words according to their first letter. This condition will be the key step in proving the following theorem.

Theorem 1.6.1 *Let $w(n)$ denote the number of words of length n from the alphabet $\{a,b\}$ that have no two adjacent a's. This sequence satisfies $w(0) = 1$, $w(1) = 2$, and*

$$w(n) = w(n-1) + w(n-2), \quad \textit{for all} \quad n \geq 2.$$

Proof By above, we have $w(0) = 1$ and $w(1) = 2$. We claim that for all $n \geq 2$:

$$w(n) = w(n-1) + w(n-2). \tag{1.1}$$

By definition, the left side of Eq. (1.1) counts the number of words of length n from the alphabet {a,b} that have no two adjacent a's.

The right side of Eq. (1.1) also counts this by counting the elements of two disjoint, exhaustive sets:

(i) The set of all words that begin with 'b.' The remaining letters form a word of length $n-1$ from the alphabet {a,b} that have no adjacent a's. This set is counted by $w(n-1)$ by definition.

(ii) The set of all words that begin with 'a.' Note that this 'a' must immediately be followed by a 'b.' Thus the remaining letters form a word of length $n-2$ from the alphabet {a,b} that have no adjacent a's. This set is counted by $w(n-2)$ by definition.

■

The relationship $w(n) = w(n-1)+w(n-2)$ is a specific example of a *recurrence relation*. A recurrence relation relates the value of a function at n to the value(s) of the function at other, smaller values. The values $w(0) = 1$ and $w(1) = 2$ are called the *initial values* for the recurrence relation. You may feel that the recurrence relation is a bit unsatisfying, preferring instead a closed form solution for $w(n)$. However, in Chap. 6, we will show that:

$$w(n) = \frac{1}{\sqrt{5}}\left[\left(\frac{1+\sqrt{5}}{2}\right)^{n+1} - \left(\frac{1-\sqrt{5}}{2}\right)^{n+1}\right].$$

For more information on combinatorial proofs, see *Proofs that really count* by Benjamin and Quinn [8].

A second method of combinatorial proof is to show that any arbitrary element is counted the same number of times on both sides of the equality. To illustrate this we revisit the Principle of Inclusion and Exclusion.

Theorem 1.6.2 *(Principle of Inclusion and Exclusion) For any sets A and B we have*

$$|A \cup B| = |A| + |B| - |A \cap B|.$$

Proof Let x be an element in the universal set. We note that if $x \notin A \cup B$ then by DeMorgan's Law, $x \notin A$ and $x \notin B$. Hence x is counted zero times in $|A \cup B|$ and zero times in each of $|A|$, $|B|$, and $|A \cap B|$. Thus we may assume that $x \in A \cup B$.

(i) Suppose that x is in exactly one of A or B. Without loss of generality, assume $x \in A$ and $x \notin B$. Thus x is counted once in $|A \cup B|$ and once in $|A|$. Further, x is counted zero times $|B|$ and zero times $|A \cap B|$. Therefore, x is counted once on each side of the equation.

(ii) Suppose that x is in both A and B. It follows that x is counted once in $|A \cup B|$. Further x is counted once in $|A|$, once in $|B|$, and negative one times in $-|A \cap B|$. Therefore, x is counted once in $|A| + |B| - |A \cap B|$.

■

Exercise 1.6.3 A *binary* string is composed entirely of zeros and ones. Find a recurrence relation and initial values for $b(n)$, the number of binary strings of length n that have no adjacent ones.

Exercise 1.6.4 In chess, a king is *attacking* another piece if it is on an adjacent square (either horizontally, vertically, or diagonally). Find a recurrence relation and initial values for $k(n)$, the number of ways of placing non-attacking kings on a $n \times 1$ chessboard.

Exercise 1.6.5 Find a recurrence relation and initial values for $s(n)$, the number of ways of writing n as an ordered sum where the parts come from $\{1, 2\}$.

Exercise 1.6.6 A *tiling* of a $n \times m$ grid is an arrangement of pieces such that every square on the grid is covered and no two pieces overlap. Find a recurrence relation and initial values for $t(n)$, the number of tilings on a $n \times 1$ grid using 1×1 squares and 2×1 dominoes.

Exercise 1.6.7 Find a recurrence relation and initial values for $W(n)$, the number of words of length n from the alphabet {a,b,c} with no adjacent a's.

Exercise 1.6.8 Find a recurrence relation and initial values for $\tau(n)$, the number of tilings on a $n \times 1$ grid using 1×1 squares and 3×1 triominoes.

Chapter 2
Basic Counting

2.1 The Multiplication Principle

Suppose that we are ordering dinner at a small restaurant. We must first order our drink, the choices being Soda, Tea, Water, Coffee, and Wine (respectively S, T, W, C, and I). Then, we order our appetizer, either Soup or Salad (respectively O and A). Next we order our entree from Beef, Chicken, Fish, and Vegetarian. Finally, we order dessert from Pie, Cake, and Ice Cream.

When ordering dinner, we can think of each choice as a separate independent event. In other words, our choice of an appetizer does not limit our choice of an entree. In reality, we may prefer certain choices of drink and appetizer with our choice of entree. However, we do not limit ourselves to these considerations for this example. So, our possibilities for drink and appetizer are:

$$\begin{array}{ccccc} SO & TO & WO & CO & IO \\ SA & TA & WA & CA & IA \end{array}$$

Note that regardless of our choice of drink, we still have two choices for an appetizer. Hence, we have $5(2) = 10$ choices for drink and appetizer. For each of these ten choices, we then have four possibilities for our entree (Beef, Chicken, Fish, and Vegetarian). This gives us a total of $5(2)(4) = 40$ possibilities. Finally, for each of these 40 possibilities, we have three possibilities for dessert (Pie, Cake, and Ice Cream). Thus we have a total of $5(2)(4)(3) = 120$ possibilities for dinner at this restaurant. The Multiplication Principle is a generalization of the above example.

Theorem 2.1.1 *(The Multiplication Principle) Suppose that there are n sets denoted $A_1,...,A_n$. If elements can be selected from each set independently, then the number of ways to select one element from each set is given by $|A_1|\cdots|A_n|$.*

Proof Note that this problem is equivalent to selecting an element from the set $A_1 \times \cdots \times A_n$. The cardinality of this set is $|A_1|\cdots|A_n|$ by Proposition 1.3.10. ■

Example 2.1.2 Suppose that we wish to go shopping. There are shopping districts in the north, east, west, and south side of town. We can take a car, bus, or train to any one of these destinations. Further, we may choose to take a scenic or a direct route.

R. A. Beeler, *How to Count*, DOI 10.1007/978-3-319-13844-2_2

While in the shopping district, we may shop for any one of clothing, groceries, or movies. While we are out, we may either go to the park, a restaurant, or neither. How many different shopping trips are possible?

Solution Let A_1 be the set of directions we can travel (N, E, W, S). Let A_2 denote the set of transportation options (car, bus, or train). We let A_3 be the set {Scenic, Direct}. A_4 will denote the shopping options (clothes, groceries, or movies). Finally, A_5 will be the set of "side trips" (park, restaurant, or no side trip). Thus the number of different shopping trips is given by $|A_1||A_2||A_3||A_4||A_5| = 4(3)(2)(3)(3) = 216$ by the Multiplication Principle. □

Suppose that we want to make a string of n colored beads. Each bead may be one of m colors and we have unlimited beads of each color. As usual, we want to determine the number of visually distinct strings. Usually, we would only be interested in visually distinct strings, in other words, those that cannot be obtained from another by flipping the string. So, if we are using the colors 1, 2, and 3, then 1223 and 3221 would be considered the same string. However, this is a more difficult problem and will wait until Chap. 8. For now, we will consider reflections to be different. In this case, we can think of the string as an n-tuple where the entries come from the set $[m]$.

Example 2.1.3 How many ways are there to make an n-tuple from the elements of $[m]$ in such a way that no two adjacent elements of the n-tuple are the same?

Solution There are m choices for the first element of the n-tuple. The ith element of the n-tuple ($i = 2, ..., n$) can be anything other than what was used for $(i-1)$th element. Thus the number of acceptable strings is

$$m \prod_{i=2}^{n} (m-1) = m(m-1)^{n-1}.$$ □

You may notice that the particular sets change for each of the choices in Example 2.1.3. For instance, suppose that $m = 3$ and we select 1 for the first entry in the n-tuple. Our set of options for the second entry is $\{2, 3\}$. However, if we select 2 as the first entry, then our set of options for the second entry is $\{1, 3\}$. While these sets are different, the cardinality of each set *is* the same. For this reason, the Multiplication Principle still applies. Further note that in Example 2.1.3, exponentiation appears as part of our solution. This is also the case in the following corollary.

Corollary 2.1.4 *Suppose that we have k distinguishable trophies to distribute to n people (who are by definition distinguishable). Each person may receive more than one trophy and some people may not receive a trophy. The number of ways to distribute the trophies is given by* n^k.

Proof Let A_i be the set of ways to distribute the ith trophy. Since there are n people, it follows that $|A_i| = n$ for all i. Since there are k trophies, there are n^k ways to distribute them by the Multiplication Principle. ■

Recall that in algebra 0^0 is undefined. Similarly, the limit form 0^0 is considered an *indeterminant form* in calculus. However, in this book we will *always* assume that $0^0 = 1$. By Corollary 2.1.4, there are $0^0 = 1$ ways to distribute 0 trophies to 0 people, the way in which no one gets anything.

The case of 0^0 is a special case of an *empty product*. In an empty product, no terms are being multiplied. Because an empty product should not change the value of a product, algebraically an empty product should equal one. Analogously, an *empty sum* should be zero.

Corollary 2.1.4 is often referred to as *sampling with replacement*. Consider the problem of selecting 5 cards from a standard poker deck of 52 cards. Each time a card is selected, it is placed back into the deck. Thus there are 52^5 possible ways to sample five cards from the deck with replacement, if the order of the cards is important.

Example 2.1.5 The Henry Classification System for fingerprints classifies the print for each of the ten digits. The possible classifications are plain arch, tented arch, radial loop, ulnar loop, plain whorl, accidental whorl, double loop whorl, peacock's eye whorl, composite whorl, and central packet whorl. How many possible finger print patterns are possible? Based on this, should we believe that no two people have the same fingerprints?

Solution There are ten classifications for each of the ten fingers. Hence there are 10^{10} possible patterns by Corollary 2.1.4. This means that there are 10 billion possible patterns under this classification system. As this exceeds the almost 7 billion people currently alive, it is plausible to assume that no two individuals have the same pattern. ■

A necessary condition for a challenging or stimulating game is that there is a large number of different games. If there is a relatively small number of possible games, then eventually the player has seen every possibility. For this reason, Tic-tac-toe (or Naughts and crosses) which has only 26830 possible different games (up to symmetries on the board) has little appeal except to school children. We will consider chess. While an exact computation of the number of games of chess is untractable, we will be satisfied by an approximation. Our approximation will be based on the estimates used by Shannon [38] and improved upon by Allis [2].

Example 2.1.6 Suppose that on average there are 80 moves made in chess (in other words, both players make 40 moves). For each player's first move, there are 20 possibilities (namely, each of the eight pawns can move either one or two spaces and each of the two knights can jump to the left or the right). For the remaining moves, each player has an average of 35 choices available at each move. Approximate the number of possible games of chess.

Solution Each player has 20 moves on their first turn. Thus there are $20^2 = 400$ possibilities for the first two moves. Assuming 35 moves for each of the 78 remaining moves, there are 35^{78} possibilities for the remaining turns by Corollary 2.1.4. Thus, there are $400 * 35^{78}$ possible games of chess by the Multiplication Principle. ■

Note that $400 * 35^{78}$ is approximately 10^{123}. For dedicated chess players, this may be very comforting. If only one in 10^{23} games is a "good game," then there are over 10^{100} "good games" of chess. Therefore, it is very likely that many "good games" of chess are yet to be played. To put this number in perspective, there are only 5×10^{20} possible games of checkers and less than 10^{81} atoms in the universe.

When using the Multiplication Principle, it is important to consider any restrictions on our choices.

Example 2.1.7 At a particular company, any valid password consists of six lower-case letters. Further, any valid password must end in a vowel (in other words, 'a,' 'e,' 'i,' 'o,' or 'u') and cannot contain the same letter twice. Find the number of valid passwords.

Solution As a first attempt at a solution, we might try to put the letters in order from left to right. It is easy to see we have 26 possibilities for the first letter, 25 for the second (anything but the first letter used), and so on. However, placing the last letter is more difficult, as we do not know how many vowels have been used.

To find a solution, we begin with the most restrictive selection we have to make. Namely, we begin by selecting the last letter. There are five ways of doing this. There are then 25 ways to select the first letter (anything but the letter used in the last slot). Similarly, there are 24 ways to select the second letter, 23 ways to select the third, 22 ways for the forth letter, and 21 ways for the fifth letter. Hence the number of valid passwords is given by $5(25)(24)(23)(22)(21) = 31878000$. □

Example 2.1.8 Areas must contain a large number of valid telephone numbers in order to accommodate their customer base. Further, it is common to restrict which telephone numbers are assigned to customers. How many seven-digit telephone numbers:

(i) Are there?
(ii) Do not start with 0 (such a number would dial the operator)?
(iii) Do not contain 911 (such a number would immediately dial emergency services)?
(iv) Do not contain 911 nor do they start with 0?

Solution

(i) Note that there are 10 possible digits that can be used. Hence there 10^7 possible telephone numbers by Corollary 2.1.4.
(ii) First find how many telephone numbers start with 0. There are six remaining numbers, so there are 10^6 possibilities. Thus there are $10^7 - 10^6$ telephone numbers that do not start with 0 by the Subtraction Principle.
Alternatively, there are nine choices for the first digit, namely anything but zero. There are then 10^6 choices for the remaining digits. Hence, there are $9 * 10^6 = 9000000$ telephone numbers that do not start with zero.
(iii) If a number contains 911, then there are 10^4 choices for the remaining digits. There are five ways to place the sequence 911 within the number, so $5 * 10^4$ telephone numbers contain 911. Thus by the Subtraction Principle there are $10^7 - 5 * 10^4 = 9950000$ acceptable numbers.
(iv) We must find the numbers that start with 0 and contain 911. If a number contains both, then there are 10^3 choices for the remaining digits. There are four choices as to where to place the 911. Hence there are $4 * 10^3$ numbers that start with

0 and contain 911. By the Principle of Inclusion and Exclusion, the numbers of ways that a telephone number can start with 0 and contain 911 is given by $10^6 + 5 * 10^4 - 4 * 10^3$. Hence by the Subtraction Principle, the number of telephone numbers that do not start with 0 nor contain 911 is given by:

$$10^7 - (10^6 + 5 * 10^4 - 4 * 10^3) = 10^7 - 10^6 - 5 * 10^4 + 4 * 10^3 = 8954000.$$

□

A different kind of restriction is one that considers certain configurations to be identical. For example, suppose that we want to label the faces of a six -sided die with distinct elements of [6]. Since a die will be rolled around on a surface, we will consider two labelings to be identical if one can be obtained from another by rotating or rolling the die.

Example 2.1.9 Find the number of distinguishable ways to label a six-sided die with the elements of [6].

Solution Label the top face with 1. This breaks one of the symmetries on the faces of the cube. The remaining faces are labeled as follows:

(i) Choose one of the remaining numbers to place on the bottom face. There are 5 possibilities.
(ii) Place one of the remaining numbers on the front face. This breaks the final symmetry.
(iii) Label the right face. There are 3 possibilities.
(iv) Label the back face. There are 2 possibilities.
(v) The final number is the forced on the left face.

Thus by the Multiplication Principle, there are $5 * 3 * 2 = 30$ possibilities.

Proposition 2.1.10 *The cardinality of the power set of A is $2^{|A|}$.*

Proof For each element x of A, x must be included in a subset or excluded from the subset. Thus there are two choices for each element of A. Since there are $|A|$ such elements, there are $2^{|A|}$ subsets of A by the Multiplication Principle. ■

Example 2.1.11 A particular deli offers sandwiches with a choice of 5 meats, 3 cheeses, 12 vegetables, and 4 condiments. A sandwich may consist of any combination of these "toppings." How many possible different sandwiches does the deli offer?

Solution Let A be the set of sandwich toppings. We note that $|A| = 24$. The contents of the sandwich can be thought of as a subset of the available toppings (the empty set can be thought of a sandwich consisting of only bread). The number of subsets of A is $2^{24} = 16777216$. Hence, there are 1677216 distinct sandwiches. □

An effective tool in examining problems involving the Multiplication Principle is the *tree diagram*. In a tree diagram, each "branch" of the "tree" corresponds to a choice or event taking place. For example, consider flipping a coin (either heads or tails), drawing a suit from a deck of cards (either spades, hearts, clubs, or diamonds), and rolling a six-sided die. The tree diagram corresponding to this chain of events is given in Fig. 2.1. At the first branch, the coin lands either heads or tails. For each of these outcomes, the diagram branches further. In particular, for each outcome of the coin, either a heart, spade, club, or diamond is drawn from the deck. Each of these branches into six possible leaves, one for each of the possible roll of the die.

We can also use these principles to prove other, more surprising results.

Theorem 2.1.12 *If $F_0 = 1$, $F_1 = 2$, and $F_n = F_{n-1} + F_{n-2}$, then*

$$F_{m+n+1} = F_m F_n + F_{m-1} F_{n-1}.$$

Proof Note that under this indexing, F_{m+n+1} counts the number of words of length $m+n+1$ from the alphabet {a,b} that have no adjacent a's (see Theorem 1.6.1). The right side also counts this by counting two disjoint, exhaustive sets:

(i) The set of words in which the $(m+1)$st letter is 'b.' The first m letters comprise a word of length m from the alphabet {a,b} in which there are no adjacent a's. There are F_m such words. Similarly, the remaining n letters comprise a word of length n from the alphabet {a,b} with no adjacent a's. There are F_n such words. Thus by the Multiplication Principle, there are $F_m F_n$ words in the first set.
(ii) The set of words in which the $(m+1)$st letter is 'a.' Note that the mth letter and the $(m+2)$nd letter must both be 'b.' The first $m-1$ letters comprise a word of length $m-1$ from the alphabet {a,b} in which there are no adjacent a's. There are F_{m-1} such words. Similarly, the remaining n letters comprise a word of length $n-1$ from the alphabet {a,b}. There are F_{n-1} such words. Thus by the Multiplication Principle there are $F_{m-1}F_{n-1}$ words in the second set.

The result then follows by the Addition Principle. ■

The above theorem gives an efficient method for computing large Fibonacci numbers. This method is actually more efficient than using the closed form discussed in the previous chapter.

Recall that in linear algebra, an $n \times n$ matrix is invertible if and only if

(i) No row is composed entirely of zeros *and*
(ii) The rows are *linearly independent*. In other words, if $r_1,...,r_n$ are the rows of this matrix and $a_1r_1 + \cdots + a_nr_n = 0$, then $a_1 = \cdots = a_n = 0$ [5].

Example 2.1.13 Find the number of invertible $n \times n$ matrices in which the entries are from the field of order q.

Solution Note that if the entries are from the field of order q, then there are q^n possible rows by the Multiplication Principle. However, one of these rows is the zero row, hence there are $q^n - 1$ choices for the first row.

Further note that there are q rows that are linearly dependent on this one. Hence there are $q^n - q$ choices for the second row.

In general, if we determined the first k rows, any linearly dependent $(k+1)$st row will be of the form:

$$r_{k+1} = a_1r_1 + \cdots + a_kr_k,$$

Fig. 2.1 A tree diagram

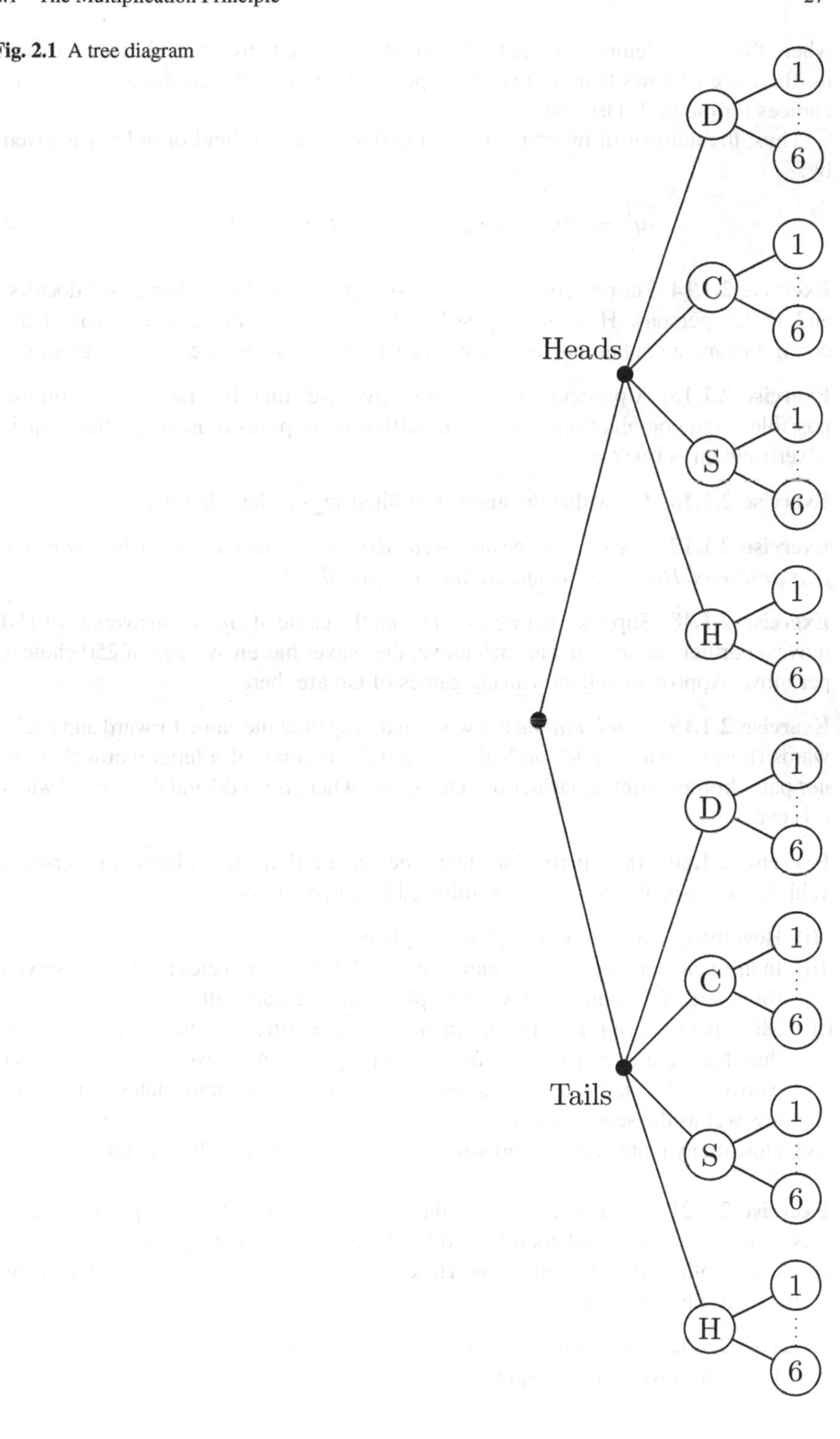

where the a_i are elements of the field of order q. Since there are q choices for each a_i, there are q^k rows that are linearly dependent on $r_1,..., r_k$. So there are $q^n - q^k$ choices for the $(k+1)$st row.

Thus, the number of invertible $n \times n$ matrices from the field of order q is given by:

$$(q^n - 1)(q^n - q)(q^n - q^2) \cdots (q^n - q^{n-1}). \quad \square$$

Exercise 2.1.14 The city council contains 8 clergy, 4 scientists, 5 lawyers, 3 doctors, and 10 lay persons. How many possible five person committees are there if the committee must contain one person from each group and no one is in two groups?

Exercise 2.1.15 A particular pizza parlor advertises that they have over a million possible pizza combinations. How many different toppings must they offer if their advertisement is true?

Exercise 2.1.16 Show that the number of bit strings of length n is 2^n.

Exercise 2.1.17 *The* snake cube *puzzle consists of 17 joints, each of which can take four positions. How many configurations are possible?*

Exercise 2.1.18 Suppose that we assume that the game of *Go* has an average of 150 moves. Further suppose that at each move, the player has an average of 250 choices per move. Approximately how many games of Go are there?

Exercise 2.1.19 A *palindrome* is a word that is spelled the same forward and backwards (for example, "civic" or "radar"). Find the number of n letter words that *are not* palindromes. Hint: Consider two cases, one where n is odd and the second when n is even.

Exercise 2.1.20 In a particular state, the standard license plates for personal vehicles consists of three numbers followed by three letters.

(i) How many possible license plates are there?
(ii) In this state, no license plate may begin with 9 as these plates are to be reserved for emergency vehicles. How many plates do not start with 9?
(iii) Many people complain that certain plates contain offensive words. To minimize this, the state is considering discontinuing plates that have a vowel (in other words, 'a,' 'e,' 'i,' 'o,' and 'u') as the second letter. How many plates do not have a vowel as the second letter?
(iv) How many plates do not start with 9 nor have a vowel as the second letter?

Exercise 2.1.21 In a particular state, the standard license plates for personal vehicles consists of three numbers followed by three letters. Specialty plates consists of two letters followed by four numbers. How many specialty plates are possible? How many license plates are possible?

Exercise 2.1.22 At a particular company, any valid password starts with six lower case letters followed by two digits.

(i) How many valid passwords are there?
(ii) How many valid passwords do not start with a?
(iii) How many valid passwords do not end with 88?
(iv) How many valid passwords do not start with a nor end with 88?

Exercise 2.1.23 A five digit number cannot start with zero.

(i) Find the number of five digit numbers.
(ii) Find the number of five digit numbers in which no digit appears twice.
(iii) Find the number of five digit odd numbers in which no digit appears twice.
(iv) Find the number of five digit even numbers in which no digit appears twice.

Exercise 2.1.24 Let $F_0 = 1$, $F_1 = 1$, and $F_n = F_{n-1} + F_{n-2}$. Show that $F_{m+n} = F_m F_n + F_{m-1} F_{n-1}$. Hint: See Exercise 1.6.5, Exercise 1.6.6, and Theorem 2.1.12.

2.2 The Addition Principle

We have already looked at two rules dealing with the cardinality of the union of two sets. In particular, Proposition 1.3.6 states that if A and B are disjoint sets, then $|A \cup B| = |A| + |B|$. As a generalization of this, Theorem 1.3.8 states that if A and B are any sets then $|A \cup B| = |A| + |B| - |A \cap B|$. In this section, we expand Proposition 1.3.6 to an arbitrary number of sets. We will not generalize Theorem 1.3.8 until Chap. 7.

Throughout the remainder of this book, we will use the notation

$$\cup_{i=1}^{n} A_i = A_1 \cup \cdots \cup A_n$$

to denote the union of the sets $A_1, ..., A_n$.

Theorem 2.2.1 *(The Generalized Addition Principle) If $A_1, ..., A_n$ are mutually disjoint sets, then:*

$$| \cup_{i=1}^{n} A_i| = \sum_{i=1}^{n} |A_i|.$$

Proof We proceed by induction on n. If $n = 1$, then the claim is obvious. Assume that for some n, the mutually disjoint sets $A_1,...,A_n$ satisfy:

$$| \cup_{i=1}^{n} A_i| = \sum_{i=1}^{n} |A_i|.$$

Let $A_1,...,A_n$, and A_{n+1} be mutually disjoint sets. Thus, $\cup_{i=1}^{n} A_i$ and A_{n+1} are disjoint sets. Hence, by Proposition 1.3.6, we have that

$$| \cup_{i=1}^{n+1} A_i| = | \cup_{i=1}^{n} A_i| + |A_{n+1}|.$$

Applying the inductive hypothesis yields

$$
\begin{aligned}
|\cup_{i=1}^{n+1} A_i| &= |\cup_{i=1}^{n} A_i| + |A_{n+1}| \\
&= \sum_{i=1}^{n} |A_i| + |A_{n+1}| = \sum_{i=1}^{n+1} |A_i|.
\end{aligned}
$$

Alternatively, if $x \in A_i$, then $x \notin A_j$ for $j \neq i$. Therefore, x is counted once by $|\cup_{i=1}^{n} A_i|$ and once by $\sum_{i=1}^{n} |A_i|$. Similarly, if $x \notin A_i$ for all i, then x is counted zero times by both sides of the equation. ■

Theorem 2.2.1 can be summarized as follows: Suppose that there are n events. The ith event can occur in a_i ways. If no two events can occur simultaneously (in other words, they are disjoint), then there are $a_i + \cdots + a_n$ ways that exactly one of the events can occur.

Example 2.2.2 Suppose that you can go to one of five small restaurants for dinner. No two restaurants offer the same menu items. The first restaurant offers 8 menu items, the second restaurant offers 6 items, the third restaurant offers 11 items, the fourth restaurant offers 4 items, and the final restaurant offers 20 items. How many meals are possible?

Solution Let A_i be the set of all menu choices in the ith restaurant. Since no two restaurants offer the same menu items, these sets are disjoint. So by the Addition Principle, we have

$$
\begin{aligned}
|A_1 \cup A_2 \cup A_3 \cup A_4 \cup A_5| &= |A_1| + |A_2| + |A_3| + |A_4| + |A_5| \\
&= 8 + 6 + 11 + 4 + 20 = 49.
\end{aligned}
$$

□

Example 2.2.3 Find the number of non-negative integer solutions to $x + y \leq 6$.

Solution Let A_i denote the set of non-negative integer solutions to $x + y = i$, where $i = 0, 1, ..., 6$. Note that the A_i are disjoint sets. So we have:

$$
\begin{aligned}
A_0 &= \{(0,0)\}. \\
A_1 &= \{(1,0),(0,1)\}, \\
A_2 &= \{(2,0),(1,1),(0,2)\}, \\
A_3 &= \{(3,0),(2,1),(1,2),(0,3)\}, \\
A_4 &= \{(4,0),(3,1),(2,2),(1,3),(0,4)\}, \\
A_5 &= \{(5,0),(4,1),(3,2),(2,3),(1,4),(0,5)\}, \\
A_6 &= \{(6,0),(5,1),(4,2),(3,3),(2,4),(1,5),(0,6)\}.
\end{aligned}
$$

Thus by the Addition Principle, we have:

$$|A_0 \cup A_1 \cup A_2 \cup A_3 \cup A_4 \cup A_5 \cup A_6|$$

$$= |A_0| + |A_1| + |A_2| + |A_3| + |A_4| + |A_5| + |A_6|$$
$$= 1 + 2 + 3 + 4 + 5 + 6 + 7 = 28. \qquad \square$$

We note that the above problem could have been done by simply listing all of the possibilities. However, this approach would not illustrate the Addition Principle. Further, in a more complicated function, it would be difficult to determine if you had examined all of the possibilities. This danger is compounded when dealing with more than two variables.

Suppose that we were to consider the number of non-negative integer solutions to $x + y \leq N$, where N is large. In this case, it is impractical to even list the individual elements of A_i, the set of all non-negative solutions to $x + y = i$. However, we can observe that $A_i = \{(k, i-k) : k = 0, 1, ..., i\}$. Thus $|A_i| = i+1$ for all $i = 0, 1, ..., N$.

You may be curious as to when to use the Addition Principle instead of the Multiplication Principle. A general rule of thumb is to use the Addition Principle when the problem uses the word "or." If the problem uses the word "and," then you should use the Multiplication Principle. In most cases, we have to combine the two rules, as is the case in our next two examples.

Example 2.2.4 Find the number of bit strings of length three, four, or five.

Solution The number of bit strings of length n is given by 2^n (see Exercise 2.1.16). Thus, there are $2^3 = 8$ bit strings of length three, $2^4 = 16$ bit strings of length four, and $2^5 = 32$ bit strings of length five. Since no bit string can be of two different lengths, it follows that there are:

$$2^3 + 2^4 + 2^5 = 8 + 16 + 32 = 66$$

bit strings of length three, four, or five. $\square$

Example 2.2.5 Suppose that you have the choice of four restaurants for lunch. The first has four choices for appetizer, ten entree choices, and two choices for dessert. The second has three choices for appetizer, eight entree choices, and five choices for dessert. The third has 5 choices for appetizer, 12 entree choices, and 3 choices for dessert. Finally, the last restaurant has 6 choices for appetizer, 20 entree choices, and 8 choices for dessert. No two restaurants have the same menu. Find the total number of different possible lunches.

Solution Let A_i be the set of possible lunches at the ith restaurant. By the Multiplication Principle, we have:

$$|A_1| = 4(10)(2) = 80;$$
$$|A_2| = 3(8)(5) = 120;$$
$$|A_3| = 5(12)(3) = 180;$$
$$|A_4| = 6(20)(8) = 960.$$

Since none of the restaurants have the same menu, the Addition Principle gives us

$$|A_1 \cup A_2 \cup A_3 \cup A_4| = |A_1| + |A_2| + |A_3| + |A_4|$$

$$= 80 + 120 + 180 + 960 = 1340. \qquad \square$$

The Addition Principle can also be used to prove various identities.

Theorem 2.2.6 *Let $F_0 = 1$, $F_1 = 2$, and $F_n = F_{n-1} + F_{n-2}$ for all $n \geq 2$. For $n \geq 0$ we have*

$$2 + \sum_{k=2}^{n+2} F_{k-2} = F_{n+2}.$$

Proof Equivalently, we prove that

$$1 + \sum_{k=2}^{n+2} F_{k-2} = F_{n+2} - 1. \tag{2.1}$$

Recall that under this indexing, the Fibonacci sequence counts the number of words of length n from the alphabet {a,b} that have no adjacent a's (see Theorem 1.6.1). It suffices to show that both sides of Eq. (2.1) count the same thing.

By Theorem 1.6.1, F_{n+2} counts the number of words of length $n+2$ from the alphabet {a,b} that have no adjacent a's. Further, there is exactly one acceptable word of length $n+2$ that has no a's. It follows that the right side of Eq. (2.1) counts the acceptable words of length $n+2$ with at least one 'a.'

The left side also counts this by counting $n+2$ disjoint, exhaustive sets. The kth set ($k = 1, \ldots, n+2$) is the set of all acceptable words in which the last 'a' occurs in the kth position. If $k = 1$, then there is one such word, namely 'a' followed by $n+1$ b's. For $k \geq 2$, the 'a' in the kth position must be immediately preceded by a 'b.' The first $k-2$ letters is a word of length $k-2$ from the alphabet {a,b} in which there are no adjacent a's. The number of such words is given by F_{k-2}. Ergo, Eq. (2.1) holds by the Addition Principle. ■

Exercise 2.2.7 Suppose that you can go to one of four small restaurants for lunch. The Chinese restaurant offers six menu items, the Mexican restaurant offers seven items, the Italian restaurant offers ten items, and the American restaurant offers eight items. How many different choices do you have for lunch if you can only get one item and no two restaurants serve the same menu?

Exercise 2.2.8 Find the number of seven letter words that start with a or do not contain a.

Exercise 2.2.9 Find the number of six letter words that either start with *a* or do not contain *b*.

Exercise 2.2.10 Find the number of non-negative integer solutions to $2x + 3y \leq 7$.

Exercise 2.2.11 Suppose that you have the choice of five restaurants for dinner. The first has 5 choices for appetizer, 11 entree choices, and 4 choices for dessert.

The second has five choices for appetizer, nine entree choices, and six choices for dessert. The third has four choices for appetizer, ten entree choices, and eight choices for dessert. The fourth restaurant has two choices for appetizer, seven entree choices, and five choices for dessert. Finally, the last restaurant has 7 choices for appetizer, 30 entree choices, and 9 choices for dessert. No two restaurants have the same menu. Find the total number of different possible dinners.

Exercise 2.2.12 Let $F_0 = F_1 = 1$ and $F_n = F_{n-1} + F_{n-2}$ for $n \geq 2$. Prove that for $n \geq 2$,

$$\sum_{k=0}^{n} F_k = F_{n+2} - 1.$$

Hint: Refer to Exercise 1.6.5, Exercise 1.6.6, and Theorem 2.2.6.

Exercise 2.2.13 In Exercise 1.6.6, we defined $t(n)$ to be the number of ways to tile a $n \times 1$ grid with 2×1 dominoes and 1×1 squares. Define $d(n)$ to be the number of ways to tile a $n \times 2$ grid with squares and dominoes. Prove that $d(0) = 1$, and for $n \geq 1$:

$$d(n) = t(n)^2 + \sum_{k=1}^{n} t(k-1)^2 d(n-k).$$

2.3 Permutations

An arrangement of the elements of $[n]$ is called a *permutation* on n. In particular, we are interested in how many permutations on $[n]$ there are. This is equivalent to the number of ways of lining up n individuals. Further, this is equivalent to the number of bijections from $[n]$ to itself.

As usual, for small values of n, we can determine the number of permutations simply by listing all the possibilities. For instance, the permutations on [4] are:

1234	1243	1324	1342	1423	1432
2134	2143	2314	2341	2413	2431
3124	3142	3214	3241	3412	3421
4123	4132	4213	4231	4312	4321

Note that we use a more compact notation for the function at hand. For instance:

$$\begin{pmatrix} 1 & 2 & 3 & 4 \\ 2 & 4 & 1 & 3 \end{pmatrix} := 2413.$$

There is an alternate "one line" notation for permutations that we will discuss in Sect. 2.7.

Table 2.1 Values of $n!$ for small n

n	0	1	2	3	4	5	6	7	8	9
$n!$	1	1	2	6	24	120	720	5040	40320	362880

If f is a function and $f(x) = x$, then we say that x is a *fixed point* of the function. Later, we will consider the problem of determining the number of permutations on n that contain no fixed points. Such a permutation is called a *derangement* on n.

Unfortunately, it would be difficult to list all permutations for larger values of n. For instance, there are 120 permutations on five elements. In general, the number of permutations on n grows faster than 2^{n-1}. Thus, it becomes necessary to find a general formula for the number of permutations.

To aid in this, we define the symbol $n!$ (read *n factorial*). This symbol is defined recursively as $0! = 1$ and $n! = n(n-1)!$ for $n \in \mathbb{Z}^+$.

Theorem 2.3.1 *The number of permutations on* $[n]$ *is given by* $n!$.

Proof We will count the number of bijections from $[n]$ to itself. Clearly there is one bijection of the empty set to itself. For all other n, there are n choices for $f(1)$, $n-1$ choices for $f(2)$,..., two choices for $f(n-1)$, and one choice for $f(n)$. Thus by the Multiplication Principle, the number of bijections from $[n]$ to itself is given by:

$$n(n-1)\cdots(2)(1) = n!.$$

■

Hence, the number of ways to order a n-element set is given by $n!$. Values of $n!$ for small n are given in Table 2.3. A tree diagram illustrating the permutations on [4] is given in Fig. 2.2.

For large values of n, it is more useful to use *Stirling's approximation*. Namely,

$$n! \approx \sqrt{2\pi n}\left(\frac{n}{e}\right)^n.$$

A full proof of Stirling's approximation is beyond the scope of this text.

We now consider a table setting problem. In a table setting problem, we find the number of visually distinct ways of sitting a group of people around a circular table (with seats labeled 1, ..., n) according to a given set of restrictions. The reason that the seats are labeled is to ensure that the rotations and reflections are visually distinct. Suppose that Alice, Bob, Chad, and Diane (denoted A, B, C, and D, respectively) are to be seated around a circular dinner table. The table settings ABCD and ADCB (see Fig. 2.3) look similar as each person is seated next to the same individuals. However, in the first setting Alice has Bob on her left, while in the second Alice has Diane on her left. If things are being passed clockwise, then this determines the order in which items are passed. Similarly, ABCD and DABC are simply rotations of the same table setting. However, if you were to enter the dining room, you would agree that these are different as a different person would be seated across from the door. Note that if you *do not* consider these distinct, then you would simply divide by n to account for the rotations and divide by two to account for the flips.

Fig. 2.2 Permutations on [4]

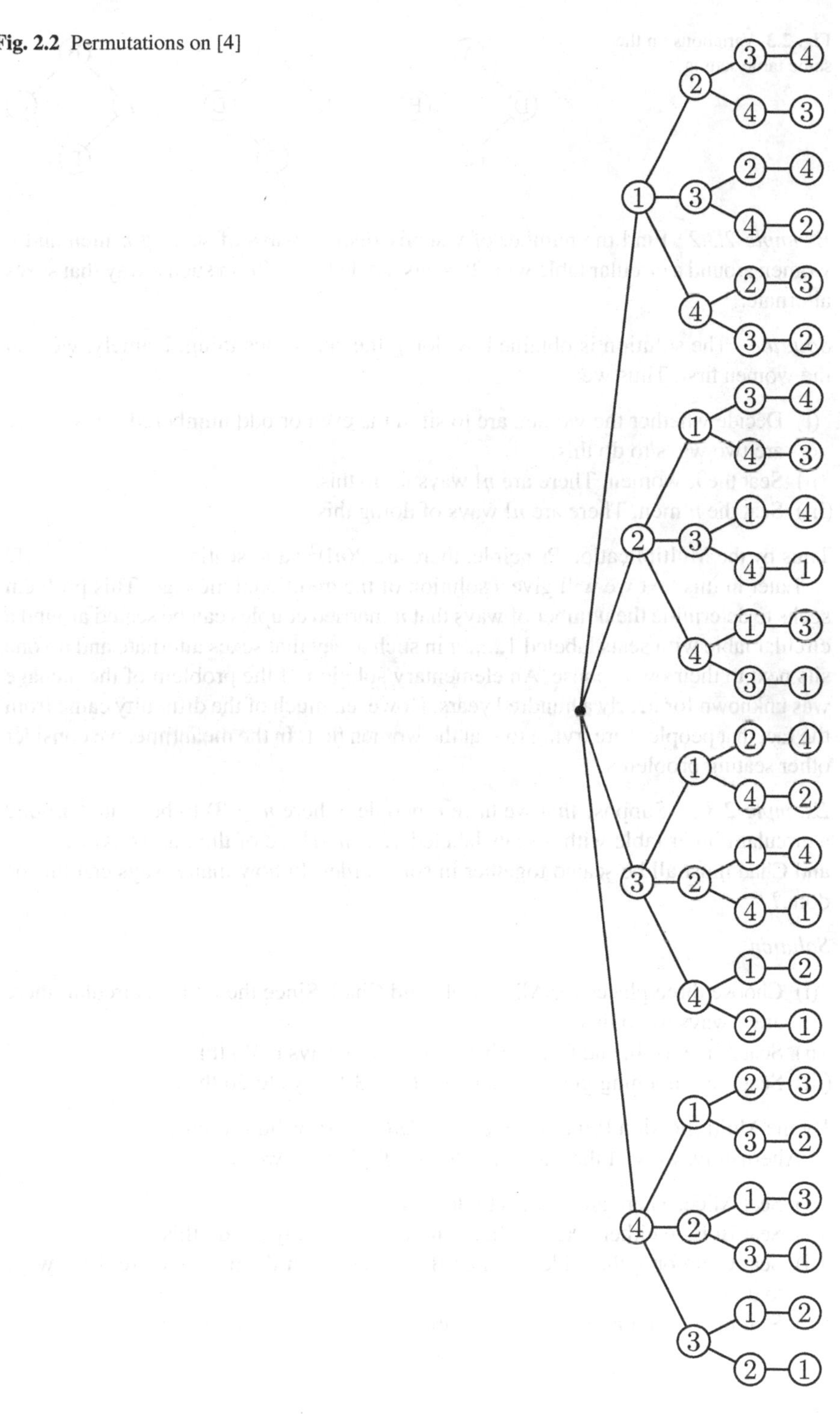

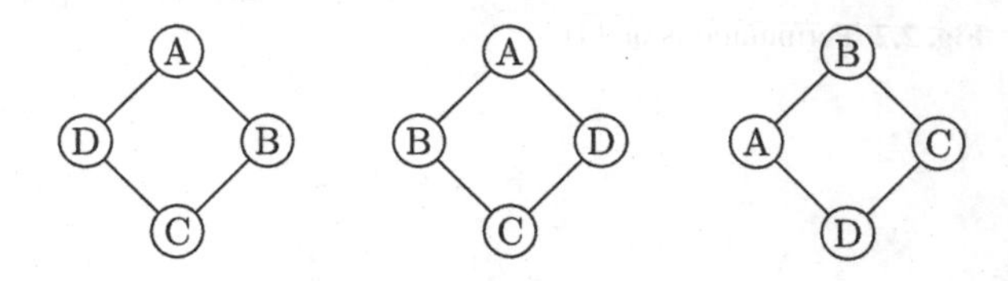

Fig. 2.3 Variations on the same table setting

Example 2.3.2 Find the number of visually distinct ways of seating n men and n women around a circular table with $2n$ seats labeled $1, \ldots, 2n$ in such a way that sexes alternate.

Solution The solution is obtained by doing the chivalrous thing. Namely, we seat the women first. Thus we:

(i) Decide whether the women are to sit in the even or odd numbered seats. There are two ways to do this.
(ii) Seat the n women. There are $n!$ ways to do this.
(iii) Seat the n men. There are $n!$ ways of doing this.

Thus by the Multiplication Principle, there are $2(n!)^2$ valid seatings. □

Later in this text we will give a solution of the traditional ménage. This problem seeks to determine the number of ways that n married couples can be seated around a circular table with seats labeled $1,\ldots,2n$ in such a way that sexes alternate and no one sits next to their own spouse. An elementary solution of the problem of the ménage was unknown for nearly a hundred years. However, much of the difficulty came from the fact that people were trying to seat the women first. In the meantime, we consider other seating problems.

Example 2.3.3 Suppose that we have n people (where $n \geq 3$) to be seated around a circular dinner table with n seats labeled $1, \ldots, n$. Three of the guests, Alice, Bob, and Chad must all be seated together in some order. In how many ways can this be done?

Solution

(i) Choose three places for Alice, Bob, and Chad. Since the table is circular, there are n ways to do this.
(ii) Seat Alice, Bob, and Chad. There are $3! = 6$ ways to do this.
(iii) Seat the remaining guests. There are $(n-3)!$ ways to do this.

By the Multiplication Principle, there are $6n(n-3)!$ valid seatings.

Alternately, we seat the trio individually. To do this, we:

(i) Seat Alice. There are n ways to do this.
(ii) Seat Bob on either side of Alice. There are two ways to do this.
(ii) Seat Chad on either side of Alice/Bob or between them. There are three ways to do this.
(iv) Seat the remaining guests. There are $(n-3)!$ ways to do this.

Again, the Multiplication Principle yields $6n(n-3)!$ valid seatings. □

Note that if Alice, Bob, and Chad refuse to be seated together, this problem becomes somewhat more difficult. The more general problem of when none of the three can sit next to each other will wait until later in the text. For now, we satisfy ourselves with a slightly easier case.

Example 2.3.4 Suppose that we have n people (where $n \geq 3$) to be seated around a circular dinner table with n seats. Bob refuses to be seated next to either Alice or Chad. Find the number of valid settings if Bob's restriction must be honored.

Solution Let U be the set of all ways to seat n guests around a circular dinner table with n seats labeled $1, \ldots, n$. Let A be the set of all settings in which Alice and Bob are seated together. Let B be the set of all settings in which Bob and Chad are seated together. We note that:

(i) The total number of settings (no restrictions) is given by $|U| = n!$.
(ii) We compute the cardinality of A by first seating Alice and Bob. Since the table is circular, there are n places to seat them. Further, there are two ways to seat them, either Alice is on the left or on the right. Finally, there $(n-2)!$ ways to seat the remaining guests. Thus by the Multiplication Principle, $|A| = 2n(n-2)!$.
(iii) By symmetry, the number of ways to seat Bob and Chad together is the same as the number of ways to seat Alice and Bob together. Thus, $|B| = 2n(n-2)!$.
(iv) Note that $A \cap B$ is the set of all settings in which Bob is between Alice and Chad. We compute $|A \cap B|$ by first seating Alice. There are n places to seat Alice. Bob can be seated on either her left or her right, so there are two possibilities. Chad is then forced to sit in the remaining seat next to Bob. Finally, there are $(n-3)!$ ways to seat the rest of the guests. Thus, by the Multiplication Principle, $|A \cap B| = 2n(n-3)!$.

By the Principle of Inclusion and Exclusion, the number of ways in which Alice and Bob are seated together or Bob and Chad are seated together is given by

$$|A \cup B| = |A| + |B| - |A \cap B|$$
$$= 2n(n-2)! + 2n(n-2)! - 2n(n-3)!$$
$$2n(n-3)!(n-2+n-2-1) = 2n(n-3)!(2n-5).$$

Hence, the number of ways in which Alice and Bob are not seated together nor Bob and Chad are seated together is given by

$$|U| - |A \cup B| = n! - 2n(n-3)!(2n-5).$$ □

Exercise 2.3.5 List all derangements of [4].

Exercise 2.3.6 Find the number of eight character passwords if each password must contain six lower case letters and two numbers.

Exercise 2.3.7 Find the number of $n \times n$ binary matrices that have one 1 in every row and column.

Exercise 2.3.8 Show that $n! \geq 2^{n-1}$ for all n.

Exercise 2.3.9 Find the number of ways of arranging the elements of $[n]$ in such a way that even and odd numbers must alternate.

Exercise 2.3.10 Suppose that n men and n women are to be seated around a rectangular dinner table with n labeled seats on each side. If no one can sit next to or across from someone of the same sex, how many visually distinct ways can this be done?

Exercise 2.3.11 Suppose that n men and n women are to be seated around a circular dinner table (with seats labeled $1, \ldots, 2n$) in such a way that sexes alternate.

(i) Suppose that two of the guests, Alice and Bob, require that they be seated together. In how many sex-alternating seatings are Alice and Bob seated together?
(ii) In how many sex alternating seatings are Alice and Bob apart?

Exercise 2.3.12 Suppose that we have n people (where $n \geq 4$) to be seated around a circular dinner table with n seats labeled $1, \ldots, n$. Four of the guests are Alice, Bob, Chad, and Diane. Find the number of valid settings if:

(i) Alice and Bob must be seated together. Chad and Diane must be seated together. Note: Alice/Bob need not be seated next to Chad/Diane.
(ii) Alice and Bob must be seated together. Chad and Diane must not be seated together. Note: Chad and/or Diane can be seated next to Alice/Bob.
(iii) Alice and Bob must not be seated together. Chad and Diane must not be seated together.

Exercise 2.3.13 Suppose that we have n people (where $n \geq k$) to be seated around a circular dinner table with n seats labeled $1, \ldots, n$. There are k guests that must be seated together in some order. In how many ways can this be done?

Exercise 2.3.14 Repeat Example 2.3.3 assuming that there are two circular tables. The first table has k seats, the second has $n - k$ seats, and $k, n - k \geq 3$.

2.4 Application: Legendre's Theorem

Suppose that we want to know how many zeros are at the end of 100! (or, in general, $n!$). To approach this problem, consider how does one get a zero as one of the final digits of a number. The only way to get a zero as a digit at the end of a number is to multiply by ten. In other words, we have to multiply by both two and five. However, there would be far more twos in the product $n!$ than there would be fives. For this reason, it suffices to count the number of times a multiple of five appears in the product $n!$.

To do this, we will need one of the earliest results from number theory. In fact, the following result appears in *Euclid's Elements*. Because this result is so well known, we present it without proof. Interested readers are referred to any elementary text on number theory for a proof. In particular, *Elementary Number Theory* by Jones and Jones [31].

Theorem 2.4.1 *Let $n \in \mathbb{Z}^+$ and $\ell \in \mathbb{Z}$. There exists a unique pair of integers q and r such that $\ell = qn + r$ and $0 \leq r < n$.*

The following corollary is an immediate corollary of Theorem 2.4.1.

Corollary 2.4.2 *Let $n, m \in \mathbb{Z}^+$. Let A be the set of positive integers less than or equal to n that are a multiple of m. In other words, $A = \{km : k \in \mathbb{Z}^+, km \leq n\}$. The cardinality of A is*

$$|A| = \left\lfloor \frac{n}{m} \right\rfloor.$$

Proof By Theorem 2.4.1, there exists a unique pair of integers q and r such that $n = qm + r$ where $0 \leq r < m$. Thus, $A = \{m, 2m, ..., qm\}$. Hence, $|A| = q$. Further,

$$\left\lfloor \frac{n}{m} \right\rfloor = \left\lfloor \frac{qm + r}{m} \right\rfloor = \left\lfloor q + \frac{r}{m} \right\rfloor.$$

Since $0 \leq r < m$, it follows that $0 \leq r/m < 1$. Ergo,

$$\left\lfloor \frac{n}{m} \right\rfloor = \left\lfloor q + \frac{r}{m} \right\rfloor = q.$$

Thus, $|A| = q = \lfloor \frac{n}{m} \rfloor$. ■

Example 2.4.3 Find the number of positive integers less than or equal to 1000 that are divisible by 2 or 3.

Solution Let A be the set of integers less than or equal to 1000 that are divisible by 2. By Corollary 2.4.2,

$$|A| = \left\lfloor \frac{1000}{2} \right\rfloor = 500.$$

Similarly, let B be the set of integers less than or equal to 1000 that are divisible by 3. Again, Corollary 2.4.2 yields

$$|B| = \left\lfloor \frac{1000}{3} \right\rfloor = 333.$$

In order to compute $|A \cup B|$ using the Principle of Inclusion and Exclusion, we would first need $|A \cap B|$. In this case, $A \cap B$ is the set of all integers less than or equal to 1000 that are divisible by both 2 and 3. In other words, this is the set of integers less than or equal to 1000 that are divisible by 6. Hence, Corollary 2.4.2 gives

$$|A \cap B| = \left\lfloor \frac{1000}{6} \right\rfloor = 166.$$

By the Principle of Inclusion and Exclusion, we have that

$$|A \cup B| = |A| + |B| - |A \cap B| = 500 + 333 - 166 = 667. \qquad \square$$

In general, suppose that A is the set of integers less than or equal to n that are divisible by a and B is the set of integers less than or equal to n that are divisible by b. To obtain $|A \cap B|$, we would compute $\left\lfloor \frac{n}{lcm(a,b)} \right\rfloor$, where $lcm(a, b)$ is the *least common multiple* of a and b.

We are now ready to proceed with Legendre's Theorem.

Theorem 2.4.4 *(Legendre's Theorem) Let p be a prime number and let $n \in \mathbb{N}$. The number of times that p appears as a factor of $n!$ is given by*

$$\sum_{k \geq 1} \left\lfloor \frac{n}{p^k} \right\rfloor.$$

Proof Note that for each multiple of p that is less than n, then the prime p appears as a factor of $n!$. However, this will not be *all* of the factors of p that appear in $n!$. For instance, any multiple of p^2 will contribute an additional factor of p to $n!$. In general, any multiple of p^k will contribute k factors of p to $n!$. However, $k-1$ of these factors will have been accounted for by the multiples of $p, p^2, \ldots, p^{k-1}$. By Corollary 2.4.2, the number of multiples of p^k that are less than or equal to n is $\left\lfloor \frac{n}{p^k} \right\rfloor$. Therefore, the number of times that p appears as a factor of $n!$ is given by

$$\sum_{k \geq 1} \left\lfloor \frac{n}{p^k} \right\rfloor. \qquad \blacksquare$$

We are now prepared to answer the question that motivated this section.

Example 2.4.5 Find the number of zeros at the end of 100!.

Solution As discussed above, it suffices to determine the number of times that 5 appears as a factor of 100. By Legendre's Theorem, this is given by

$$\sum_{k \geq 1} \left\lfloor \frac{100}{5^k} \right\rfloor$$

$$= \left\lfloor \frac{100}{5} \right\rfloor + \left\lfloor \frac{100}{25} \right\rfloor = 20 + 4 = 24. \qquad \square$$

Exercise 2.4.6 Determine the number of positive integers less than or equal to 5000 that are divisible by 3 or 5.

Exercise 2.4.7 Determine the number of positive integers less than or equal to 8500 that are neither divisible by 4 nor 7.

Exercise 2.4.8 Determine the number of positive integers less than or equal to 10000 that are divisible by 21 or 33.

Exercise 2.4.9 How many times does 7 appear as a factor of 200!?

Exercise 2.4.10 How many zeros are at the end of 333!?

Exercise 2.4.11 Give an intuitive explanation as to why Legendre's Theorem is restricted to primes.

2.5 Ordered Subsets of $[n]$

We now return to one of the problems that we introduced in Sect. 1.1. Namely, we look at the problem of choosing an ordered k-element subset of $[n]$. Such subsets are often referred to as k-tuples, because the order is important. In Sect. 1.1, we listed out all 60 ordered 3 element subsets of {A, B, C, D, E}. However, this approach will not work for arbitrary n and k. Hence we must find a way to compute this quantity. To do this, we let $P(n, k)$, read "n place k," denote the number of k-tuples of $[n]$.

Often, $P(n, k)$ is thought of as the number of ways to *sample without replacement*. Suppose that we are playing a card game such as 5-card stud poker. In this game, cards are revealed one at a time. Thus, when trying to "bluff" your opponents, you are concerned with not only what cards appear, but the order in which they appear. Therefore, order is important. Further, cards are never returned to the deck while a round is in progress. Hence, we are sampling cards out of the deck without replacing them. Since each hand can be thought of as an 5-tuple from [52], there are $P(52, 5)$ possible hands in 5-card stud poker.

We now give a formula to compute $P(n, k)$ for any $n, k \in \mathbb{N}$.

Proposition 2.5.1 *For $n, k \in \mathbb{N}$ and $n \geq k$ we have:*

$$P(n,k) = \frac{n!}{(n-k)!}.$$

Proof We instead prove that $n! = P(n, k)(n-k)!$. The number of ways of arranging the elements of n is given by $n!$. This can also be computed by:

(i) Choosing and arranging the first k elements in the permutation. There are $P(n, k)$ ways to do this by definition.
(ii) Arrange the remaining $n - k$ elements. There are $(n-k)!$ ways to do this.

Ergo, we have:

$$n! = P(n,k)(n-k)! \Rightarrow P(n,k) = \frac{n!}{(n-k)!}.$$ ■

Table 2.2 Values of $P(n,k)$ for small n and k

$n \setminus k$	0	1	2	3	4	5	6	7	8	9
0	1									
1	1	1								
2	1	2	2							
3	1	3	6	6						
4	1	4	12	24	24					
5	1	5	20	60	120	120				
6	1	6	30	120	360	720	720			
7	1	7	42	210	840	2520	5040	5040		
8	1	8	56	336	1680	6720	20160	40320	40320	
9	1	9	72	504	3024	15120	60480	181440	362880	362880

The values of $P(n,k)$ for small n and k are given in Table 2.2.

One of the most common interpretations of $P(n,k)$ is selecting k officers (of different rank) from a club of n members. Since each officer is of a different rank, it certainly makes a difference who is selected as president, vice-president, etcetera.

Example 2.5.2 The Science club has ten members and must select three officers to go to a retreat. The Spanish club has 20 members and must select 5 officers to go to the same retreat. If no one is in both clubs, how many different combinations of officers can be at the retreat?

Solution Note there are $P(10,3) = 720$ ways for the Science club to select its officers. Similarly, there are $P(20,5) = 1860480$ ways for the Spanish club to select their officers. Hence by the Multiplication Principle, the number of combinations of officers is given by:

$$P(10,3)P(20,5) = 720(1860480) = 1339545600. \quad \blacksquare$$

Example 2.5.3 The Mathletes Club has 15 members and is selecting five officers (president, vice-president, secretary, treasurer, and sergeant-at-arms). One of the members, Alice, is quite stubborn. If she is not elected president, then she refuses to serve as another officer. How many ways can the club select its officers if Alice's demand must be met?

Solution The set of all ways that the club can elect its officers can be partitioned into two disjoint, exhaustive sets:

(i) The set of elections in which Alice is the president. This leaves four positions to select from 14 members. There are $P(14,4) = 24024$ elements in this set.

(ii) The set of elections in which Alice is not elected to any position. This leaves five positions to select from 14 members. There are $P(14,5) = 240240$ elements in this set.

By the Addition Principle, the number of ways that the club can elect officers is given by:

$$P(14,4) + P(14,5) = 24024 + 240240 = 264264. \qquad \square$$

A useful recurrence for determining the values of $P(n,k)$ is given in the following theorem.

Theorem 2.5.4 *For all $n, k \in \mathbb{N}$, we have:*

$$P(n,k) = P(n-1,k) + kP(n-1,k-1).$$

Proof We note that $P(n,k)$ counts the number of ordered k-sets from $[n]$ by definition. The right side of the equation also counts this by counting two disjoint, exhaustive sets:

(i) The element n is not in the ordered k-set. Thus we must select an ordered k-set from the remaining $n-1$ elements. There are $P(n-1,k)$ ways of doing this by definition.
(ii) The element n is in the k-set. There are k choices for the position of n. Selecting the remaining elements is equivalent to selecting an ordered $(k-1)$-set from the set $[n-1]$. There are $P(n-1,k-1)$ ways of doing this by definition. Hence, by the Multiplication Principle, there are $kP(n-1,k-1)$ elements in this set.

Therefore, the identity holds by the Addition Principle. ∎

Exercise 2.5.5 Confirm the entries in Table 2.2.

Exercise 2.5.6 In chess, a rook is *attacking* if it is in the same row or column as another piece. Find the number of ways to place k non-attacking rooks on a $n \times k$ chessboard if $n \geq k$.

Exercise 2.5.7 Evaluate each of the following:

(i) $P(16,7)$;
(ii) $P(12,3)$;
(iii) $P(15,4)$;
(iv) $P(15,11)$.

Exercise 2.5.8 Each club in the college must select a representative and an alternate to speak for them at the student council. How many possible council meetings are possible if there are five clubs, the ith club has n_i members, and no one is in two clubs?

Exercise 2.5.9 The Science Club has 20 members and needs to elect 4 officers (president, vice-president, secretary, and treasurer). Two of the members, Alice and Bob, require that at least one of the following is satisfied:

(i) Alice is elected president;
(ii) Bob is elected vice-president;

(iii) Alice and Bob are elected secretary and treasurer (in some order);
(iv) Neither Alice nor Bob have a position.

How many elections are possible if their demands must be satisfied?

Exercise 2.5.10 Give a combinatorial proof that $P(n, n) = n!$.

2.6 Application: Possible Games of Tic-tac-toe

Suppose that we wish to count the possible games of Tic-tac-toe. Trivially, this will be bounded above by 9!. In this section, we will obtain the exact number of possible games of Tic-tac-toe.

For this analysis we will assume that the symmetries on the 3×3 board result in distinct games. We will base our solution on that of Henry Bottomley [11]. To win a game of Tic-tac-toe, the first player must get three X's in a row (either vertical, horizontal, or diagonal). Similarly, the second player must get three O's in a row. For this reason, the game cannot end until the fifth turn. We will count the number of possible games by counting the sets A_5, A_6, A_7, A_8, and A_9, where A_i is the set of all games that end on turn i. Note that these sets are mutually disjoint. For this reason, $| \cup_{i=5}^{9} A_i|$ counts the number games of Tic-tac-toe.

Proposition 2.6.1 *The number of Tic-tac-toe games that end on the fifth turn is* $|A_5| = 1440$.

Proof For a game to end on the fifth turn, we must place three X's on a row and two O's elsewhere on the grid. Further note that there are three vertical rows, three horizontal rows, and two diagonal rows. Thus:

(i) There are 8 choices for which row to use;
(ii) There are 3! choices for the order in which to place the X's.
(iii) Player two must then choose two of the remaining six squares.

By the Multiplication Principle, the number of games in this set is given by $|A_5| = 8 * 3! * P(6, 2) = 1440$ games in this set. ■

The calculations for $|A_6|$, $|A_7|$, and $|A_8|$ will be slightly more complicated. However, all three calculations will be very similar.

Proposition 2.6.2 *The number of Tic-tac-toe games that end on the sixth turn is* $|A_6| = 5328$.

Proof For the game to end on the sixth turn, there are three O's and three X's on the grid such that the three O's are on a row. To count the number of such arrangements, we:

(i) Choose one of the 8 rows for player two to place their O's.
(ii) Order the three O's. There are 3! possibilities.

(iii) Choose three of the remaining six squares to place their X's. Thus, there are $P(6,3)$ choices for player one.

Hence, by the Multiplication Principle, there are $8 * 3! * P(6,3) = 5760$ ways to place three O's in a row and three X's on the grid. However, this also counts those arrangements in which there are three X's in a row, which would have resulted in a player one win on the fifth turn. If the three O's are on a diagonal, then there cannot be three X's in a row. If there are three O's in one of the six horizontal or vertical row, then there are two rows for the three X's. Thus if we want three O's in a row and three X's in a row, then:

(i) Choose rows for the O's. There are 6 possibilities;
(ii) Order the O's. There are 3! ways to order the O's;
(iii) Choose a row for the X's. There are 2 choices;
(iv) Order the X's in 3! possible ways.

So there are $6 * 3! * 2 * 3! = 432$ ways to place three O's in a row and three X's in a row by the Multiplication Principle.

Hence, $|A_6| = 5760 - 432 = 5328$ possible games in this set by the Subtraction Principle. ■

Note that in the first step of the previous proposition, we count the number of ways to arrange three X's and three O's on a 3×3 grid such that the three O's form a row. In the second step, we count the number of ways to arrange three X's and three O's on a 3×3 grid such that the X's form a row and the O's form a row. This will be the strategy for the remaining cases.

Proposition 2.6.3 *The number of Tic-tac-toe games that end on the seventh turn is* $|A_7| = 47952$.

Proof We first count the number of arrangements of four X's and three O's on the grid in such a way that three X's in a row and the first three X's do not complete a row. To do this:

(i) Choose a row for the three X's. There are 8 ways to do this.
(ii) Choose an additional square for the fourth X. There are 6 ways to do this.
(iii) Choose one square on the row to be the square that player one uses to complete the row. There are 3 ways to do this.
(iv) Place three X's on the other two squares of the row and the additional square in some order. There are 3! ways to do this.
(v) Place the three O's on the remaining five squares in some order. There are $P(5,3)$ ways to do this.

Thus, by the Multiplication Principle, there are $8 * 6 * 3 * 3! * P(5,3) = 51840$ possibilities. However, some of these possibilities allow for player two to place three O's in a row. For this reason, we must count the number of arrangements of four X's and three O's on the grid in such a way that three X's in a row, three O's in a row, and the first three X's do not complete a row. To do this:

(i) Choose a row for the three X's. Since these cannot be on a diagonal, there are 6 possibilities.
(ii) Choose an additional square for the fourth X. There are 6 ways to do this. Note that this removes one possible row to place the three O's.
(iii) Choose one square on the row to be the square that player one uses to complete the row. There are 3 ways to do this.
(iv) Place three X's on the other two squares of the row and the additional square in some order. There are 3! ways to do this.
(v) Place the three O's on the remaining row in some order. There are 3! possibilities.

Thus there are $6 * 6 * 3 * 3! * 3! = 3888$ possible arrangements. Ergo, by the Subtraction Principle, $|A_7| = 51840 - 3888 = 47952$. □

Proposition 2.6.4 *The number of games of Tic-tac-toe that end on the eighth move is* $|A_8| = 72576$.

Proof Again, we begin by determining the number of ways to arrange four O's and four X's on the grid in such a way that the first three O's do not complete a row. To do this:

(i) Choose a row for the three O's. There are 8 ways to do this.
(ii) Choose an additional square for the fourth O. There are 6 ways to do this.
(iii) Choose one square on the row to be the square that player two uses to complete the row. There are 3 ways to do this.
(iv) Place three O's on the other two squares of the row and the additional square in some order. There are 3! ways to do this.
(v) Place the four X's on the remaining five squares in some order. There are $P(5, 4)$ ways to do this.

It follows from the Multiplication Principle that there are $8 * 6 * 3 * 3! * P(5, 4) = 103680$ such arrangements.

We now determine the number of arrangements of four O's and four X's such that three O's are in a row, three X's are in a row, and the first three O's do not complete a row. To do this:

(i) Choose a row for the three O's. Since this cannot be a diagonal row, there are 6 ways to do this.
(ii) Choose an additional square for the fourth O. There are 6 ways to do this. Note that this leaves only one row for the X's.
(iii) Choose one square on the row to be the square that player two uses to complete the row. There are 3 ways to do this.
(iv) Place three O's on the other two squares of the row and the additional square in some order. There are 3! ways to do this.
(v) The row chosen in (ii) has two squares where one can place an X. Choose one of the two possible squares.
(vi) Place the four X's in any order. There are 4! possibilities.

The Multiplication Principle yields $6 * 6 * 3 * 3! * 2 * 4! = 31104$ possible arrangements. Thus by the Subtraction Principle $|A_8| = 103680 - 31104 = 72576$. ■

Finally, we will consider the possibility of a game lasting the full nine moves. This will require a bit more calculation that the previous cases. We will leave several of the smaller calculations as exercises to the reader.

Proposition 2.6.5 *The number of Tic-tac-toe games lasting nine moves is given by* $|A_9| = 127872$.

Proof We determine the cardinality of A_9 by partitioning it into four disjoint, exhaustive sets. Three of these sets will involve the possibility of a win. These sets will be denoted W_1, W_2, and W_3. The fourth set, D, will be the set of all games ending in a draw. Note that for any of the W_i, we need to ensure that there are no three O's in a row before the fifth X is placed and that the fifth X is the only one that completes a row. If this condition is not met, then the game will end sooner than the ninth move.

Let W_1 denote the set of Tic-tac-toe games in A_9 in which the win is achieved using only a single diagonal. To compute this, we

(i) Choose one of the diagonals. There are two possibilities.
(ii) Choose a square on this diagonal to place the fifth X. There are 3 possibilities.
(iii) There are $P(6, 2)/2 = 15$ ways to choose two squares from the remaining six. However, only 8 of these will satisfy the conditions above.
(iv) Place first four X's in some order. There are 4! possibilities.
(v) Place the four O's in some order. There are 4! possibilities.

So $|W_1| = 2 * 3 * 8 * 4! * 4! = 27648$. Let W_2 denote the set of Tic-tac-toe games in A_9 in which the win is achieved using only one vertical or horizontal row. This is computed by:

(i) Choose one of the rows. There are 6 possibilities.
(ii) Choose a square on this row to place the fifth X. There are 3 possibilities.
(iii) Of the 15 possible pairs of squares, only 4 of these will satisfy the conditions above.
(iv) Place first four X's in some order. There are 4! possibilities.
(v) Place the four O's in some order. There are 4! possibilities.

So $|W_2| = 6 * 3 * 4 * 4! * 4! = 41472$.

Let W_3 denote the subset of A_9 in which the final X completes two distinct rows at their intersection. This can be computed by choosing one of the 22 possible intersecting pairs of rows. The fifth X must be at this intersection. Further there is no possible way that player two can complete a row. Using a similar calculation to the above, $|W_3| = 22 * 4! * 4! = 12672$.

Finally, we will consider the number of draws. Here there are 16 arrangements of five X's and four O's in which there are no three in a row. So $|D| = 16 * 5! * 4! = 46080$.

Thus,

$$|A_9| = |W_1| + |W_2| + |W_3| + |D|$$
$$= 27648 + 41472 + 12672 + 46080 = 127872. \quad \blacksquare$$

From the above analysis, the possible games of Tic-tac-toe (including symmetries) is

$$|\cup_{i=5}^{9} A_i| = |A_5| + |A_6| + |A_7| + |A_8| + |A_9|$$
$$= 1440 + 5328 + 47952 + 72576 + 127872 = 255168.$$

Steve Schaefer [37] did a similar analysis of the game in 2002 that accounted for symmetries in the board. In his analysis, he concluded that there are 26830 possible games. However, he notes that there are games that are over before a row is completed. These games include those in which a draw is forced or in which one player is guaranteed to win because they have executed a "fork" (in other words, they can complete two different rows). If we consider a game to be over when its outcome is determined, then there are only 23129 possible games. He also notes that if players follow basic strategy (in other words, they try to win or block when possible), then the number of possible games would decrease.

For more information on games, the reader is referred to *Combinatorial Games: Tic-tac-toe Theory* by Beck [6], *Games and Mathematics: Subtle Connections* by Wells [44], or *Mathematical Recreations and Essays* by Ball and Coxeter [4].

Exercise 2.6.6 In Proposition 2.6.5, player one achieves a win by completing a diagonal. Confirm that there are eight possible pairs of squares in this case that (i) do not result in player one completing a second row and (ii) do not allow player two to complete a row.

Exercise 2.6.7 In Proposition 2.6.5, player one achieves a win by completing a single vertical or horizontal row. Confirm that there are four possible pairs of squares in this case that (i) do not result in player one completing a second row and (ii) do not allow player two to complete a row.

Exercise 2.6.8 In Proposition 2.6.5, player one achieves a win by completing two intersecting rows simultaneously. Confirm that there are 22 pairs of intersecting rows.

Exercise 2.6.9 Give an alternate proof of Proposition 2.6.5 by using the Subtraction Principle and the results of the previous propositions in this section.

Exercise 2.6.10 Consider a game of 4×4 Tic-tac-toe in which a player must get four in a row to win. Determine the number of games that end on turn 7.

Exercise 2.6.11 Consider a game of 4×4 Tic-tac-toe in which a player must get four in a row to win. Determine the number of games that end on turn 8.

Exercise 2.6.12 Consider a game of 4×4 Tic-tac-toe in which a player must get four in a row to win. Determine the number of games that end on turn 9.

2.7 Stirling Numbers of the First Kind

In Sect. 2.3, we made reference to the fact that there is an alternate, and generally preferred, single line notation used to denote permutations. In this section we delve into this in more detail. Let S_n denote the set of all permutations on $[n]$. Note that the composition of two elements of S_n is likewise an element of S_n (see Exercise 1.4.8).

As an example, consider the permutation

$$\pi = \begin{pmatrix} 1 & 2 & 3 & 4 & 5 & 6 & 7 & 8 & 9 \\ 5 & 1 & 7 & 9 & 3 & 4 & 2 & 8 & 6 \end{pmatrix}.$$

In the permutation π, 1 maps to 5. Thus, $\pi(1) = 5$. Similarly, 5 maps to 3, in other words, $\pi(5) = 3$. Using composition of functions, we can write $\pi(\pi(1)) = 3$, or more compactly, $\pi^2(1) = 3$. In generally, we let $\pi(\pi^{k-1}(x)) = \pi^k(x)$ for all $k \in \mathbb{Z}$. We say that y is in the *orbit* of x if there exists $k \in \mathbb{Z}$ such that $\pi^k(x) = y$.

Proposition 2.7.1 *For each $x \in [n]$ and $\pi \in S_n$, there exist $k \in \mathbb{Z}^+$ such that $\pi^k(x) = x$.*

Proof Consider the set $S = \{\pi^i(x) : i \in \mathbb{Z}^+\}$. If $x \in S$, then we are done. If not, then we can think of the elements of $\mathbb{Z}^+$ as pigeons and the elements of $[n] - \{x\}$ as pigeonholes. By the Pigeonhole Principle, there are at least two elements of $\mathbb{Z}^+$ that are mapped to the same element of $[n] - \{x\}$. In other words, there exists $i, j \in \mathbb{Z}^+$ such that $\pi^i(x) = \pi^j(x)$. Without loss of generality, assume that $j > i$. Since π is a permutation on $[n]$, it follows that π^i is a permutation on $[n]$ by Exercise 1.4.8. Thus, we take the inverse of π^i on both sides. Hence $\pi^{j-i}(x) = x$. Ergo, $j - i$ is the required positive integer. ■

Using the above proposition, we can write the orbit of x under π as a cycle $(x, \pi(x), ..., \pi^{k-1}(x))$. For example, in the specific permutation above, the orbit of 1 is $(1, 5, 3, 7, 2)$. We can continue this process and compute the remaining orbits of π. So the orbit of 4 is $(4, 9, 6)$ and the orbit of 8 is (8). Notice that no two cycles contain the same element. This follows from the fact that π is a bijection. If two cycles contain none of the same elements, then we say that the cycles are *disjoint*. Hence we can write π as a *product* of its orbits or cycles:

$$\pi = \begin{pmatrix} 1 & 2 & 3 & 4 & 5 & 6 & 7 & 8 & 9 \\ 5 & 1 & 7 & 9 & 3 & 4 & 2 & 8 & 6 \end{pmatrix} = (1, 5, 3, 7, 2)(4, 9, 6)(8).$$

Often times, mathematicians will omit the fixed points when writing a permutation as a product of disjoint cycles. For instance, the above permutation could be written as $\pi = (1, 5, 3, 7, 2)(4, 9, 6)$. This is appropriate if it is clear what set is being permuted. For our purposes, we will always list the fixed points in a permutation unless otherwise noted.

Above, we wrote π as a product of disjoint cycles. In fact, the ability to write a permutation as a product of disjoint cycles holds for any permutation. Thus the next theorem follows immediately from the above comments.

Theorem 2.7.2 *Every permutation on* $[n]$ *can be written as a product of disjoint cycles.*

There are two useful parameters that often appear in discussions of permutations and their cycle decompositions. The first is the number of disjoint cycles of the permutation π. This is called the *cycle index* of π. The cycle index of π is denoted $cyc(\pi)$. In the example above, $cyc(\pi) = 3$. The second, more descriptive, parameter is the *cycle type* of the permutation. This parameter not only considers the number of cycles but also the length of each cycle. For example, the cycle type of π is $[5, 3, 1]$ in the example above.

Example 2.7.3 Consider the permutation

$$\sigma = \begin{pmatrix} 1 & 2 & 3 & 4 & 5 & 6 & 7 & 8 & 9 \\ 4 & 5 & 9 & 7 & 3 & 6 & 1 & 2 & 8 \end{pmatrix}.$$

Write σ as a product of disjoint cycles. Find the cycle index and cycle type for σ.

Solution Begin by computing the orbit of each element. The orbit of 1 is $(1, 4, 7)$, the orbit of 2 is $(2, 5, 3, 9, 8)$, and the orbit of 6 is (6). Thus $\sigma = (1, 4, 7)(2, 5, 3, 9, 8)(6)$. Hence $cyc(\sigma) = 3$ and the cycle type is $[5, 3, 1]$. □

A natural combinatorial problem is to determine the number of permutations in S_n that are of a given cycle type.

Example 2.7.4 Find the number of permutations in S_{15} that have $[5, 3, 3, 2, 2]$ as their cycle type.

Solution This can be computed as follows:

(i) Select five elements from $[15]$ to place in order on the 5-cycle. There are $P(15, 5)$ ways to select and order these elements.
(ii) Since rotations in a cycle are the same, we divide by the number of rotations, namely 5.
(iii) Select three elements from the remaining ten for the first 3-cycle. There are $P(10, 3)/3$ ways to select and order these elements in a cycle.
(iv) Select three elements from the remaining seven for the second three cycle. There are $P(7, 3)/3$ ways to select and order these elements in a cycle.
(v) Since the two 3-cycles can be written in any order, we divide by the number of ways to order the two 3-cycles, namely 2.
(vi) Select two elements from the remaining four for the first 2-cycle. There are $P(4, 2)/2$ ways to select and order these elements in a cycle.
(vii) Select two elements from the remaining two for the second 2-cycle. There are $P(2, 2)/2$ ways to select and order these elements in a cycle.
(viii) Since the two 2-cycles can be written in any order, we divide by 2.

So, by the Multiplication Principle, the number of permutations in S_{15} that have $[5, 3, 3, 2, 2]$ as their cycle type is

$$\left(\frac{P(15,5)}{5}\right)\left(\frac{P(10,3)}{3}\right)\left(\frac{P(7,3)}{3}\right)\left(\frac{1}{2}\right)\left(\frac{P(4,2)}{2}\right)\left(\frac{P(2,2)}{2}\right)\left(\frac{1}{2}\right)$$

$$= 1816214400.$$

□

We can apply this idea to the notion of table settings as well. In these examples, we will assume that the seats are unlabeled. Thus, rotations of a table are equivalent. Further, if there are two tables of the same size, then we do not care which of the two tables we are seated at. We only care about who we are seated with and our relative position to them at the table. However, it *does* matter whether we seat someone on the left or the right of someone else.

Example 2.7.5 Suppose that 15 people are to attend a dinner party. The guests are to be seated at one of three circular tables each with unlabeled seats. The first table has seven seats, the second has five seats, and the third has three seats. Two of the guests, Alice and Bob, need to be seated at the same table, but not necessarily next to each other. How many valid settings are there?

Solution We count the number of table settings by considering three disjoint, exhaustive sets. The first set, A_7, will be the set of all table settings in which Alice and Bob are at the first table. The second set, A_5, will be the set of all table settings in which Alice and Bob are at the second table. The third set, A_3, will be the set of all table settings in which Alice and Bob are at the third table. We will now count the elements in each of these sets in turn.

To count A_7:

(i) Seat Alice at the first table. Note that this breaks the cycle.
(ii) Seat Bob at one of the remaining six places at this table.
(iii) Choose and order five of the remaining 13 guests to sit at the first table. There are $P(13,5)$ possibilities.
(iv) Place five of the remaining eight guests around the second table. There are $P(8,5)/5$ possibilities.
(v) Place three of the remaining three guests around the final table. There are $P(3,3)/3$ possibilities.

So, by the Multiplication Principle,

$$|A_7| = 6P(13,5)\left(\frac{P(8,5)}{5}\right)\left(\frac{P(3,3)}{3}\right) = 2490808320.$$

The remaining two sets will be counted in a similar fashion. So to count A_5:

(i) Seat Alice at the second table. Note that this breaks the cycle.
(ii) Seat Bob at one of the remaining four places at this table.
(iii) Choose and order three of the remaining 13 guests to sit at the second table. There are $P(13,2)$ possibilities.

(iv) Place seven of the remaining ten guests around the first table. There are $P(10,7)/7$ possibilities.
(v) Place three of the remaining three guests around the final table. There are $P(3,3)/3$ possibilities.

So, by the Multiplication Principle,

$$|A_5| = 4P(13,2)\left(\frac{P(10,7)}{7}\right)\left(\frac{P(3,3)}{3}\right) = 107827200.$$

Finally, to count A_3:

(i) Seat Alice at the third table. Note that this breaks the cycle.
(ii) Seat Bob at one of the two remaining places at this table.
(iii) Choose and order one of the remaining 13 guests to sit at the third table. There are $P(13,1)$ possibilities.
(iv) Place seven of the remaining 12 guests around the first table. There are $P(12,7)/7$ possibilities.
(v) Place five of the remaining five guests around the final table. There are $P(5,5)/5$ possibilities.

So, by the Multiplication Principle,

$$|A_3| = 2P(13,1)\left(\frac{P(12,7)}{7}\right)\left(\frac{P(5,5)}{5}\right) = 355829760.$$

By the Addition Principle, the number of valid settings is:

$$|A_7 \cup A_5 \cup A_3| = |A_7| + |A_5| + |A_3|$$

$$= 2490808320 + 107827200 + 355829760 = 2954465280. \qquad \square$$

Our next example will be more involved.

Example 2.7.6 Suppose that 15 people are to attend a dinner party. The guests are to be seated at one of three circular tables each with unlabeled seats. The first table has seven seats, the second has five seats, and the third has three seats. Bob refuses to be the same table as either Alice or Chad. How many valid settings are there?

Solution Let U be the set of all settings. The cardinality of this set can be computed as follows:

(i) Choose and order seven guests to sit around the large table. There are $P(15,7)/7$ ways to do this.
(ii) Choose and order five of the remaining eight guests to sit at the middle table. There are $P(8,5)/5$ ways to do this.
(iii) Choose and order the final three guests to sit around the last table. There are $P(3,3)/3$ ways to do this.

So by the Multiplication Principle,

$$|U| = \left(\frac{P(15,7)}{7}\right)\left(\frac{P(8,5)}{5}\right)\left(\frac{P(3,3)}{3}\right) = 12454041600.$$

Let A be the set of settings in which Alice and Bob are at the same table. By the previous example, we know that $|A| = 2954465280$. Similarly, let B be the set of settings in which Bob and Chad are at the same table. Analogously, $|B| = 2954465280$.

We need only compute the cardinality of $A \cap B$, that is, the set of all settings in which all three are seated at the same table. This can be done by counting three disjoint, exhaustive sets. These sets will be A_i, where $i \in \{3, 5, 7\}$. Here, A_i denotes the set of all settings in which all three are at the table with i seats.

To compute $|A_7|$:

(i) Seat Alice at the large table. This breaks the cycle.
(ii) Choose one of the six remaining seats for Bob.
(iii) Choose one of the five remaining seats for Chad.
(iv) Choose and order four of the remaining 12 guests to sit at this table. There are $P(12, 4)$ ways to do this.
(v) Choose and order five of the remaining eight guests to sit at the middle table. There are $P(8, 5)/5$ ways to do this.
(vi) Choose and order three of the remaining three guests to sit at the last table. There are $P(3, 3)/3$ ways to do this.

So by the Multiplication Principle,

$$|A_7| = 30P(12,4)\left(\frac{P(8,5)}{5}\right)\left(\frac{P(3,3)}{3}\right) = 958003200.$$

The computation for $|A_5|$ is similar:

(i) Seat Alice at the middle table. This breaks the cycle.
(ii) Choose one of the four remaining seats for Bob.
(iii) Choose one of the three remaining seats for Chad.
(iv) Choose and order two of the remaining twelve guests to sit at this table. There are $P(12, 2)$ ways to do this.
(v) Choose and order seven of the remaining ten guests to sit at the large table. There are $P(10, 7)/7$ ways to do this.
(vi) Choose and order three of the remaining three guests to sit at the last table. There are $P(3, 3)/3$ ways to do this.

So by the Multiplication Principle,

$$|A_5| = 12P(12,2)\left(\frac{P(10,7)}{7}\right)\left(\frac{P(3,3)}{3}\right) = 273715200.$$

Finally, we compute $|A_3|$:

(i) Seat Alice at the small table. This breaks the cycle.
(ii) Choose one of the two remaining seats for Bob.
(iii) Place Chad at the last place of this table.
(v) Choose and order seven of the remaining 12 guests to sit at the large table. There are $P(12,7)/7$ ways to do this.
(vi) Choose and order five of the remaining five guests to sit at the middle table. There are $P(5,5)/5$ ways to do this.

So by the Multiplication Principle,

$$|A_3| = 2\left(\frac{P(12,7)}{7}\right)\left(\frac{P(5,5)}{5}\right) = 27371520.$$

Thus,

$$|A \cap B| = |A_7| + |A_5| + |A_3|$$
$$= 958003200 + 273715200 + 27371520 = 1259089920$$

by the Addition Principle.

So by the Principle of Inclusion and Exclusion, the number of settings in which Bob is at the same table as either Alice or Chad is given by

$$|A \cup B| = |A| + |B| - |A \cap B|$$
$$= 2954465280 + 2954465280 - 1259089920 = 4649840640.$$

Hence, the number of settings in which Bob is not at the same table as Alice nor Chad is given by

$$|U| - |A \cup B| = 12454041600 - 4649840640 = 7804200960.$$ □

We now return our attention to permutations. Suppose that we want to know the number of permutations in S_n that have cycle index k. Let this be denoted $s(n,k)$. These numbers are known as *Stirling numbers of the first kind*.

Definition 2.7.7 The *Stirling numbers of the first kind*, denoted $s(n,k)$, count the number of permutations on $[n]$ in which the cycle index is k. Equivalently, this is the number of ways to seat n individuals around k circular unlabeled tables, where each table must seat at least one person.

As usual, $s(0,0) = 1$ because there is one empty seating. We will begin with additional elementary properties of these numbers.

Proposition 2.7.8 *If $k > n$, then $s(n,k) = 0$.*

Proof Here, there are more individuals than seats to sit them. Hence this is impossible by the Pigeonhole Principle. Thus, $s(n,k) = 0$. ■

Proposition 2.7.9 *For all $n \in \mathbb{N}$, $s(n,n) = 1$.*

Proof Since there are as many people as tables, each person must be at their own table. Since the tables are indistinguishable, there is one way to do this.

Table 2.3 Values of $s(n, k)$ for small n and k

$n \setminus k$	0	1	2	3	4	5	6	7	8	9
0	1									
1	0	1								
2	0	1	1							
3	0	2	3	1						
4	0	6	11	6	1					
5	0	24	50	35	10	1				
6	0	120	274	225	85	15	1			
7	0	720	1764	1624	735	175	21	1		
8	0	5040	13068	13132	6769	1960	322	28	1	
9	0	40320	109584	118124	67284	22449	4536	546	36	1

Proposition 2.7.10 *For all* $n \in \mathbb{Z}^+$, $s(n, 1) = (n-1)!$.

Proof Seat one person at the table. This breaks the cycle. Then seat the remaining $n-1$ people in the remaining $n-1$ seats. There are $(n-1)!$ ways to do this. ■

Finally, we give a recurrence for additional values of $s(n, k)$.

Theorem 2.7.11 *For all* $k, n \in \mathbb{N}$ *such that* $1 \le k < n$,

$$s(n+1, k) = ns(n, k) + s(n, k-1).$$

Proof By definition, $s(n+1, k)$ counts the number of permutations in S_{n+1} with cycle index k. The right side also counts this by counting two disjoint, exhaustive sets:

(i) The set of all permutations in S_{n+1} such that 1 is a fixed point. The remaining n elements must be placed into $k-1$ cycles. There are $s(n, k-1)$ ways to do this by definition.

(ii) The set of all permutations in S_{n+1} such that 1 is not a fixed point. Place the remaining n elements into k cycles. There are $s(n, k)$ ways to do this by definition. Place 1 within the cycle decomposition. Since 1 can be placed anywhere except the last position (which is equivalent to placing it before the first element of the last cycle), there are n places to insert 1 into the cycle decomposition. Thus there are $ns(n, k)$ elements in this set by the Multiplication Principle.

■

The above recursion can be used to generate the entries in Table 2.3. We now present an example of the Stirling numbers to a table setting problem.

Example 2.7.12 Suppose that n guests ($n \ge 2$) are to be seated around k circular dinner tables with unlabeled seats. Each table must sit at least one person. Further, two of the guests, Alice and Bob, must be seated at the same table, though not necessarily next to each other. Find the number of ways to seat the tables under this restriction.

Solution We count the set of valid table settings by counting $n-1$ disjoint, exhaustive sets of settings. The ℓth set, A_ℓ ($\ell = 2, ..., n$) is the set of all settings in which Alice and Bob are seated at a table with ℓ seats. To determine $|A_\ell|$:

(i) Seat Alice. This breaks the cycle.
(ii) Seat Bob at one of the remaining $\ell - 1$ seats at Alice's table.
(iii) Choose and order $\ell - 2$ guests from the remaining $n - 2$ guests to sit at Alice's table. There are $P(n-2, \ell-2)$ ways to do this.
(iv) Place the remaining $n - \ell$ guests around the remaining $k - 1$ tables. There are $s(n-\ell, k-1)$ ways to do this.

So, by the Multiplication Principle,

$$|A_\ell| = (\ell - 1)P(n-2, \ell-2)s(n-\ell, k-1).$$

By the Addition Principle, the number of valid settings is given by

$$\left|\cup_{\ell=2}^{n} A_\ell\right| = \sum_{\ell=2}^{n} |A_\ell| = \sum_{\ell=2}^{n} (\ell-1)P(n-2, \ell-2)s(n-\ell, k-1). \qquad \square$$

Exercise 2.7.13 Consider the permutation

$$\tau = \begin{pmatrix} 1 & 2 & 3 & 4 & 5 & 6 & 7 & 8 & 9 \\ 5 & 1 & 7 & 9 & 3 & 4 & 2 & 8 & 6 \end{pmatrix}.$$

Write τ as a product of disjoint cycles. Find the cycle index and cycle type for τ.

Exercise 2.7.14 Find the number of permutations in S_{20} that have cycle type $[7, 3, 3, 3, 2, 1, 1]$.

Exercise 2.7.15 Find the number of permutations in S_{25} that have cycle type $[8, 8, 6, 1, 1, 1]$.

Exercise 2.7.16 Suppose that 18 people are to attend a dinner party. The guests are to be seated at one of three circular tables each with unlabeled seats. The first table has eight seats, the second has six seats, and the third has four seats. Two of the guests, Alice and Bob, need to be seated at the same table but not necessarily next to each other. How many settings are there?

Exercise 2.7.17 Suppose that 20 people are to attend a dinner party. The guests are to be seated at one of four circular tables each with unlabeled seats. The first table has eight seats, the second has five seats, the third has five seats, and the fourth has two seats. Two of the guests, Alice and Bob, need to be seated at the same table but not necessarily next to each other. How many settings are there?

Exercise 2.7.18 Suppose that 20 people are to attend a dinner party. The guests are to be seated at one of four circular tables each with unlabeled seats. The first table has seven seats, the second has five seats, the third has five seats, the fourth has three seats. Two of the guests, Alice and Bob, cannot be seated at the same table. How many settings are there?

Exercise 2.7.19 Confirm the entries in Table 2.3.

Exercise 2.7.20 Give a combinatorial proof that $\sum_{k=0}^{n} s(n,k) = n!$.

Exercise 2.7.21 Suppose that n guests ($n \geq 3$) are to be seated around k circular dinner tables with unlabeled seats. Each table must sit at least one person. Further, three of the guests must be seated at the same table, though not necessarily next to each other. Find the number of ways to seat the tables.

Exercise 2.7.22 Suppose that n guests ($n \geq 2$) are to be seated around k circular dinner tables with unlabeled seats. Each table must sit at least one person. Further, two of the guests cannot be seated together, though they may be at the same table. Find the number of ways to seat the tables.

Exercise 2.7.23 Suppose that n guests ($n \geq 3$) are to be seated around k circular dinner tables with unlabeled seats. Each table must sit at least one person. Further, Bob refuses to be at the same table as either Alice or Chad. Find the number of ways to seat the tables.

Chapter 3
The Binomial Coefficient

We now turn our attention to one of the most fundamental and useful notions in all of combinatorics, *the binomial coefficient*. You may recall the binomial coefficient from high-school algebra class. However, we will give several other interpretations for this concept.

3.1 Unordered Subsets of $[n]$

By definition, the order of elements in a subset is unimportant. However, we emphasize "unordered" here to distinguish the sets studied here from those of Sect. 2.5. The difference can be described as follows. Suppose that a club with n members wishes to form a committee of k to discuss important business. By definition, the members of the committee are all equal. Hence, we must select a k-subset from $[n]$ instead of a k-tuple from $[n]$. This being the case, $P(n,k)$ is not appropriate. However, if we wish to select k officers (of different rank), then $P(n,k)$ is appropriate.

The *binomial coefficient*, denoted $\binom{n}{k}$, counts the number of unordered k-subsets of $[n]$. The quantity $\binom{n}{k}$ is read "n choose k." We now compute the binomial coefficient.

Proposition 3.1.1 *For $n,k \in \mathbb{N}$ and $n \geq k$, we have:*

$$\binom{n}{k} = \frac{P(n,k)}{k!} = \frac{n!}{k!(n-k)!}.$$

Proof It suffices to prove $\binom{n}{k}k! = P(n,k)$. By definition, $P(n,k)$ counts the number of ordered k-element subsets of $[n]$. The left side also counts this as follows:

(i) First, choose an unordered k-element subset of $[n]$. There are $\binom{n}{k}$ ways to do this.
(ii) Assign an order to this subset. There are $k!$ ways to order a k-element set by definition.

Thus, by the Multiplication Principle, we have that $P(n,k) = \binom{n}{k}k!$. Since $P(n,k) = \frac{n!}{(n-k)!}$ by Proposition 2.5.1, we have that $\binom{n}{k} = \frac{n!}{k!(n-k)!}$. ■

R. A. Beeler, *How to Count*, DOI 10.1007/978-3-319-13844-2_3

Table 3.1 Values of the binomial coefficient for small n and k

$n \setminus k$	0	1	2	3	4	5	6	7	8	9
0	1									
1	1	1								
2	1	2	1							
3	1	3	3	1						
4	1	4	6	4	1					
5	1	5	10	10	5	1				
6	1	6	15	20	15	6	1			
7	1	7	21	35	35	21	7	1		
8	1	8	28	56	70	56	28	8	1	
9	1	9	36	84	126	126	84	36	9	1

By convention, if $k > n$, then $\binom{n}{k} = 0$. The values for the binomial coefficient for small values of n and k ($n \geq k$) are given in Table 3.1.

You may recognize the values in Table 3.1 as *Pascal's Triangle*. In fact, we can show that the binomial coefficients satisfy the Pascal Recurrence. This recurrence can easily be shown using a combination of elementary algebra and Proposition 3.1.1. However, we prefer a combinatorial approach.

Theorem 3.1.2 *(Pascal's Recurrence) For $n, k \in \mathbb{N}$ such that $n \geq k$, we have:*

$$\binom{n}{k} = \binom{n-1}{k} + \binom{n-1}{k-1}. \tag{3.1}$$

Proof The left side of Eq. 3.1 counts the number of unordered k-element subsets of $[n]$ by definition. The right side of Eq. 3.1 also counts this by counting two disjoint, exhaustive classes:

(i) The class of subsets that do not contain the element n. There are $\binom{n-1}{k}$ ways of selecting the k objects from the remaining $n - 1$ elements.
(ii) The class of subsets that contain the element n. There are $\binom{n-1}{k-1}$ ways of selecting the remaining $k - 1$ objects from the remaining $n - 1$ elements.

■

The second observation that can be gained from Pascal's Triangle is that it is symmetric, in other words, $\binom{n}{k} = \binom{n}{n-k}$ for all $n, k \in \mathbb{N}$. Again, this can be proven readily using algebra, however, we prefer the combinatorial approach.

Theorem 3.1.3 *For $n, k \in \mathbb{N}$ such that $n \geq k$, we have $\binom{n}{k} = \binom{n}{n-k}$.*

Proof By definition, there are $\binom{n}{k}$ ways to select a committee of k people from a club with n members. Equivalently, there are $\binom{n}{n-k}$ ways to select $n - k$ people from a club with n members to *not* be part of the committee. ■

We now point out a more subtle observation regarding the binomial coefficients. Begin summing up the diagonals of Pascal's Triangle:

$$\binom{0}{0} = 1, \quad \binom{1}{0} = 1,$$

$$\binom{1}{1} + \binom{2}{0} = 2, \quad \binom{2}{1} + \binom{3}{0} = 3,$$

$$\binom{2}{2} + \binom{3}{1} + \binom{4}{0} = 5, \quad \binom{3}{2} + \binom{4}{1} + \binom{5}{0} = 8, \ldots$$

It appears that the sum of the diagonals on Pascal's Triangle is a Fibonacci number (under the right indexing). In fact, if $F_0 = 1$, $F_1 = 1$, and $F_n = F_{n-1} + F_{n-2}$, then:

$$\sum_{k=0}^{n} \binom{n-k}{k} = F_n.$$

This can also be proven using a combinatorial interpretation of the Fibonacci sequence. However, the necessary machinery will not be developed until Sect. 3.5.

Other identities involving the binomial coefficient will be developed in Sect. 3.4. For now, we will be satisfied with some additional examples.

Example 3.1.4 The Science club has ten members and must select three members to go to a retreat. The Spanish club has 20 members and must select five members to go to the same retreat. If no one is in both clubs, how many different combinations of members can be at the retreat?

Solution Note there are $\binom{10}{3} = 120$ ways for the Science club to select its members. Similarly, there are $\binom{20}{5} = 15504$ ways for the Spanish club to select their members. Therefore, by the Multiplication Principle yields the number of combinations as

$$\binom{10}{3}\binom{20}{5} = 120(15504) = 1860480.$$

□

Example 3.1.5 The Mathletes Club has 15 members and is selecting a chaired committee of five individuals (in a chaired committee, all members except the chair are of the same rank). One of the members, Alice, is quite stubborn. If she is not the chair of the committee, then she refuses to serve on the committee. How many ways can the club select its committee if Alice's demand must be met?

Solution The set of all ways that the club can elect its committee can be partitioned into two disjoint, exhaustive sets:

(i) The set of elections in which Alice is the chair. There are $\binom{14}{4} = 1001$ ways for the club to select the remainder of the committee.
(ii) The set of elections in which Alice is not on the committee. There are 14 choices for the chair of the committee. This leaves 13 members to select the remaining

4 members. There are $\binom{13}{4} = 715$ ways to do this. Thus by the Multiplication Principle, there are $14\binom{13}{4} = 14(715) = 10010$ ways to do this.

Hence, by the Addition Principle, the number of ways that the club can select its committee is given by:

$$\binom{14}{4} + 14\binom{13}{4} = 1001 + 10010 = 11011.$$

□

Example 3.1.6 The Burger Joint restaurant offers three types of bun (white, sesame, or wheat), four patties (hamburger, turkey, portebello, or veggie), seven cheeses (American, swiss, colby, pepper jack, cheddar, mozzarella, and goat cheese), five vegetables (pickle, lettuce, tomato, onion, and green peppers), and ten condiments (mustard, mayonnaise, ketchup, horseradish, steak sauce, salsa, teriyaki sauce, tartar sauce, guacamole, and barbecue sauce). Suppose that you must have a bun and a patty, you may have up to two cheeses, up to three condiments, and any number of vegetables. How many burgers are possible?

Solution There are three choices for bun and four choices for patty. The number of ways to select cheeses can be counted using three disjoint, exhaustive sets:

(i) The set of all choices in which you select zero cheeses. There are $\binom{7}{0}$ ways to do this.
(ii) The set of all choices in which you select one cheese. There are $\binom{7}{1}$ ways to do this.
(iii) The set of all choices in which you select two cheeses. There are $\binom{7}{2}$ ways to do this.

Thus by the Addition Principle, the number of ways to select cheeses is given by

$$\binom{7}{0} + \binom{7}{1} + \binom{7}{2}.$$

Similarly, the number of ways to select condiments can be counted using four disjoint, exhaustive sets. In this case, the ith set ($i = 0, 1, 2, 3$) is the set of all selections in which i condiments are selected. Hence there are $\binom{10}{i}$ elements in the ith set. By the Addition Principle, the number of ways to select condiments is given by:

$$\binom{10}{0} + \binom{10}{1} + \binom{10}{2} + \binom{10}{3}.$$

Finally, since any number of vegetables can be selected, the number of ways to select vegetables is given by the cardinality of the power set of $A = \{\text{pickle, lettuce, tomato, onion, green peppers}\}$. Since $|A| = 5$, it follows that $|P(A)| = 2^5$. Thus there are $2^5 = 32$ ways to select vegetables.

Hence, by the Multiplication Principle, the number of possible burgers at the Burger Joint is:

$$3(4)\left(\binom{7}{0}+\binom{7}{1}+\binom{7}{2}\right)\left(\binom{10}{0}+\binom{10}{1}+\binom{10}{2}+\binom{10}{3}\right)(2^5).$$

□

Example 3.1.7 At a particular company, employees have four digit pin numbers. Two of the digits must come from $\{1,5,7\}$. The remaining digits must come from $\{2,3,9\}$. Repetition of digits is allowed.

(a) How many pin numbers are there?
(b) Suppose that the company has 500 employees. Will two employees have the same pin number?
(c) Suppose that pin numbers cannot start with 1 nor end with 3. How many pin numbers are possible with this restriction?

Solution *(a)* The number of pin numbers can be counted as follows:

(i) Choose two of the positions for $\{1,5,7\}$. There are $\binom{4}{2}=6$ ways to do this.
(ii) Assign numbers from the set $\{1,5,7\}$ to these positions. Since repetition is allowed, there are $3^2=9$ ways to do this by Corolloar 2.1.4.
(iii) Assign numbers from the set $\{2,3,9\}$ to the remaining positions. Since repetition is allowed, there are 3^2 to do this by Corollary 2.1.4.

By the Multiplication Principle, the number of pin numbers is given by

$$\binom{4}{2}3^2 * 3^2 = 486.$$

(b) We can think of the 500 employees as "pigeons." The 486 possible pin numbers are the "pigeonholes." Since there are more employees than pin numbers, at least two employees must have the same pin number by the Pigeonhole Principle.

(c) Let S be the total number of possible passwords. By (a), $|S|=486$. Let A be the set of passwords that begin with 1. Let B be the set of passwords that end with 3.

To compute $|A|$:

(i) Place 1 in the first position. Choose one of the remaining positions to be from the set $\{1,5,7\}$. There are $\binom{3}{1}=3$ ways to do this.
(ii) Assign one of the numbers from the set $\{1,5,7\}$ to this position. There are three possibilities.
(iii) Assign numbers from the set $\{2,3,9\}$ to the remaining positions. Since repetition is allowed, there are 3^2 to do this by Corollary 2.1.4.

By the Multiplication Principle,

$$|A| = \binom{3}{1} * 3 * 3^2 = 81.$$

Similarly, we compute $|B|$:

(i) Place 3 in the last position. There are now $\binom{3}{2} = 3$ ways to select positions for $\{1, 5, 7\}$.
(ii) Assign numbers from the set $\{1, 5, 7\}$ to these positions. Since repetition are allowed, there are $3^2 = 9$ ways to do this by Corollary 2.1.4.
(iii) Assign a number from the set $\{2, 3, 9\}$ to the remaining position. There are three ways to do this.

By the Multiplication Principle,

$$|B| = \binom{3}{2} * 3^2 * 3 = 81.$$

Finally, we compute $|A \cap B|$:

(i) Place 1 in the first position and 3 in the last position. There are now $\binom{2}{1} = 2$ ways to select positions for $\{1, 5, 7\}$.
(ii) Assign a number from the set $\{1, 5, 7\}$ to this position. There are 3 possibilities.
(iii) Assign a number from the set $\{2, 3, 9\}$ to the remaining position. There are 3 possibilities.

By the Multiplication Principle,

$$|A \cap B| = \binom{2}{1} * 3 * 3 = 18.$$

Thus, the number of pin numbers that neither start with 1 nor end with 3 is

$$|S| - |A| - |B| + |A \cap B|$$
$$= 486 - 81 - 81 + 18 = 342,$$

by the Principle of Inclusion and Exclusion. □

In the next example, we use the binomial coefficient and the Pigeonhole Principle.

Proposition 3.1.8 *In any set of 12 distinct elements of* [50] *there are 4 distinct elements, a, b, c, and d such that* $a + b = c + d$.

Proof It suffices to find two pairs of elements that have the same difference. Note that for distinct $a, b \in [50]$, there are 49 possibilities for the difference $b - a$, namely, 1, ..., 49. These will be the "pigeonholes." Further, in any 12-set, there are $\binom{12}{2} = 66$ possible pairs of distinct elements a and b. However, some of these pairs will contain the same element. For example, $\{3, 5\}$ and $\{5, 17\}$ both contain 5. Two pairs *overlap at a* if they both contain a and have the same difference. In other words, (a_i, a) and (a, a_j) overlap if $a_i - a = a - a_j$.

Suppose that there exists distinct a_i, a_j, a_k, a_ℓ such that $a_i - a = a - a_j$ and $a_k - a = a - a_\ell$. It follows that $a_i + a_j = 2a = a_k + a_\ell$, thus affirming the claim. Hence, without loss of generality, we may suppose that there exists at most two pairs that overlap at a. Similarly, it follows that for each a_i, at most two pairs can overlap at a_i. For each set of two pairs that overlap at a_i, we remove one such pair. Thus, we have removed 12 pairs from the 66 available. The remaining 54 pairs will be our "pigeons." Thus by the Pigeonhole Principle, there exists distinct pairs (a_i, a_j) and (a_k, a_ℓ) such that $a_i - a_j = a_k - a_\ell$. From this it follows that $a_i + a_\ell = a_k + a_j$. ■

Exercise 3.1.9 Confirm the entries in Table 3.1.

Exercise 3.1.10 Evaluate each of the following:

(i)
$$\binom{16}{7};$$

(ii)
$$\binom{12}{3};$$

(iii)
$$\binom{15}{4};$$

(iv)
$$\binom{15}{11}.$$

Exercise 3.1.11 Each club in the college must select two senators to represent them at council. How many possible council meetings are possible if there are five clubs, the ith club has n_i members, and no one is in two clubs.

Exercise 3.1.12 The Science Club, Math Club, and English club have 15, 20, and 25 members, respectively. Alice is in all three clubs. No one else is in more than one club. In how many ways can the clubs select committees of three individuals if Alice cannot be on more than one committee.

Exercise 3.1.13 The United Nations has five senior members (the United States, Russia, China, Germany, and France) and 100 junior members. Find the number of five nation committees if the committee must contain two senior members and three junior members.

Exercise 3.1.14 The Burgero Magnifico restaurant offers five types of bun, eight patties, ten cheeses, eight vegetables, and thirteen condiments. Each patty can be cooked rare, medium rare, medium, medium well, or well done. Suppose that you must have a bun and a patty, you may have up to three cheeses, up to four condiments, and any number of vegetables. How many burgers are possible?

Exercise 3.1.15 At a particular company, employees have an six digit security code. Three of the digits must come from $\{2, 3, 5, 9\}$. The remaining digits must come from $\{1, 4, 6, 7\}$. Repeated digits are allowed.

(i) How many security codes are there?
(ii) How many of these begin with 1?
(iii) How many of these end with 5?
(iv) How many of these neither begin with 1 nor end with 5?

Exercise 3.1.16 At a particular company, employees have an seven digit security code. Five of the digits must come from $\{1, 4, 5, 7\}$ with repetition allowed. The remaining digits must come from $\{2, 3, 9\}$ with no repetition.

(i) How many security codes are possible?
(ii) Suppose that this company has 150,000 employees. Should they be concerned that two employees have the same security code?

Exercise 3.1.17 Show that for any eight distinct elements of [20], there exists distinct elements a, b, c, d such that $a + b = c + d$.

Exercise 3.1.18 Show that for any 20 distinct elements of [150], there exists distinct elements a, b, c, d such that $a + b = c + d$.

Exercise 3.1.19 Find the smallest n such that for any 15 distinct elements of $[n]$, there exists distinct elements a, b, c, d such that $a + b = c + d$.

Exercise 3.1.20 Find the smallest n such that for any k distinct elements of $[n]$, there exists distinct elements a, b, c, d such that $a + b = c + d$.

3.2 Application: Hands in Poker

We now examine one of the most interesting applications of combinatorics, the game of poker. There are many variations on poker. However, we restrict our attention to five-card draw with no wild cards. It is not necessary to understand all the subtleties of the game for our exhibition. Nonetheless, we will outline some of the basics of the game as we proceed.

The standard poker deck has 13 *values* or *ranks*. These values are Ace (A), 2, 3, 4, 5, 6, 7, 8, 9, 10, Jack (J), Queen (Q), and King (K). Each value is represented exactly once in each of four suits: Spades, Clubs, Hearts, and Diamonds. This gives a total of 52 cards.

In five-card draw poker, we are dealt a hand of five cards. Since we can rearrange the cards in our hand, the order in which they are dealt is not important. Hence, our hand can be thought of as choosing an unordered 5-element subset from a set of 52. Thus there are $\binom{52}{5}$ possible hands. We now count specific hands in poker.

Example 3.2.1 How many ways are there are to be dealt "three of a kind?"

Solution A three of a kind hand must consist of three distinct values: the triple and two cards with different ranks from each other and the triple. Note that if they are the same value, the hand is called a "full house." If one of the two values matches the value of the triple, the hand is called "four of a kind." So the number of ways of drawing a three of a kind can be computed as follows:

(i) Choose the three values, there are $\binom{13}{3}$ ways to do this;
(ii) Choose one of the three values to triple, there are $\binom{3}{1}$ ways to do this;
(iii) Choose the suits for the triple, there are $\binom{4}{3}$ ways to do this;
(iv) Choose suits for the remaining two cards, there are 4^2 ways to do this.

Thus, by the Multiplication Principle, the number of ways of being dealt three of a kind is given by:

$$\binom{13}{3}\binom{3}{1}\binom{4}{3}4^2 = 54912.$$

□

Example 3.2.2 Find the number of ways of drawing a pair.

Solution Similarly, a "pair" consists of four distinct values: the pair and three cards with different ranks from each other and the pair. Thus the number of ways of drawing a pair can be computed as follows:

(i) Choose the four values, there are $\binom{13}{4}$ ways to do this;
(ii) Choose one of the four values to double, there are $\binom{4}{1}$ ways to do this;
(iii) Choose the suits for the double, there are $\binom{4}{2}$ ways to do this;
(iv) Choose suits for the remaining three cards, there are 4^3 ways to do this.

Thus by the Multiplication Principle, the number of ways of being dealt a pair is given by:

$$\binom{13}{4}\binom{4}{1}\binom{4}{2}4^3 = 1098240.$$

□

Example 3.2.3 A "straight" consists of five cards in sequence. Find the number of ways of drawing a straight if the Ace may be played high (above the King) or low (below the two).

Solution Note that there are ten sequences that give straights:

$$\{A, 2, 3, 4, 5\}, \{2, 3, 4, 5, 6\}, \{3, 4, 5, 6, 7\}, \{4, 5, 6, 7, 8\},$$
$$\{5, 6, 7, 8, 9\}, \{6, 7, 8, 9, 10\}, \{7, 8, 9, 10, J\}, \{8, 9, 10, J, Q\},$$
$$\{9, 10, J, Q, K\}, \quad \text{and} \quad \{10, J, Q, K, A\}.$$

Therefore, there are ten choices for the sequence and 4^5 choices for the suits. Hence there are $10*4^5 = 10240$ ways to draw a straight by the Multiplication Principle. □

Example 3.2.4 A "flush" consists of five cards that are the same suit. Find the number of ways to draw a flush.

Solution Note that a flush necessarily consists of five different values. Hence there are $\binom{13}{5}$ ways to choose the values and four ways to choose the suit. So there are $\binom{13}{5} * 4 = 5148$ ways to draw a flush. □

Example 3.2.5 A hand is considered to be "junk" if it is five cards of different values that is neither a straight nor a flush. Find the number of "junk" hands.

Solution The number of junk hands can be computed as follows:

(i) Compute the number of hands that contain five different values. There are $\binom{13}{5}$ ways to choose the values and 4^5 ways to choose the suits. Hence there are $\binom{13}{5} * 4^5 = 1317888$ ways of drawing a hand with five different values.
(ii) Compute the number of straights. There are $10 * 4^5 = 10240$ ways to do this by Example 3.2.3.
(iii) Compute the number of flushes. There are $\binom{13}{5} * 4 = 5148$ ways to do this by Example 3.2.4.
(iv) Some hands can be both a straight and a flush, in other words, a "straight flush." There are ten choices for the sequence and four choices for the suit. Hence there are 40 possibilities for a straight flush.
(v) By the Principle of Inclusion and Exclusion, the number of hands that are either a straight or a flush is given by:

$$10 * 4^5 + \binom{13}{5} * 4 - 40 = 15348.$$

(vi) It follows that the number of hands that are junk is given by $\binom{13}{5}4^5 - 15348 = 1302540$.

□

Exercise 3.2.6 Find the number of ways to draw "two pair," in other words, two distinct pairs of values with a third unmatched value.

Exercise 3.2.7 Find the number of ways to draw a "full house," in other words, three cards of one value and a pair of a different value.

Exercise 3.2.8 Find the number of ways of drawing "four of a kind," in other words, four cards of one value and a second unmatched value.

The imaginary game of Combo is played with a standard poker deck and six card hands.

Exercise 3.2.9 Find the number of ways of drawing "three of a kind" in Combo.

Exercise 3.2.10 Find the number of ways of drawing a "straight" (six cards in sequence) in Combo.

Exercise 3.2.11 Find the number of ways of drawing a "flush" (six cards of the same suit) in Combo.

Exercise 3.2.12 A Combo hand is called "super" if it contains a pair and four cards in sequence. Find the number of hands that are super.

Exercise 3.2.13 Find the number of "junk" hands in Combo, in other words, hands that contain no pairs and are neither straights nor flushes.

3.3 The Binomial Theorem

In your algebra classes, you probably expanded out simple binomials such as $(x+y)^n$ for small values of n. You may have even proven the binomial theorem in your introductory proof class using induction. In this section, we will give a different proof for the binomial expansion. We will also give several implications.

Theorem 3.3.1 *(The Binomial Theorem) For all $n \in \mathbb{N}$, we have:*

$$(x+y)^n = \sum_{k=0}^{n} \binom{n}{k} x^k y^{n-k}.$$

Proof It suffices to show that the coefficient of $x^k y^{n-k}$ in $(x+y)^n$ is $\binom{n}{k}$. Note that by definition,

$$(x+y)^n = \underbrace{(x+y)\cdots(x+y)}_{n \text{ times}}.$$

To obtain x^k, we must use the x in k copies of $x+y$. There are $\binom{n}{k}$ ways to select the terms that will provide this x. The remaining $n-k$ terms must all contribute a y, thus giving us a y^{n-k} in the product. Thus the coefficient of $x^k y^{n-k}$ is $\binom{n}{k}$. ∎

Example 3.3.2 Find the coefficient of x^4 in each of the following (treat y as an unknown fixed constant):

(i) $(x+1)^7$; (ii) $(x+5)^6$;
(iii) $(3x+7)^5$; (iv) $(3x+5y+7)^6$.

Solution

(i) By Theorem 3.3.1, the coefficient of x^4 is $\binom{7}{4} = 35$.
(ii) By Theorem 3.3.1, the coefficient of x^4 in $(x+5)^6$ is $\binom{6}{4}5^2 = 375$.
(iii) By Theorem 3.3.1, the coefficient of x^4 in $(3x+7)^5$ is $\binom{5}{4}3^4 7^1 = 2835$.
(iv) Note that $(3x+5y+7)^6 = (3x+(5y+7))^6$. Thus, by the Binomial Theorem, we have:

$$(3x+5y+7)^6 = (3x+(5y+7))^6 = \sum_{k=0}^{6} \binom{6}{k} 3^k x^k (5y+7)^{6-k}.$$

Hence the coefficient on x^4 is:

$$\binom{6}{4}3^4(5y+7)^2.$$

□

In Sect. 3.6, we will determine how to compute the coefficient of $x_1^{k_1}...x_\ell^{k_\ell}$ in the expansion of $(x_1 + ... + x_\ell)^n$.

Corollary 3.3.3 *For all $n \in \mathbb{N}$,*

$$\sum_{k=0}^{n}\binom{n}{k} = 2^n.$$

Proof Let $x = y = 1$ in the Binomial Theorem. ■

We will give a combinatorial proof of the above identity in the next section.

Corollary 3.3.4 *For all $n \in \mathbb{N}$,*

$$\binom{n}{0}+\binom{n}{2}+\cdots = \binom{n}{1}+\binom{n}{3}+\cdots.$$

Proof Let $x = -1$ and $y = 1$. Note that $x + y = 0$. Thus by the Binomial Theorem, we have:

$$\begin{aligned} 0 &= \sum_{k=0}^{n}\binom{n}{k}(-1)^k 1^{n-k} = \sum_{k=0}^{n}\binom{n}{k}(-1)^k \\ &= \binom{n}{0}-\binom{n}{1}+\binom{n}{2}-\binom{n}{3}\pm\cdots \\ &\Rightarrow \binom{n}{0}+\binom{n}{2}+\cdots = \binom{n}{1}+\binom{n}{3}+\cdots. \end{aligned}$$

■

Corollary 3.3.4 may seem esoteric, however, it actually suggests something fairly subtle. Namely, it shows that the number of subsets of $[n]$ of odd cardinality is the same as the number of subsets of even cardinality. In certain cases, this is relatively easy to see. For instance, if n and k are both odd, then $n - k$ is even. Further, $\binom{n}{k} = \binom{n}{n-k}$, implying the result. The more general bijection is left as an exercise.

Using the Binomial Theorem, we can also discover additional combinatorial identities.

Corollary 3.3.5 *For all $n \in \mathbb{N}$,*

$$n2^{n-1} = \sum_{k=0}^{n} k\binom{n}{k}.$$

Proof From the Binomial Theorem, we know that

$$(1+x)^n = \sum_{k=0}^{n} \binom{n}{k} x^k.$$

Differentiating both sides with respect to x yields

$$n(1+x)^{n-1} = \sum_{k=0}^{n} k\binom{n}{k} x^{k-1}.$$

Letting $x = 1$ in this equation gives us the desired result. ■

Using the same technique as in Corollary 3.3.5, we can show

$$P(n,m)2^{n-m} = \sum_{k=0}^{n} P(k,m)\binom{n}{k}.$$

This result will be left as an exercise to the reader.

Exercise 3.3.6 Find the coefficient of x^3 in each of the following:

(i) $(x+y)^6$; (ii) $(x+3)^7$;
(iii) $(2x+5)^4$; (iv) $(2x+3y+5)^7$.

Exercise 3.3.7 Prove that for all $n, m \in \mathbb{N}$,

$$P(n,m)2^{n-m} = \sum_{k=0}^{n} P(k,m)\binom{n}{k}.$$

Exercise 3.3.8 Give an explicit bijection between the subsets of $[n]$ with even cardinality and the subsets of $[n]$ with odd cardinality.

3.4 Identities Involving the Binomial Coefficient

In this section, we prove several identities involving binomial coefficients. Several of these identities may seem esoteric. In many cases, these identities can be proven using induction. However, we prefer to use combinatorial proofs (see Sect. 1.6). We begin by giving the sum of binomial coefficients in various ways. Our first example was previously proven as Corollary 3.3.3.

Theorem 3.4.1 *For all $n \in \mathbb{N}$,*

$$\sum_{k=0}^{n} \binom{n}{k} = 2^n.$$

Proof By Example 2.1.10, 2^n is the cardinality of the power set of $[n]$. In other words, 2^n is the number of subsets of $[n]$. The left side also counts this in $n+1$ disjoint, exhaustive classes.

The kth class ($k = 0, ..., n$) counts the number of k-element subsets of n. This is given by $\binom{n}{k}$ by definition.

It follows that the result holds by the Addition Principle. ■

In the next theorem, we look at summing the "columns" of Pascal's triangle. Surprisingly enough, these sums are also binomial coefficients.

Theorem 3.4.2 *For all $n, k \in \mathbb{N}$,*

$$\sum_{j=0}^{n} \binom{j}{k} = \binom{n+1}{k+1}.$$

Proof Consider the problem of choosing $(k+1)$-element subsets of $[n+1]$. The number of ways to do this is given by $\binom{n+1}{k+1}$. The left side also counts this in $n+1$ disjoint, exhaustive classes.

The jth class ($j = 0, ..., n$) counts the number of $(k+1)$-subsets in which the largest element is $j+1$. This problem reduces down to choosing a k-element subset from $[j]$. There are $\binom{j}{k}$ ways to do this by definition.

Hence, the theorem holds by the Addition Principle. ■

Corollary 3.4.3 *For all $n, k \in \mathbb{N}$,*

$$\sum_{i=0}^{k} \binom{n+i}{i} = \binom{n+k+1}{k}.$$

Proof Note that $\binom{n+j}{j} = \binom{n+j}{n}$ by Theorem 3.1.3.

Hence,

$$\sum_{i=0}^{k} \binom{n+i}{i} = \sum_{i=0}^{k} \binom{n+i}{n}.$$

We now do a change of variables, namely $j = n + i$. Note that if $i = 0$ then $j = n$. Further, if $i = k$, then $j = n + k$. Changing the bounds of the summation to match the new variable, we have:

$$\sum_{i=0}^{k} \binom{n+i}{n} = \sum_{j=n}^{n+k} \binom{j}{n}.$$

Recall that $\binom{j}{n} = 0$ for all $j < n$. Thus,

$$\sum_{j=n}^{n+k} \binom{j}{n} = \sum_{j=0}^{n+k} \binom{j}{n}.$$

Applying Theorem 3.4.2 yields

$$\sum_{j=0}^{n+k} \binom{j}{n} = \binom{n+k+1}{n+1}.$$

Applying Theorem 3.1.3 again gives us

$$\sum_{i=0}^{k} \binom{n+i}{i} = \binom{n+k+1}{k}.$$

■

Another useful identity is known as Vandermonde's Sum. This is also known as the hypergeometric identity.

Theorem 3.4.4 *(Vandermonde's Sum) For all $n, m, k \in \mathbb{N}$, we have:*

$$\binom{n+m}{k} = \sum_{i=0}^{k} \binom{n}{i}\binom{m}{k-i}.$$

Proof Consider the problem of selecting a committee of k individuals from a club with n men and m women. The number of ways of doing this is given by $\binom{n+m}{k}$.

The set of all committees of k from a club with n men and m women can also be counted in $k+1$ disjoint sets.

The ith set ($i = 0, ..., k$) is the set of all committees that contain i men. There are $\binom{n}{i}$ ways to select the men for this committee. The remaining $k-i$ members of the committee must all be women. There are $\binom{m}{k-i}$ ways of selecting the women for the committee. Thus, by the Multiplication Principle, the number of ways of selecting a committee with i men is given by:

$$\binom{n}{i}\binom{m}{k-i}.$$

The identity holds by the Addition Principle. ■

Theorem 3.4.5 *For all $n, k, p \in \mathbb{N}$, we have:*

$$\binom{n+k}{n+p}\binom{n+p}{p} = \binom{n+k}{n}\binom{k}{p}.$$

Proof Consider the problem of selecting p employees from a pool of $n+k$ applicants. The left side counts this by:

(i) Choosing a "short list" of $n+p$ applicants. There are $\binom{n+k}{n+p}$ ways to do this.
(ii) From the "short list" of $n+p$ applicants, choose the p employees. There are $\binom{n+p}{p}$ ways to do this.

Hence, by the Multiplication Principle, the number of ways of selecting the employees is given by:

$$\binom{n+k}{n+p}\binom{n+p}{p}.$$

The right side also counts this by:

(i) Selecting n candidates to reject. There are $\binom{n+k}{n}$ ways to do this. Note that this leaves us k applicants in the pool.
(ii) From the remaining k applicants, choose p employees. There are $\binom{k}{p}$ ways to do this.

So by the Multiplication Principle, the number of ways of selecting the employees is given by:

$$\binom{n+k}{n}\binom{k}{p}.$$

■

Exercise 3.4.6 Prove combinatorially that

$$n\binom{n-1}{k-1} = \binom{n}{k}k.$$

Exercise 3.4.7 Prove combinatorially that

$$\sum_{k=0}^{n} k\binom{n}{k} = n2^{n-1}.$$

Exercise 3.4.8 Prove combinatorially that

$$\sum_{k=0}^{n} P(k,m)\binom{n}{k} = P(n,m)2^{n-m}.$$

Exercise 3.4.9 Prove combinatorially that

$$\binom{n+m}{2} = \binom{n}{2} + \binom{m}{2} + nm.$$

Exercise 3.4.10 Prove combinatorially that $1 + \cdots + (n-1) = \binom{n}{2}$.

*|******||****|**|***||*

|**|**|***|***|**|*|

*||***|**|****||******|*

*****************|||||||

|*|**|***|**|***|**|

Fig. 3.1 Various arrangements of 17 stars and 7 bars

Exercise 3.4.11 Prove that:

$$\sum_{t \geq 0} \binom{n}{j+t}\binom{m}{k+t} = \binom{n+m}{n+k-j}.$$

Exercise 3.4.12 Prove combinatorially that

$$\binom{n+m+\ell}{p} = \binom{n+m+\ell}{m+\ell}\binom{m+\ell}{\ell}\binom{\ell}{p}.$$

Exercise 3.4.13 Prove combinatorially that

$$\binom{n+m+\ell}{k} = \sum_{i=0}^{k}\sum_{j=0}^{k-i}\binom{n}{i}\binom{m}{j}\binom{\ell}{k-i-j}.$$

3.5 Stars and Bars

While the name of this section may seem innocent, perhaps even silly, the principle of Stars and Bars is actually one of the most useful notions in combinatorics. Consider the problem of determining the number of ways of arranging n (indistinguishable) stars and k (indistinguishable) bars. Note that if we could distinguish the stars and distinguish the bars, then the problem reduces down to that of arranging $n+k$ distinct objects. There are $(n+k)!$ ways to do this. Hence, the problem of stars and bars is that of arranging n indistinguishable objects of the first type and k indistinguishable objects of the second type. When we refer to stars and bars, we will *always* assume that they are indistinguishable. Several arrangements of seventeen stars and seven bars are given in Fig. 3.1.

As a first attempt to solve this problem, we might observe that we have two choices for the first element in the sequence, two choices for the second element in the sequence, and so on. While this approach will work for the first k (or n) slots, we do not know how many stars were placed in these slots. Thus, we have no idea how to proceed with the remaining slots.

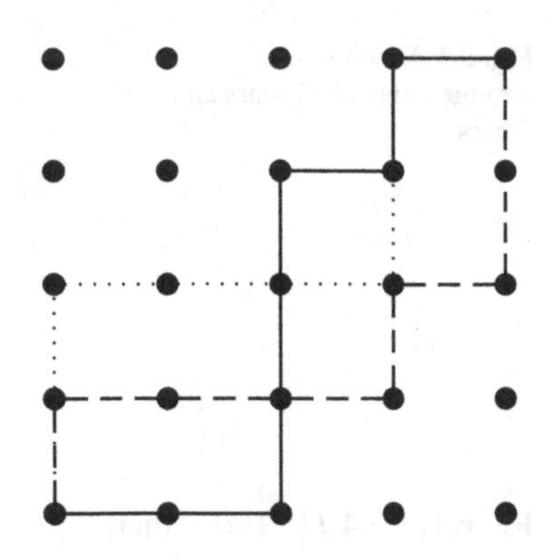

Fig. 3.2 Three lattice paths

Theorem 3.5.1 *(Stars and Bars) The number of ways of arranging n (indistinguishable) stars and k (indistinguishable) bars is* $\binom{n+k}{k}$.

Proof Since we are arranging n stars and k bars, there are a total of $n + k$ slots to place them. Out of these $n + k$ slots, we choose k of them to place the bars. There are $\binom{n+k}{k}$ ways to do this. The stars are then forced into the remaining n slots.

One of the key applications of Stars and Bars is to lattice path problems. A *lattice path* from $(0, 0)$ to (n, k) is a path through the integer points of the Cartesian grid in which only moves to the north and to the east are allowed. Figure 3.2 gives three examples of lattice paths from $(0, 0)$ to $(4, 4)$.

Corollary 3.5.2 *The number of lattice paths from* $(0, 0)$ *to* (n, k) *is given by* $\binom{n+k}{k}$.

Proof Any lattice path from $(0, 0)$ to (n, k) uses only moves to the north and moves to the east. Thus, a lattice path is equivalent to an arrangement of n E's (east moves) and k N's (north moves). Hence, by Stars and Bars, the number of lattice paths from $(0, 0)$ to (n, k) is given by $\binom{n+k}{k}$. ∎

Example 3.5.3 Find the number of lattice paths from:

(i) $(-1, 2)$ to $(7, 10)$;
(ii) $(-3, -4)$ to $(-1, 2)$;
(iii) $(-3, -4)$ to $(7, 10)$;
(iv) $(-3, -4)$ to $(7, 10)$ that pass through $(-1, 2)$;
(v) $(-3, -4)$ to $(7, 10)$ that do not pass through $(0, 0)$;
(vi) $(-3, -4)$ to $(7, 10)$ that pass through $(0, 0)$ or $(-1, 2)$.

Solution

(i) Note that a lattice path from $(-1, 2)$ to $(7, 10)$ is equivalent to a lattice path from $(0, 0)$ to $(8, 8)$. So the number of paths is given by $\binom{16}{8} = 12870$.
(ii) Similarly, the number of paths from $(-3, 4)$ to $(-1, 2)$ is equivalent to the number of paths from $(0, 0)$ to $(-1 + 3, 2 + 4) = (2, 6)$. Hence the number of paths is given by $\binom{2+6}{2} = \binom{8}{2} = 28$.
(iii) By the comments above, $\binom{7+3+(10+4)}{7+3} = \binom{24}{10} = 1961256$.

(iv) Any such path must go from $(-3,-4)$ to $(-1,2)$ and then from $(-1,2)$ to $(7,10)$. The number of paths from $(-3,-4)$ to $(-1,2)$ is given by $\binom{8}{2}$. The number of paths from $(-1,-2)$ to $(7,10)$ is given by $\binom{20}{8}$. Hence, by the Multiplication Principle, the number of acceptable paths is given by $\binom{8}{2}\binom{20}{8} = 3527160$.

(v) The number of paths from $(-3,-4)$ to $(0,0)$ is given by $\binom{7}{3}$. The number of paths from $(0,0)$ to $(7,10)$ is given by $\binom{17}{7}$. Thus the number of paths from $(-3,-4)$ to $(7,10)$ that pass through $(0,0)$ is given by $\binom{7}{3}\binom{17}{7}$. Since the number of paths from $(-3,-4)$ to $(7,10)$ is given by $\binom{24}{10}$, it follows that the number of acceptable paths is given by $\binom{24}{10} - \binom{7}{3}\binom{17}{7} = 1280576$.

(vi) The number of paths from $(-3,-4)$ to $(7,10)$ that pass through $(0,0)$ is given by $\binom{7}{3}\binom{17}{7}$ (see (v)). The number of paths from $(-3,-4)$ to $(7,10)$ that pass through $(-1,2)$ is given by $\binom{8}{2}\binom{20}{8}$ (see (iv)). Since no path can pass through both $(-1,2)$ and $(0,0)$, the total number of acceptable paths is given by:

$$\binom{7}{3}\binom{17}{7} + \binom{8}{2}\binom{20}{8} = 4207840.$$

□

As a more difficult problem, we now consider the task of determining the number of lattice paths from $(0,0)$ to (n,k) (where $n > k$) that stay strictly below the diagonal $y = x$, except at the point $(0,0)$.

Theorem 3.5.4 *Let $n > k$. The number of lattice paths from $(0,0)$ to (n,k) that stay below the diagonal $y = x$ (except at $(0,0)$) is given by:*

$$\frac{n-k}{n+k}\binom{n+k}{k}.$$

Proof Note that any lattice path from $(0,0)$ to (n,k) that stays below the diagonal, must begin by moving east to $(1,0)$. Thus, this problem is equivalent to finding the number of lattice paths from $(1,0)$ to (n,k) that stay below the diagonal. By Theorem 3.5.1, we know that the total number of lattice paths from $(1,0)$ to (n,k) is given by $\binom{n+k-1}{n-1}$. It suffices to compute the number of lattice paths from $(1,0)$ to (n,k) that touch or cross the diagonal and to subtract the two numbers.

We claim that the number of lattice paths from $(1,0)$ to (n,k) that touch or cross the diagonal is equal to the number of lattice paths from $(0,1)$ to (n,k). This equivalence can be established be giving a bijection between the two sets.

Take any lattice path from $(1,0)$ to (n,k) that touches or crosses the diagonal. In this path, there is an initial point (p,q) at which it touches or crosses the diagonal. We "reflect" the steps of this path from $(0,0)$ to (p,q) across the diagonal $y = x$ (see Fig. 3.3). Hence, north moves on the original lattice path become east moves on the reflected path. Similarly, east moves on the original lattice path become north moves on the reflected path. This gives a path from $(0,1)$ to (n,k). We need only establish that this mapping is a bijection.

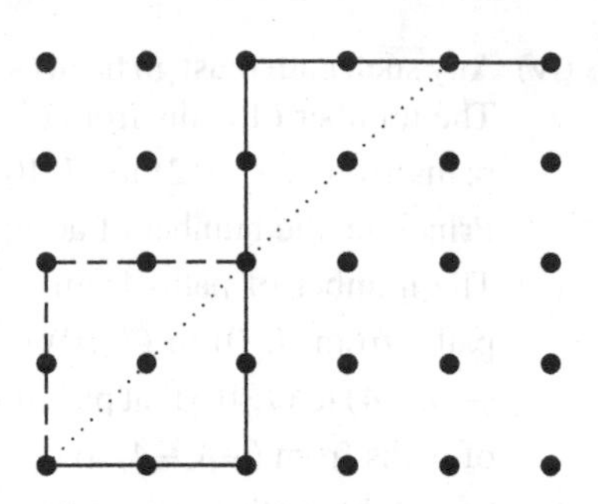

Fig. 3.3 The reflection principle

(i) Any lattice path consists of a unique sequence of N's and E's (corresponding to north and east moves, respectively). Thus the reflection will likewise consist of a unique sequence of N's and E's. It follows that no two distinct lattice paths will map to the same reflected path. Hence, the mapping is injective.

(ii) Since $n > k$, it follows that (n, k) lies below the diagonal. Then any lattice path from $(0, 1)$ to (n, k) must cross the diagonal at some point. Therefore it is a reflection of a path from $(1, 0)$ to (n, k) that touches or crosses the diagonal.

Note that the number of lattice paths from $(0, 1)$ to (n, k) is given by

$$\binom{n+k-1}{k-1} = \binom{n+k-1}{n}.$$

As shown above, there is a bijection between the set of lattice paths from $(0, 1)$ to (n, k) and the set of lattice paths from $(1, 0)$ to (n, k) that touch or cross the diagonal. Hence, both sets have the same cardinality. From this it follows that, the number of lattice paths from $(1, 0)$ to (n, k) that touch or cross the diagonal is also given by $\binom{n+k-1}{n}$. Hence the number of lattice paths from $(0, 0)$ to (n, k) that stay below the diagonal (except at the origin) can be obtained by the Subtraction Principle:

$$\begin{aligned}
\binom{n+k-1}{n-1} - \binom{n+k-1}{n} &= \frac{(n+k-1)!}{(n-1)!k!} - \frac{(n+k-1)!}{n!(k-1)!} \\
&= \frac{n+k}{n+k}\left(\frac{(n+k-1)!}{(n-1)!k!} - \frac{(n+k-1)!}{n!(k-1)!}\right) \\
&= \frac{1}{n+k}\left(\frac{(n+k)!}{(n-1)!k!} - \frac{(n+k)!}{n!(k-1)!}\right) \\
&= \frac{1}{n+k}\left(\frac{n}{n}\left(\frac{(n+k)!}{(n-1)!k!}\right) - \frac{k}{k}\left(\frac{(n+k)!}{n!(k-1)!}\right)\right) \\
&= \frac{1}{n+k}\left(n\frac{(n+k)!}{n!k!} - k\frac{(n+k)!}{n!k!}\right) \\
&= \frac{1}{n+k}\left(n\binom{n+k}{k} - k\binom{n+k}{k}\right) \\
&= \frac{1}{n+k}\binom{n+k}{k}(n-k) = \frac{n-k}{n+k}\binom{n+k}{k}. \qquad \blacksquare
\end{aligned}$$

A closely related concept is that of the *Catalan numbers*. The Catalan numbers count the number of lattice paths from $(0, 0)$ to (n, n) that do not cross the diagonal (they are allowed to touch the diagonal). Using a similar technique as above, the number of such paths is given by

$$\frac{1}{n+1}\binom{2n}{n}.$$

Exercise 3.5.5 Find the number of binary strings of length n in which there are exactly k ones.

Exercise 3.5.6 Show that the number of ways of placing k non-overlapping indistinguishable 2×1 dominoes on a $m \times 1$ grid is given by $\binom{m-k}{k}$.

Exercise 3.5.7 Find the number of arrangements of n stars and k bars where:

(i) $n = 5, k = 7$;
(ii) $n = 6, k = 15$;
(iii) $n = 3, k = 9$;
(iv) $k = 2n$.

Exercise 3.5.8 Find the number of lattice paths from

(i) $(-3, -2)$ to $(-1, 0)$;
(ii) $(-3, -2)$ to $(1, 2)$;
(iii) $(-3, -2)$ to $(4, 5)$;
(iv) $(-1, 0)$ to $(4, 5)$;
(v) $(1, 2)$ to $(4, 5)$;
(vi) $(-1, 0)$ to $(1, 2)$;
(vii) $(-3, -2)$ to $(4, 5)$ that pass through $(-1, 0)$;
(viii) $(-3, -2)$ to $(4, 5)$ that pass through $(1, 2)$;
(ix) $(-3, -2)$ to $(4, 5)$ that pass through neither $(-1, 0)$ nor $(1, 2)$.

Exercise 3.5.9 Find the number of lattice paths from $(0, 0)$ to $(8, 5)$ that:

(i) Stay below the diagonal $y = x$ except at $(0, 0)$;
(ii) Pass through the point $(4, 4)$;
(iii) Pass through the point $(4, 4)$ and stay below the diagonal $y = x$ except at $(0, 0)$ and $(4, 4)$;
(iv) Pass through the point $(4, 4)$ or stay below the diagonal (except at $(0, 0)$ and $(4, 4)$).

Exercise 3.5.10 Find the number of lattice paths from $(0, 0)$ to $(13, 12)$ that pass through neither $(4, 5)$ nor $(9, 9)$.

Exercise 3.5.11

(i) Find the number of ways to arrange n left parentheses and k right parentheses;
(ii) An arrangement of left and right parentheses is *legal* if, when counting left to right, the number of left parentheses is at least the number of right parentheses. Find the number of legal ways to arrange n left parentheses and k right parentheses.

Exercise 3.5.12 Give a combinatorial proof for Corollary 3.4.3.

Exercise 3.5.13 Let $F_0 = F_1 = 1$ and $F_n = F_{n-1} + F_{n-2}$. Give a combinatorial proof that:

$$\sum_{k=0}^{n} \binom{n-k}{k} = F_n.$$

Exercise 3.5.14 Show that the number of lattice paths from $(0, 0)$ to (n, n), that do not cross the diagonal, is given by

$$\frac{1}{n+1}\binom{2n}{n}.$$

Note that such paths are allowed to touch the diagonal.

3.6 The Multinomial Coefficient

The binomial theorem can be generalized to the expansion of the *multinomial*, $(x_1 + \cdots + x_k)^n$.

If non-negative integers $\lambda_1,\ldots,\lambda_k$ sum to n, in other words, $\lambda_1 + \cdots + \lambda_k = n$, then $(\lambda_1,\ldots,\lambda_k)$ is an *ordered weak partition of n into k parts*. Note that order here implies that $(1, 2, 3)$ is different from $(3, 2, 1)$. While this makes no difference in computing the multinomial coefficients, it does impact some computations later. The notion of a weak partition here implies that we allow the λ_i to be zero. This is opposed to a *strong partition* in which only positive values of λ_i are allowed.

Definition 3.6.1 Let $n \in \mathbb{N}$ and take $(\lambda_1,\ldots,\lambda_k)$ to be an ordered weak partition of n. The number of ways to select a λ_1-subset from $[n]$, followed by a λ_2-subset from the remaining elements,..., followed by a λ_k-subset from the remaining elements is denoted:

$$\binom{n}{\lambda_1, \ldots, \lambda_k}.$$

This is referred to as the *multinomial coefficient*.

The multinomial coefficient is read "n choose $\lambda_1,\ldots,\lambda_k$." We will now see how to compute multinomial coefficients.

A bit of clarification is in order. Namely, each of the subsets is distinguishable, even if they are the same size. This is different from the examples used in Sect. 2.7. In that section, tables of the same size were considered to be indistinguishable. In this case, it does matter which set the objects are placed into. This is analogous to assigning children into pre-named teams. In this case, the children not only care who is on their team, but what the team is named. We will deal with the case where sets of the same size are indistinguishable later in this section.

Theorem 3.6.2 *Let $n \in \mathbb{N}$ and take $(\lambda_1,\dots,\lambda_k)$ to be an ordered weak partition of n. The multinomial coefficient $\binom{n}{\lambda_1,\dots,\lambda_k}$ can be computed as:*

$$\binom{n}{\lambda_1,\dots,\lambda_k} = \binom{n}{\lambda_1}\binom{n-\lambda_1}{\lambda_2}\binom{n-\lambda_1-\lambda_2}{\lambda_3}\cdots\binom{n-\lambda_1-\cdots-\lambda_{k-1}}{\lambda_k}$$

$$= \frac{n!}{\lambda_1!\cdots\lambda_k!}.$$

Proof It suffices to show that both sides count the same set. By definition, $\binom{n}{\lambda_1,\dots,\lambda_k}$ counts the number of ways of selecting a λ_1-subset from n, followed by a λ_2-subset from the remaining elements,..., followed by a λ_k-subset from the remaining elements. This can also be counted as follows:

(i) Select the λ_1 elements for the first subset, there are $\binom{n}{\lambda_1}$.
(ii) Choose λ_2 elements from the remaining $n-\lambda_1$ elements. There are $\binom{n-\lambda_1}{\lambda_2}$ ways to do this.
(iii) Choose λ_3 elements from the remaining $n-\lambda_1-\lambda_2$ elements.
(⋮)
(Ω) Choose λ_k elements from the remaining $n-\lambda_1-\cdots-\lambda_{k-1}$ elements. There are $\binom{n-\lambda_1-\cdots-\lambda_{k-1}}{\lambda_k}$ ways to do this.

Thus, by the Multiplication Principle,

$$\binom{n}{\lambda_1,\dots,\lambda_k} = \binom{n}{\lambda_1}\binom{n-\lambda_1}{\lambda_2}\binom{n-\lambda_1-\lambda_2}{\lambda_3}\cdots\binom{n-\lambda_1-\cdots-\lambda_{k-1}}{\lambda_k}.$$

Since $\binom{n}{k} = \frac{n!}{k!(n-k)!}$, it follows that:

$$\binom{n}{\lambda_1,\dots,\lambda_k} = \binom{n}{\lambda_1}\binom{n-\lambda_1}{\lambda_2}\binom{n-\lambda_1-\lambda_2}{\lambda_3}\cdots\binom{n-\lambda_1-\cdots-\lambda_{k-1}}{\lambda_k}$$

$$= \frac{n!}{\lambda_1!(n-\lambda_1)!}\,\frac{(n-\lambda_1)!}{\lambda_2!(n-\lambda_1-\lambda_2)!}\cdots\frac{(n-\lambda_1-\cdots-\lambda_{k-1})!}{\lambda_k!(n-\lambda_1-\cdots-\lambda_k)!}$$

$$= \frac{n!}{\lambda_1!\cdots\lambda_k!}.$$

□

Example 3.6.3 Evaluate each of the following multinomial coefficients:

(i)
$$\binom{12}{4,3,5};$$

(ii)
$$\binom{25}{7,6,5,7};$$

(iii)
$$\binom{60}{7,10,9,8,19,7};$$

(iv)
$$\binom{25}{5,5,5,5,5}.$$

Solution

(i)
$$\binom{12}{4,3,5} = \frac{12!}{4!3!5!} = 27720;$$

(ii)
$$\binom{25}{7,6,5,7} = \frac{25!}{7!6!5!7!} = 7067582121600;$$

(iii)
$$\binom{60}{7,10,9,8,19,7} = \frac{60!}{7!10!9!8!19!7!};$$

(iv)
$$\binom{25}{5,5,5,5,5} = 623360743125120.$$

□

We now generalize the Binomial Theorem to the expansion of multinomials.

Theorem 3.6.4 *(The Multinomial Theorem) For all $n \in \mathbb{N}$, we have:*

$$(x_1 + \cdots + x_k)^n = \sum_{\lambda_k=0}^{n-\lambda_1-\cdots-\lambda_{k-1}} \cdots \sum_{\lambda_2=0}^{n-\lambda_1} \sum_{\lambda_1=0}^{n} \binom{n}{\lambda_1, \cdots, \lambda_k} x_1^{\lambda_1} \cdots x_k^{\lambda_k}$$
$$= \sum_{\substack{\lambda_1 + \cdots + \lambda_k = n \\ \lambda_1, \ldots, \lambda_k \geq 0}} \binom{n}{\lambda_1, \cdots, \lambda_k} x_1^{\lambda_1} \cdots x_k^{\lambda_k}.$$

Proof By definition,

$$(x_1 + \cdots + x_k)^n =$$

$$\underbrace{(x_1 + \cdots + x_k) \cdots (x_1 + \cdots + x_k)}_{n \text{ times}}.$$

To achieve $x_1^{\lambda_1} \cdots x_k^{\lambda_k}$ in the expansion, we:

(i) Choose λ_1 terms from this product to take our x_1 term;
(ii) Choose λ_2 of the remaining terms to take x_2;

($\vdots$)
(Ω) Choose λ_k of the remaining terms to take x_k.

There are $\binom{n}{\lambda_1, \ldots, \lambda_k}$ ways to do this by definition. ■

Summations such as the one in Theorem 3.6.4 are often referred to as *k-fold summations*. While the notation:

$$\sum_{\lambda_k=0}^{n-\lambda_1-\cdots-\lambda_{k-1}} \cdots \sum_{\lambda_2=0}^{n-\lambda_1} \sum_{\lambda_1=0}^{n}$$

is perhaps technically more precise, we prefer the more compact notation,

$$\sum_{\substack{\lambda_1 + \cdots + \lambda_k = n \\ \lambda_1, \ldots, \lambda_k \geq 0}}.$$

We will use the second notation throughout the remainder of this book.

Example 3.6.5 Expand $(x + y + z)^4$.

Solution First we list all ordered weak partitions of four into three parts. The required partitions are:

$$(0, 0, 4), (0, 1, 3), (1, 0, 3), (0, 2, 2), (2, 0, 2), (1, 1, 2),$$

$$(0, 3, 1), (3, 0, 1), (1, 2, 1), (2, 1, 1), (0, 4, 0), (4, 0, 0).$$

Hence by the Multinomial Theorem,

$$(x + y + z)^4 = \sum_{\substack{\lambda_1 + \lambda_2 + \lambda_3 = 4 \\ \lambda_1, \lambda_2, \lambda_3 \geq 0}} \binom{4}{\lambda_1, \lambda_2, \lambda_3} x^{\lambda_1} y^{\lambda_2} z^{\lambda_3}$$

$$= \binom{4}{0,0,4}x^0y^0z^4 + \binom{4}{0,1,3}x^0y^1z^3 + \binom{4}{1,0,3}x^1y^0z^3 + \binom{4}{0,2,2}x^0y^2z^2$$
$$+\binom{4}{2,0,2}x^2y^0z^2 + \binom{4}{1,1,2}x^1y^1z^2 + \binom{4}{0,3,1}x^0y^3z^1 + \binom{4}{3,0,1}x^3y^0z^1$$
$$+\binom{4}{1,2,1}x^1y^2z^1 + \binom{4}{2,1,1}x^2y^1z^1 + \binom{4}{0,4,0}x^0y^4z^0 + \binom{4}{4,0,0}x^4y^0z^0$$
$$= z^4 + 4yz^3 + 4xz^3 + 6y^2z^2 + 6x^2z^2 + 12xyz^2$$
$$+4y^3z + 4x^3z + 12xy^2z + 12x^2yz + y^4 + x^4.$$

□

Example 3.6.6 Find the coefficient of $x^2y^4z^3$ in each of the following:

(i) $(x + y + z)^9$;
(ii) $(1 + x + y + z)^{13}$;
(iii) $(1 + 2x + 3y + 4z)^{14}$;
(iv) $(4 + x + 2y + 5z)^{17}$.

Solution

(i) By the Multinomial Theorem, the coefficient on $x^2y^4z^3$ in $(x + y + z)^9$ is given by $\binom{9}{2,4,3}$.
(ii) Here, we must find the coefficient of $1^4x^2y^4z^3$ in $(1 + x + y + z)^{13}$. By the Multinomial Theorem, this is given by $\binom{13}{4,2,4,3}$.
(iii) Similarly, the coefficient of $1^5(2x)^2(3y)^4(4z)^3$ is $2^23^44^3\binom{14}{4,2,4,3}$.
(iv) By the same logic as above, the coefficient is $4^81^22^45^3\binom{17}{7,2,4,3}$.

□

Corollary 3.6.7 *For all $n, k \in \mathbb{N}$, we have:*

$$\sum_{\substack{\lambda_1 + \cdots + \lambda_k = n \\ \lambda_1, \cdots, \lambda_k \geq 0}} \binom{n}{\lambda_1, \cdots, \lambda_k} = k^n.$$

Proof Let $x_1 = \cdots = x_k = 1$ in Theorem 3.6.4. ■

We will later be able to give a combinatorial proof of the above Corollary.

Theorem 3.6.8 *(Generalized Stars and Bars) Let $k \in \mathbb{N}$. The number of ways to arrange λ_i stars labeled i, for $i = 1, ..., k$, is given by*

$$\binom{\lambda_1 + \cdots + \lambda_k}{\lambda_1, ..., \lambda_k}.$$

Proof Let $n = \lambda_1 + \cdots + \lambda_k$. Note that there are:

(i) $\binom{n}{\lambda_1}$ ways to place the λ_1 stars labeled 1 in the n available slots;
(ii) $\binom{n-\lambda_1}{\lambda_2}$ ways to place the λ_2 stars labeled 2 in the $n - \lambda_1$ remaining slots;
(iii) $\binom{n-\lambda_1-\lambda_2}{\lambda_3}$ ways to place the λ_3 stars labeled 3 in the $n - \lambda_1 - \lambda_2$ remaining slots;

($\vdots$)
(Ω) $\binom{n-\lambda_1-\cdots-\lambda_{k-1}}{\lambda_k}$ ways to place the λ_k stars labeled k in the $n - \lambda_1 - \cdots - \lambda_{k-1}$ remaining slots.

Thus, by the Multiplication Principle, the number of arrangements is given by

$$\binom{n}{\lambda_1}\binom{n-\lambda_1}{\lambda_2}\binom{n-\lambda_1-\lambda_2}{\lambda_3}\cdots\binom{n-\lambda_1-\cdots-\lambda_{k-1}}{\lambda_k}$$
$$= \binom{n}{\lambda_1, \ldots, \lambda_k} = \binom{\lambda_1 + \cdots + \lambda_k}{\lambda_1, \ldots, \lambda_k}.$$

■

Recall that an *anagram* for a word contains the same letters as the original word with each letter occurring the same number of times as the original word. For example, "era" and "ear" are anagrams. As usual, we will not assume that the resulting "word" has any meaning.

Example 3.6.9 How many anagrams are there for the word "MATHEMATICS?"

Solution Note that the word "MATHEMATICS" contains two m's, two a's, two t's, one h, one e, one i, one c, and one s. Each distinct letter can be thought of a star labeled with the appropriate letter. By the generalization of stars and bars, the number of anagrams is given by

$$\binom{11}{2, 2, 2, 1, 1, 1, 1, 1} = 4989600.$$

□

The idea of a lattice path can be extended to any number of dimensions. In this case, we look for paths from $(0, \ldots, 0)$ to $(\lambda_1, \ldots, \lambda_k)$ in which only steps of the form $n_i = (0, \ldots, 0, 1, 0, \ldots, 0)$, where '1' is in the ith coordinate, can be taken. This can be thought of as arranging λ_i indistinguishable copies of n_i, for $i = 1, \ldots, k$. This being the case, the number of lattice paths from $(0, \ldots, 0)$ to $(\lambda_1, \ldots, \lambda_k)$ is given by

$$\binom{\lambda_1 + \cdots + \lambda_k}{\lambda_1, \ldots, \lambda_k}.$$

Example 3.6.10 Find the number of lattice paths from $(0, 0, 0, 0)$ to $(5, 4, 8, 7)$ that pass through neither the point $(2, 2, 4, 4)$ nor the point $(4, 3, 5, 5)$.

Solution Let S denote the set of all lattice paths from $(0, 0, 0, 0)$ to $(5, 4, 8, 7)$. Let A denote the set of all lattice paths from $(0, 0, 0, 0)$ to $(5, 4, 8, 7)$ that pass through the point $(2, 2, 4, 4)$. Let B denote the set of all lattice paths from $(0, 0, 0, 0)$ to $(5, 4, 8, 7)$ that pass through the point $(4, 3, 5, 5)$. We now determine the cardinality of each set involved.

(i)

$$|S| = \binom{5+4+8+7}{5,4,8,7} = \binom{24}{5,4,8,7};$$

(ii) The number of paths from $(0, 0, 0, 0)$ to $(2, 2, 4, 4)$ is given by $\binom{12}{2,2,4,4}$. The number of paths from $(2, 2, 4, 4)$ to $(5, 4, 8, 7)$ is given by

$$\binom{5+4+8+7-(2+2+4+4)}{5-2, 4-2, 8-4, 7-4} = \binom{12}{3,2,4,3}.$$

Thus by the Multiplication Principle,

$$|A| = \binom{12}{2,2,4,4}\binom{12}{3,2,4,3}.$$

(iii) The number of paths from $(0, 0, 0, 0)$ to $(4, 3, 5, 5)$ is given by $\binom{17}{4,3,5,5}$. The number of paths from $(4, 3, 5, 5)$ to $(5, 4, 8, 7)$ is given by

$$\binom{5+4+8+7-(4+3+5+5)}{5-4, 4-3, 8-5, 7-5} = \binom{7}{1,1,3,2}.$$

Thus by the Multiplication Principle,

$$|B| = \binom{17}{4,3,5,5}\binom{7}{1,1,3,2}.$$

(iv) The number of paths from $(0, 0, 0, 0)$ to $(2, 2, 4, 4)$ is given by $\binom{12}{2,2,4,4}$. The number of paths from $(2, 2, 4, 4)$ to $(4, 3, 5, 5)$ is given by

$$\binom{4+3+5+5-(2+2+4+4)}{4-2, 3-2, 5-4, 5-4} = \binom{5}{2,1,1,1}.$$

The number of paths from $(4, 3, 5, 5)$ to $(5, 4, 8, 7)$ is given by $\binom{7}{1,1,3,2}$. By the Multiplication Principle, the number of lattice paths passing through both $(2, 2, 4, 4)$ and $(4, 3, 5, 5)$ is given by

$$|A \cap B| = \binom{12}{2,2,4,4}\binom{5}{2,1,1,1}\binom{7}{1,1,3,2}.$$

Then, by the Principle of Inclusion and Exclusion, the number of paths passing through either $(2, 2, 4, 4)$ or $(4, 3, 5, 5)$ is given by

$$|A \cup B| = |A| + |B| - |A \cap B|$$

$$= \binom{12}{2,2,4,4}\binom{12}{3,2,4,3} + \binom{17}{4,3,5,5}\binom{7}{1,1,3,2}$$
$$- \binom{12}{2,2,4,4}\binom{5}{2,1,1,1}\binom{7}{1,1,3,2}.$$

By the Subtraction Principle, the number of lattice paths that pass through neither $(2,2,4,4)$ nor $(4,3,5,5)$ is given by

$$|(A \cup B)^c| = |S| - |A \cup B|$$
$$= \binom{24}{5,4,8,7} - \left(\binom{12}{2,2,4,4}\binom{12}{3,2,4,3}\right.$$
$$\left. + \binom{17}{4,3,5,5}\binom{7}{1,1,3,2} - \binom{12}{2,2,4,4}\binom{5}{2,1,1,1}\binom{7}{1,1,3,2}\right).$$

□

In our previous examples, the sets were distinguishable, in the next example, we will consider sets of the same size to be indistinguishable.

Example 3.6.11 Suppose that 20 children are to be placed into six unlabeled groups. We are to have two groups of four and four groups of three. How many ways can this be done?

Solution There are $\binom{20}{4,4,3,3,3,3}$ ways to place the children into groups. However, as the two groups of four are indistinguishable, we divide by 2. Further, the four groups of three are indistinguishable. Hence, we divide by 4!. Therefore, the number of ways to place the children into groups is given by

$$\frac{1}{2(4!)}\binom{20}{4,4,3,3,3,3} = 67897830000.$$

□

Exercise 3.6.12 Evaluate each of the following multinomial coefficients:

(i)

$$\binom{11}{3,2,6};$$

(ii)

$$\binom{27}{8,7,3,9};$$

(iii)

$$\binom{55}{5,11,8,9,15,7};$$

(iv)

$$\binom{40}{8,8,8,8,8}.$$

Exercise 3.6.13 Expand $(x+y+z)^5$.

Exercise 3.6.14 Find the coefficient of $x^3y^4z^5$ in each of the following:

(i) $(x+y+z)^{12}$;
(ii) $(1+x+y+z)^{17}$;
(iii) $(1+2x+3y+4z)^{15}$;
(iv) $(4+x+2y+5z)^{20}$.

Exercise 3.6.15 At a particular company, security codes must be ten digit numbers. Three of the digits must come from $\{1,4,5,7\}$, three of the digits must come from $\{2,3,6\}$, and four of the digits must come from $\{0,8,9\}$. Repetition of digits is allowed.

(i) How many security codes are there?
(ii) How many of these security codes neither start with 1 nor end with 9?

Exercise 3.6.16 Find the number of lattice paths from $(0,0)$ to (n,n) if we allow diagonal moves (in other words, north-east moves).

Exercise 3.6.17 Find the number of lattice paths from $(0,0)$ to (n,k) if we allow diagonal moves (in other words, north-east moves). Hint: You may assume without loss of generality that $n \geq k$.

Exercise 3.6.18

(i) Find the number of lattice paths from $(0,0)$ to $(n,2n)$ if we allow moves of the form $(1,2)$.
(ii) Find the number of lattice paths from $(0,0)$ to $(n,3n)$ if we allow moves of the form $(1,3)$.
(iii) What can you say about the number of lattice paths from $(0,0)$ to (n,kn) if we allow moves of the form $(1,k)$?

Exercise 3.6.19 How many anagrams are there for the word "COMBINATORICS?"

Exercise 3.6.20 How many anagrams are there for the word "CONTINUOUS?"

Exercise 3.6.21 Find the number of lattice paths from $(0,0,0,0,0)$ to $(7,5,6,8,10)$ that do not pass through the point $(3,2,4,6,5)$ nor the point $(5,4,5,7,8)$.

Exercise 3.6.22 Find the number of lattice paths from $(0,0,0,0,0,0)$ to $(8,7,9,4,5,8)$ that pass through either the point $(3,4,5,2,2,5)$ or the point $(5,5,7,3,3,7)$.

Exercise 3.6.23 Suppose that 30 children are to be placed into eight unlabeled groups. We are to have three groups of five and five groups of three. How many ways can this be done?

Exercise 3.6.24 Prove the following identities:

(i)

$$\sum_{\substack{\lambda_1+\lambda_2+\lambda_3=n \\ \lambda_1,\lambda_2,\lambda_3\geq 0}} \binom{n}{\lambda_1,\lambda_2,\lambda_3} = 3^n,$$

(ii)

$$\sum_{\substack{\lambda_1+\lambda_2+\lambda_3=n \\ \lambda_1,\lambda_2,\lambda_3\geq 0}} \binom{n}{\lambda_1,\lambda_2,\lambda_3}\lambda_1 = n3^{n-1},$$

(iii)

$$\sum_{\substack{\lambda_1+\lambda_2+\lambda_3=n \\ \lambda_1,\lambda_2,\lambda_3\geq 0}} \binom{n}{\lambda_1,\lambda_2,\lambda_3}\lambda_1\lambda_2 = n(n-1)3^{n-2}.$$

Exercise 3.6.25 Generalize the results of Exercise 3.6.24 as much as possible.

3.7 Application: Cryptosystems and the Enigma

Suppose that Alice needs to transmit a message to Bob. However, there is always a danger that the message will be intercepted in transit by a third party, Oscar. Thus, Alice and Bob seek to disguise the message in such a way that it is very difficult, or at least impractical, for Oscar to read the message. In this case, the original message is called the *plaintext*, while the disguised message is called the *ciphertext*.

As an example, we consider the *Caesar cipher*. In the Caesar cipher, each letter in the alphabet is assigned a non-negative integer. So 'A' is assigned 0, 'B' is assigned 1,..., and 'Z' is assigned 25. Alice and Bob then select an integer $k \in [26]$. To send the word 'STAY,' they convert the plaintext to numbers, in this case 18, 19, 0, 24. For each number x in the plaintext, Alice then computes $x + k \pmod{26}$. So if $k = 3$, then the ciphertext is 21, 22, 3, 1 or 'VWDB.' This is then transmitted to Bob.

In general, we can assume that Oscar has received the disguised message. We can also assume that Oscar knows the general method that Alice and Bob are using to disguise their messages. So, much of the security of the message is dependent on a *message key*. A message key is the specific method used within the system to disguise the message. So for instance, the specific k used within the Caesar cipher is the message key for that cipher.

Definition 3.7.1 A *cryptosystem* consists of all possible plaintexts, all possible ciphertexts, all possible message keys, an encryption rule corresponding to each message key, and its inverse function.

We will primarily be concerned with the set of all possible message keys, which we will denote K. A necessary condition for a secure cryptosystem is that $|K|$ is large. The reason for this is simple. If $|K|$ is small, then Oscar could break the code by trying all possible message keys. For example, the Caesar cipher is not very secure as there are only 26 possible message keys. However, it is not the case that a large number of possible message keys is a guarantee of a secure cryptosystem.

A generalization of the Caesar cipher is the *substitution cipher*. In the substitution cipher, each letter in the alphabet is replaced with another. For this reason, we can think of the substitution cipher as a permutation on the alphabet. Thus there are $26! \approx 4 \times 10^{26}$ possible message keys.

Numerically, the substitution cipher seems much more secure. However, there are linguistic weaknesses to the substitution cipher. Namely, the English language (or any human language) has patterns in it. Namely, certain letters appear more frequently than others. For instance, the letter 'E' is most common, accounting for 12.7 % of all letters. The next most common letters are 'T' (at 9.1 %), 'A' (at 8.2 %), 'O' (at 7.5 %), and 'S' (at 6.3 %). If someone was looking to break a substitution cipher, they can also look at pairs of letters (the most common being 'TH,' 'HE,' 'IN,' 'ER,' and 'AN') or sequences of three letters (the most common being 'THE,' 'ING,' 'AND,' 'HER,' and 'ERE'). These frequencies will be preserved in the ciphertext. Therefore, the most frequently used letter in the ciphertext is likely the one that corresponds to 'E.'

One way to improve a substitution cipher is to use a *polyalphabetic substitution*. In a polyalphabetic substitution, multiple substitution alphabets are used. This can mask the frequencies in the plaintext. Perhaps the most famous example of a polyalphabetic cipher is the *Vigenère cipher*. In the Vigenère cipher, a keyword is selected and written as a sequence of numbers as with the Caesar cipher. Suppose that we select 'MATH' (12, 0, 19, 7) as our key word. Since our keyword is of length four, we break our plaintext into blocks of length four, writing each block as a sequence of numbers. Suppose that $x_1,...,x_4$ are the numbers in the first block of the plaintext. To obtain the ciphertext, we add each of the corresponding numbers in the keyword, reducing modulo 26. So if our plaintext is 'COMBINATORICS,' then our first block is (2, 14, 12, 1). The corresponding block in the ciphertext is $(2 + 12 \pmod{26}, 14 + 0 \pmod{26}, 12 + 19 \pmod{26}, 1 + 7 \pmod{26}) = (14, 14, 5, 8)$.

Repeating this process with remaining blocks, yields a ciphertext of 'OOFIUZTAARBJE' $(14, 14, 5, 8, 20, 25, 19, 0, 0, 17, 1, 9, 4)$.

Example 3.7.2 If the plaintext has n characters, then find the number of possible message keys in the Vigenère cipher.

Solution The length of the message key can be at most the length of the message itself. Thus even if the keyword is four letters, say 'MATH' and $n = 10$, then the message key would be 'MATHMATHMA.' Hence there are at most 26^n possible message keys. □

Friedrich Kasiski was the first to publish a general method for breaking the Vigenère cipher. Kasiski's method involves first determining the length of the keyword (or the *period* of the cipher), then using a frequency analysis to determine the specific key. For a more detailed description of the cryptanalysis of the Vigenère cipher (particularly in the case where the keyword is as long as itself), the reader is encouraged to look at many of the excellent sources on cryptography. In particular, *The Code Book* by Simon Singh [39] and *Cryptography Theory and Practice* by Douglas Stinson [42]. We now turn our attention to the *Enigma machine* made famous for its use by the German military during World War II. The Enigma was invented by Arthur Scherbius in 1918. The Enigma made use of both electrical and mechanical components to create a polyalphabetic cipher. There were several versions of the Enigma used by the German military, however we will concentrate our efforts on the version used by the Army. Further, a description of the cryptanalysis of the Enigma is too detailed and would be out of place in this text. Interested readers are referred to the many excellent books written on the Enigma including *Enigma: How the Poles Broke the Nazi Code* by Kozaczuk and Straszak [33] as well as *The German Enigma Cipher Machine: Beginnings, Success, and Ultimate Failure* edited by Winkel, Deavours, Kahn, and Kruh [47].

For our purposes, we will be content with a description of the components of the Enigma (see Fig. 3.4) and the number of message keys. The Enigma consisted of a keyboard, where plaintext was entered, and a lampboard, where the ciphertext was displayed. The front panel of the Enigma had a plugboard that allowed the operator to connect pairs of letters using cables. The effect of the plugboard was to switch connected letters after input and before display. Suppose that 'A' and 'B' were connected on the plugboard. When 'A' is pressed, it is switched to 'B' before being sent into the next step of the Enigma cipher. Similarly, if the inner workings of the Enigma sent a 'B,' it would be switched to 'A' before it was displayed on the lampboard.

Proposition 3.7.3 *If the Enigma operator uses* k $(k = 0, \ldots, 13)$ *cables, then the number of plugboard combinations is given by*

$$\frac{26!}{2^k(26-2k)!k!}.$$

Proof For each of the k cables, the operator chooses a pair of letters to connect. The cables are indistinguishable and there is no order to each pair of letters. Thus, this is given by

$$\binom{26}{2, \ldots, 2, 26-2k}\frac{1}{k!} = \frac{26!}{2^k(26-2k)!k!}.$$

■

Fig. 3.4 The Enigma machine

The internal workings of the Enigma consisted of three rotors (see Fig. 3.5), each of which would perform a different substitution cipher. Each day, the German military would specify which rotors would be used, the order they were to be placed in the assembly, and their initial position for each message. The assembly also featured a reflector. When a letter passed through the three rotors, it would then pass through the reflector. The reflector would switch pairs of letters before sending the signal back through the rotor assembly.

The Enigma featured a mechanical component that would turn the rotors every time a key was pressed. The first rotor would turn after every letter press. The second rotor would rotate every 26 keystrokes. The final rotor would rotate every 676 keystrokes. Thus, the substitution being used would change with every letter. Further, it would take many keystrokes before the rotors would return to their original positions. The final variable component of the Enigma was two rings. The first ring was between the first and second rotor. This allowed the operator to alter when the second rotor would be turned. So for instance, if the operator turned this ring to position 7, then the second rotor would turn when the first rotor passed through 7. The second ring was between the second and third rotors and performed an analogous function.

Fig. 3.5 The rotor assembly

Example 3.7.4 Suppose that the wiring of the rotors and the reflector are unknown. Further suppose that it is unknown how many plugboard cables are being used. Determine the number of possible configurations for the Enigma.

Solution We count the number of possible configurations by considering 14 disjoint, exhaustive sets. The kth set, A_k ($k = 0, \ldots, 13$), will be the set of all configurations in which k plugboard cables are being used. The cardinality of this set can be determined by:

(i) Counting the ways in which the plugboard can be configured. There are $\frac{26!}{2^k(26-2k)!k!}$ ways to do this by Proposition 3.7.3.
(ii) Counting the ways to wire the three rotors. There are 26! possibilities for the first rotor. The second rotor can have any wiring other than that which was used on the first rotor. So there are $26! - 1$ ways to wire the second rotor. Similarly, there are $26! - 2$ ways to wire the third rotor. So, by the Multiplication Principle, the number of ways to wire the three rotors is given by

$$26!(26! - 1)(26! - 2).$$

(iii) Counting the possible ways to set the two rings. Each ring can have any one of 26 positions. Thus, there are 26^2 possibilities.
(iv) Counting the ways to wire the reflector. This is equivalent to selecting 13 indistinguishable 2-subsets from [26]. There are $\frac{26!}{2^{13}13!}$ ways to do this.

Thus, by the Multiplication Principle,

$$|A_k| = \left(\frac{26!}{2^k(26-2k)!k!}\right)(26!(26! - 1)(26! - 2))(26^2)\left(\frac{26!}{2^{13}13!}\right).$$

So, by the Addition Principle, the total number of configurations is given by

$$\sum_{k=0}^{13} |A_k| = \sum_{k=0}^{13} \left(\frac{26!}{2^k(26-2k)!k!} \right) (26!(26!-1)(26!-2))(26^2) \left(\frac{26!}{2^{13}13!} \right)$$
$$\approx 3 \times 10^{114}.$$

□

One of the key steps in the cryptanalysis of the Enigma was to determine the wiring on the rotors and the reflectors. This was done by Marion Rejewski of the Polish Cipher Bureau in 1932. Intelligence documents revealed that the German procedure at the time was to use only three rotors (which could be placed in any order) and exactly six plugboard cables.

Example 3.7.5 Suppose that we know the wiring of the rotors and the reflector. Further, we know that the Germans were using three rotors (which could be placed in any order) and exactly six plugboard cables. Determine the possible number of configurations for the Enigma.

Solution This can be done by:

(i) Counting the number of configurations of the plugboard. There are $\frac{26!}{2^6 14!6!}$ configurations by Proposition 3.7.3.
(ii) Counting the number of ways to arrange the three rotors. There are $3! = 6$ ways to order the rotors.
(iii) Counting the number of ways to set the two rings. As above, there are 26^2 ways to do this.

By the Multiplication Principle, the number of settings is given by

$$\left(\frac{26!}{2^6 14!6!} \right) * 6 * 26^2 \approx 4 \times 10^{14}.$$

□

Exercise 3.7.6 Suppose that we want to design a substitution cipher that is capable of being spoken. For this to be possible, we substitute vowels for vowels (in this case, the vowels are 'a,' 'e,' 'i,' 'o,' 'u,' and 'y') and consonants with consonants. How many possible message keys are there?

Exercise 3.7.7 Later in the war, the Germans began using exactly ten plugboard cables. They also began selecting three rotors from a pool of five possible rotors. However, the wiring of the rotors and the reflector were still known. Find the number of possible configurations of the Enigma.

Chapter 4
Distribution Problems

4.1 Introduction

In this chapter, we examine *the problem of occupancy*. The problem of occupancy is a *distribution problem*. In a distribution problem, we are to place a set of objects, called "balls" into a set of containers, called "urns." In this chapter, we will always assume that there are n balls and k urns.

Some natural questions occur:

(i) *Must each urn receive at least one ball?* Note that if we require that each urn must receive at least one ball, then we must have $n \geq k$ by the Pigeonhole Principle. This requirement can usually be accomplished by assigning one ball into each urn. We can then assign the remaining balls into urns assuming no such restriction.

(ii) *Can any urn receive more than one ball?* If each urn can receive at most one ball, then by the Pigeonhole Principle, we must have $k \geq n$. If we assume that each urn must receive exactly one ball, then $n = k$.

(iii) *Can we distinguish the balls?* If we can distinguish the balls, then we assume that the balls are *labeled* with the numbers $1, ..., n$. If the balls are unlabeled, then we are only concerned with how many balls are placed into each urn, not which balls are grouped together.

(iv) *Can we distinguish the urns?* If we can distinguish the urns, then we assume that the urns are *labeled* with the numbers $1, ..., k$. In which case, it matters where the balls are placed.

As an example of the differences created by these scenarios, consider Table 4.1. This table illustrates all possible different distributions of three balls into two urns. Note that when the urns are unlabeled, we need not consider all possible permutations of the urns. Similarly, if the balls are unlabeled, then we need only consider the number of balls that have been placed in each urn.

We now begin to determine the number of ways to distribute n balls into k urns. The easiest case is when we require that each urn receive exactly one ball.

R. A. Beeler, *How to Count*, DOI 10.1007/978-3-319-13844-2_4

Table 4.1 Distributions of three balls into two urns

	Labeled	urns	Unlabeled	urns
Labeled balls	123			
	12	3		
	13	2	123	
	23	1	12	3
	1	23	13	2
	2	13	23	1
	3	12		
		123		
Unlabeled balls	***			
	**	*	***	
	*	**	**	*
		***		

Proposition 4.1.1 *The number of distributions of n balls into k urns in which each urn must receive exactly one ball is given by:*

(i) 0 *when* $n \neq k$*;*
(ii) 1 *when* $n = k$ *and either the balls or the urns are unlabeled;*
(iii) $n!$ *when* $n = k$ *and both the balls and the urns are labeled.*

Proof

(i) If $k > n$, then there is at least one urn which receives no ball. If $n > k$, then there is at least one urn which receives more than one ball by the Pigeonhole Principle. As either of these cases is not allowed, there is no way to accomplish the task.
(ii) If $n = k$ and either the balls or urns are unlabeled, then there is exactly one distribution. This is the distribution that assigns exactly one ball to each urn. Since the balls or urns are unlabeled, we are not concerned with the permutations on the set of urns or the set of balls.
(iii) This is equivalent to the number of bijections between $[n]$ and itself. This is given by $n!$. ■

We note that the "balls" and "urns" may not be "balls" and "urns" in the traditional sense as we will see in the next few examples.

Suppose that we have ten cherry gumdrops to be given to four children. In this case, the children can be considered as "labeled urns." Gumdrops, provided they are all the same flavor, can be thought of as "unlabeled balls." Any child would throw a fit if they were not given a gumdrop when the other children were given gumdrops. This being the case, we would be wise to restrict ourself to assignments in which each child receives at least one gumdrop.

A variation on the above example would be to consider the case where we are are offering the children gumdrops before dinner. Again, the gumdrops can be considered "unlabeled balls" and the children can be considered "labeled urns." However, if we are concerned about spoiling the children's dinner, we would be wise to give each child at most one gumdrop.

Note that children (and people in general) are not always to be considered as "urns." Suppose that a teacher wishes to place their students in groups to work on a project. In terms of groups, children are usually only concerned with the other children in the group, not the name of the group (if such a name exists). Thus, we think of the groups as "unlabeled urns" and the children as "labeled balls." By definition, each group must contain at least one child.

As a final example, we consider a scheduling problem. At a college or university, no two classes can meet in the same room at the same time. This being the case, we think of each room during a specific block of time as a "labeled urn." For instance, Room 1 at 1 PM, Room 2 at 1 PM, and Room 1 at 2 PM would be considered as three different urns. Each class can be thought of as a "labeled balls." As no two classes may be in the same room at the same time, we assume that the urns can have at most one ball.

In practice, the above scheduling problem is not entirely straight forward. For instance, suppose that the same instructor teaches two sections of the same course. This being the case, we may consider the two sections to be indistinguishable. However, we would still be able to distinguish the sections from other courses by different instructors. Further, the instructor can not be expected to teach two classes at the same time. The scheduling problem is further complicated by the requirements of the individual classes. For instance, certain classes may require a room with computer access. Other classes may require more seats to accommodate a larger number of students.

In even a small university, there may be dozens of such restrictions. As the number of restrictions escalates, our concern shifts from determining the number of possibilities. Our concern quickly becomes determining if satisfying all of the restrictions is even possible.

4.2 The Solution of Certain Distribution Problems

The goal of this section is to present the solution to certain distribution problems. Specifically, we will solve those distribution problems in which the necessary machinery has already been developed. In some cases, these solutions are simply a translation of the vocabulary of "balls and urns" into the vocabulary of previous sections. These cases should be considered a review of previous material. In other cases, we will be extending these results to include those cases not dealt with previously.

Generally the cases in which we require that no urn receives more than one ball are the easiest. Hence, we begin with these cases.

Proposition 4.2.1 *The number of ways to distribute n balls to k unlabeled urns in such a way that no urn receives more than one ball is given by:*

(i) 0 if $n > k$;
(ii) 1 otherwise.

Proof If $n > k$, then at least one urn will receive more than one ball by the Pigeonhole Principle. Thus, there is no way to do this.

If $n \leq k$, then there is exactly one distribution. This is the distribution that places exactly one ball in n of the k urns. As the urns are unlabeled, permutations on the set of urns are irrelevant. So there is exactly one way to do this. ■

Now we deal with the case where we distribute n balls to k labeled urns in such a way that no urn receives more than one ball.

Proposition 4.2.2 *The number of ways to distribute n balls to k labeled urns such that no urn receives more than one ball is given by:*

(i) $\binom{k}{n}$ when the balls are unlabeled;
(ii) $P(k, n)$ when the balls are labeled.

Proof

(i) Suppose that the balls are not labeled. As each of the k labeled urns may receive at most one ball, we must simply choose n of the k urns to receive one of the n balls. There are $\binom{k}{n}$ ways to do this by definition.
(ii) Begin by removing the labels from the balls. There are then $\binom{k}{n}$ ways to distribute the now unlabeled balls into the labeled urns. We then assign labels to the balls. There are $n!$ ways to do this. Hence there are $\binom{k}{n}n!$ ways to distribute the balls. This is equivalent to $P(k, n)$ by Proposition 3.1.1. ■

We can also consider distributions with no restrictions, as we will see.

Proposition 4.2.3 *The number of ways to distribute n labeled balls to k labeled urns is given by k^n.*

Proof There are k choices for each of the n balls. Hence, by the Multiplication Principle, there are k^n such distributions. ■

Example 4.2.4 Find the number of distributions of n_1 unlabeled red balls and n_2 labeled white balls into k labeled urns such that no urn receives more than one red ball.

Solution The number of distributions of n_1 unlabeled (red) balls into k urns such that no urn receives more than one ball is given by $\binom{k}{n_1}$. The number of distributions of n_2 labeled (white) balls into k urns is given by k^{n_2}. Thus by the Multiplication Principle, the number of distributions is

$$\binom{k}{n_1}k^{n_2}.$$

□

Fig. 4.1 A distribution of 15 unlabeled balls into 6 labeled urns

$$\underbrace{*}_{U1}|\underbrace{*****}_{U2}|\underbrace{\ }_{U3}|\underbrace{****}_{U4}|\underbrace{**}_{U5}|\underbrace{***}_{U6}|$$

In some cases, it is desirable to look at distributions in which each urn receives a specified number of balls. This will be considered in the next proposition.

Proposition 4.2.5 *The number of distributions of $n = \lambda_1 + \cdots + \lambda_k$ labeled balls into k labeled urns in which the ith urn receives λ_i balls is*

$$\binom{n}{\lambda_1, \ldots, \lambda_k}.$$

Proof Follows immediately from the definition of the multinomial coefficient. ■

Now, we solve the problem of distributing n unlabeled balls into k labeled urns. To do this, think of the collection of urns as a set of dividers or bars. The balls will be considered as a collection of unlabeled stars. A distribution of balls into urns is then an arrangement of n stars and k bars, where the contents of the urn labeled i is considered to be the stars (balls) to the left of the ith bar and to the right of the $(i-1)$st bar. For example, in Fig. 4.1, we see that the first urn (U1) contains one ball, the second urn (U2) contains five balls, urn three is empty, the fourth urn contains four balls, there are two balls in the fifth urn, and three balls in the final urn. Notice that the final bar will always be in the last position in such an arrangement. This being the case, we could simply ignore the last bar and proceed.

Theorem 4.2.6 *The number of ways to distribute n unlabeled balls into k labeled urns is given by $\binom{n+k-1}{k-1}$.*

Proof Consider the urns as a collection of dividers or bars and the collection of balls as unlabeled stars. The contents of the urn labeled i is considered to be the stars (balls) to the left of the ith bar and to the right of the $(i-1)$st bar. Notice that the final bar is forced to be in the last position. Hence, this problem is equivalent to the arrangement of n stars and $k-1$ bars. There are $\binom{n+k-1}{k-1}$ ways to do this by Stars and Bars. ■

Proposition 4.2.7 *The number of ways to distribute n unlabeled balls into k labeled urns in such a way that no urn is empty is given by $\binom{n-1}{k-1}$.*

Proof Begin by placing one ball into each of the urns. This problem reduces to placing $n-k$ unlabeled balls into k labeled urns with no restriction. By Theorem 4.2.6, the number of distributions is

$$\binom{n-k+k-1}{k-1} = \binom{n-1}{k-1}.$$

■

Example 4.2.8

(i) Find the number of distributions of n_1 unlabeled red balls and n_2 unlabeled white balls into k labeled urns.
(ii) Find the number of distributions of n_1 unlabeled red balls and n_2 unlabeled white balls into k labeled urns if each urn must receive at least one white ball.
(iii) Find the number of distributions of n_1 unlabeled red balls and n_2 unlabeled white balls into k labeled urns if each urn must receive at least one ball.

Solution

(i) There are $\binom{n_1+k-1}{k-1}$ distributions of n_1 unlabeled red balls into k labeled urns by Theorem 4.2.6. Similarly, there are $\binom{n_2+k-1}{k-1}$ distributions of n_2 unlabeled white balls into k labeled urns. Thus by the Multiplication Principle, the number of distributions is given by

$$\binom{n_1+k-1}{k-1}\binom{n_2+k-1}{k-1}.$$

(ii) Again, there are $\binom{n_1+k-1}{k-1}$ distributions of n_1 unlabeled red balls into k labeled urns by Theorem 4.2.6. Since each urn must contain at least one white ball, there are $\binom{n_2-1}{k-1}$ distributions of the white balls by Proposition 4.2.7. By the Multiplication Principle, the number of distributions is

$$\binom{n_1+k-1}{k-1}\binom{n_2-1}{k-1}.$$

(iii) We solve this problem by considering k disjoint, exhaustive sets. Let A_i (for $i = 1, ..., k$) be the set of all distributions in which exactly i of the urns contain white balls. The cardinality of A_i may be computed as follows:
(a) Choose i of the labeled urns to receive white balls. There are $\binom{k}{i}$ ways to do this.
(b) Distribute the n_2 white balls into the i selected urns in such a way that no urn is left empty. There are $\binom{n_2-1}{i-1}$ ways to do this by Proposition 4.2.7.
(c) Place one red ball into each of the $k - i$ urns that were not selected in (a).
(d) Distribute the remaining n_1-k+i red balls into the k urns. As the restriction has been fulfilled, we need not consider the possibility of empty urns. By Theorem 4.2.6, the number of ways to distribute the remaining red balls is given by

$$\binom{n_1-k+i+(k-1)}{k-1} = \binom{n_1+i-1}{k-1}.$$

Thus, by the Multiplication Principle,

$$|A_i| = \binom{k}{i}\binom{n_2-1}{i-1}\binom{n_1+i-1}{k-1}.$$

Since the A_i are disjoint sets, it follows that the number of distributions is given by

$$\sum_{i=1}^{k} |A_i| = \sum_{i=1}^{k} \binom{k}{i}\binom{n_2-1}{i-1}\binom{n_1+i-1}{k-1}.$$

□

Note that in the previous example, the roles of the red balls and white balls can be reversed. In other words, the above derivation could have been accomplished by distributing the red balls first. This leads to the combinatorial identity:

$$\sum_{i=1}^{k} \binom{k}{i}\binom{n_2-1}{i-1}\binom{n_1+i-1}{k-1} = \sum_{i=1}^{k} \binom{k}{i}\binom{n_1-1}{i-1}\binom{n_2+i-1}{k-1}.$$

We now proceed with another example.

Example 4.2.9 Find the number of distributions of n unlabeled balls into k_1 labeled red urns and k_2 labeled white urns so that either the red urns or the white urns receive at least one ball.

Solution Let A be the set of all distributions in which the red urns receive at least one ball. Let B be the set of distributions in which the white urns receive at least one ball. We begin by counting the elements of A by considering n disjoint, exhaustive sets. The ith set A_i is the set of all distributions in which i balls are distributed to the red urns in such a way that each of the red urns receive at least one ball. To compute $|A_i|$:

(i) Distribute i of the unlabeled balls into the k_1 labeled red urns in such a way that the none of the red urns are left empty. There are $\binom{i-1}{k_1-1}$ ways to do this by Proposition 4.2.7.
(ii) Distribute the remaining $n-i$ unlabeled balls into the k_2 labeled white urns. The number of ways to do this is $\binom{n-i+k_2-1}{k_2-1}$ by Theorem 4.2.6.

The Multiplication Principle yields

$$|A_i| = \binom{i-1}{k_1-1}\binom{n-i+k_2-1}{k_2-1}.$$

By the Addition Principle, we have that

$$|A| = \sum_{i=1}^{n} |A_i| = \sum_{i=1}^{n} \binom{i-1}{k_1-1}\binom{n-i+k_2-1}{k_2-1}.$$

Reversing the roles of k_1 and k_2 gives the cardinality of B as

$$|B| = \sum_{i=1}^{n} \binom{i-1}{k_2-1}\binom{n-i+k_1-1}{k_1-1}.$$

However, some distributions may be in both A and B. Note that $A \cap B$ is the set of all distributions in which every urn receives at least one ball. In this case, we can simply ignore the colors. Hence, this is a distribution of n unlabeled balls into k_1+k_2 labeled urns such that no urn is empty. Proposition 4.2.7 gives the number of such distributions as

$$|A \cap B| = \binom{n-1}{k_1+k_2-1}.$$

Thus, the number of appropriate distributions is

$$\begin{aligned}|A \cup B| &= |A| + |B| - |A \cap B| \\ &= \sum_{\ell=0}^{1}\left(\sum_{i=1}^{n}\binom{i-1}{k_{\ell+1}-1}\binom{n-i+k_{2+(-1)^\ell}-1}{k_{2+(-1)^\ell}-1}\right) - \binom{n-1}{k_1+k_2-1}\end{aligned}$$

by the Principle of Inclusion and Exclusion. □

The problem of distributing balls into urns can be extended to other applications as well.

Example 4.2.10

(i) Find the number of non-negative integer solutions to $x_1 + x_2 + x_3 + x_4 = 20$.
(ii) Find the number of non-negative integer solutions to $x_1 + x_2 + x_3 + x_4 = 20$ if $x_1 \geq 3$, $x_2 \geq 5$, $x_3 \geq 6$, and $x_4 \geq 4$.

Solution

(i) We consider each variable x_i as a "labeled urn." Since only the specific values of the variables are relevant, we consider the sum of 20 to represent 20 unlabeled balls. Therefore this problem reduces down to that of distributing 20 unlabeled balls into 4 labeled urns. By Theorem 4.2.6, the number of non-negative integer solutions to $x_1 + x_2 + x_3 + x_4 = 20$ is

$$\binom{20+4-1}{4-1} = \binom{23}{3} = 1771.$$

(ii) We begin by making sure that each of the restrictions are satisfied. This is equivalent to placing three balls into the first urn, five balls into second urn, six into the third, and four into the final urn. Thus we have distributed 18 of the 20 balls. Hence this problem is equivalent to distributing two unlabeled balls into four unlabeled urns. By Theorem 4.2.6, the number of distributions is given by

$$\binom{2+4-1}{4-1} = \binom{5}{3} = 10.$$

□

A more complicated variation of the above example would be to consider cases where the x_i were bounded above. In other words, determine the number of non-negative integer solutions to $x_1 + x_2 + x_3 + x_4 = 20$, where $x_1 \leq 7$, $x_2 \leq 10$, $x_3 \leq 6$, and $x_4 \leq 13$. Unfortunately, the machinery necessary to solve this problem has not been developed at this time. We will revisit this problem in the next chapter.

Exercise 4.2.11 In 1907, Albert Einstein proposed a model for solids that consisted of a collection of k oscillators with identical properties (they did however differ in their spatial coordinates)and n indistinguishable units of energy. How many ways can the units of energy collect in the oscillators?

Exercise 4.2.12 Find the number of distinct k-tuples of positive integers whose sum is n.

Exercise 4.2.13

(i) Describe what the following multinomial coefficient counts in terms of a distribution of balls into urns:

$$\binom{11}{4,5,2}.$$

(ii) Give a combinatorial proof for the following identity:

$$\binom{11}{4,5,2} = \binom{10}{3,5,2} + \binom{10}{4,4,2} + \binom{10}{4,5,1}.$$

(iii) Generalize the result in (ii).

Exercise 4.2.14 Suppose that we have 15 $ bills to give to 10 children. If each child must receive at least 1 $, how many ways can this be done?

Exercise 4.2.15 Find the number of distributions of 10 balls into 6 urns if:

(i) Both the balls and the urns are labeled;
(ii) The balls are unlabeled and the urns are labeled;
(iii) The balls are unlabeled, the urns are labeled, and each urn receives at least one ball.

Exercise 4.2.16 Find the number of distributions of n unlabeled balls into k labeled urns if each urn must receive at least four balls.

Exercise 4.2.17 Twenty-five (indistinguishable) gumdrops must be distributed among five children: Alice, Bob, Chad, Diane, and Edward.

(i) How many distributions are possible?
(ii) How many distributions are possible if each child must receive at least one gumdrop?
(iii) How many distributions are possible if Alice must receive at least four gumdrops, Bob must receive at least three gumdrops, Chad must receive at least two gumdrops, Diane must receive at least three gumdrops, and Edward must receive at least five gumdrops?

Exercise 4.2.18 Find the number of distributions of n_1 labeled red balls and n_2 unlabeled white balls into k labeled urns if each urn must receive at least one white ball. What if each urn must receive both a red and a white ball?

Exercise 4.2.19 Give a combinatorial proof that:

$$\sum_{\substack{\lambda_1+\cdots+\lambda_k=n \\ \lambda_1,\ldots,\lambda_k \geq 0}} \binom{n}{\lambda_1,\ldots,\lambda_k} = k^n.$$

4.3 Partition Numbers and Stirling Numbers of the Second Kind

In the last section, we gave solutions to several distribution problems. However, there are several distribution problems that have not been addressed. In particular, in this section we explore the following problems:

(i) Distributions of unlabeled balls into unlabeled urns;
(ii) Distributions of unlabeled balls into unlabeled urns with no empty urns allowed;
(iii) Distributions of labeled balls into unlabeled urns;
(iv) Distributions of labeled balls into unlabeled urns with no empty urns allowed;
(v) Distributions of labeled balls into labeled urns with no empty urns allowed.

We begin with the problem of distributing n unlabeled balls into k unlabeled urns with no empty urns allowed. For motivation, we consider the following example.

Example 4.3.1 Find all distributions of 10 unlabeled balls into 4 unlabeled urns such that there are no empty urns.

Solution Since the urns are unlabeled, we arrange the urns in order of how many balls they contain. These distributions are given in Table 4.2. As we see, there are nine distributions of ten unlabeled balls into four unlabeled urns such that no urn is empty. □

Note that in the above example, we only need to know how many balls are placed in each urn. Thus these distributions can be listed more compactly in the following way:

$$\{1,1,1,7\} \quad \{1,1,2,6\} \quad \{1,1,3,5\}$$

$$\{1,1,4,4\} \quad \{1,2,2,5\} \quad \{1,2,3,4\}$$

$$\{1,3,3,3\} \quad \{2,2,2,4\} \quad \{2,2,3,3\}.$$

Notice that in each case there are four (not necessarily distinct) elements in each of the multisets. Moreover, the sum of the elements is ten. This notion is generalized in the following definition.

Table 4.2 All distributions of 10 unlabeled balls into 4 unlabeled urns such that no urn is empty

* * * *******	* * ** ******	* * *** *****
* * **** ****	* ** ** *****	* ** *** ****
* *** *** ***	** ** ** ****	** ** *** ***

Table 4.3 The number of partitions of n into k parts

$n \setminus k$	1	2	3	4	5	6	7	8	9	10
1	1									
2	1	1								
3	1	1	1							
4	1	2	1	1						
5	1	2	2	1	1					
6	1	3	3	2	1	1				
7	1	3	4	3	2	1	1			
8	1	4	5	5	3	2	1	1		
9	1	4	7	6	5	3	2	1	1	
10	1	5	8	9	7	5	3	2	1	1

Definition 4.3.2 A *partition of n into k parts* is a collection of k (not necessarily distinct) positive integers which sum to n. Let $p(n, k)$ denote the number of partitions of n into k parts.

Table 4.3 gives the value of $p(n, k)$ for small n and k. Note that $p(0, 0) = 1$, $p(n, 0) = 0$ for $n > 1$, and $p(n, k) = 0$ for $k > n$.

These partition numbers give us a way of counting the number of distributions of n unlabeled balls into k unlabeled urns such that no urn is left empty.

Proposition 4.3.3 *The number of distributions of n unlabeled balls into k unlabeled urns such that no urn is empty is given by* $p(n, k)$.

Proof It suffices to exhibit the required bijection. Consider the distribution in which λ_i unlabeled balls are placed in the ith urn for $i = 1, ..., k$. Since each urn must contain at least one ball, $\lambda_i > 0$ for $i = 1, ..., k$. Since there are n balls, we must have $\lambda_1 + \cdots + \lambda_k = n$. Thus $\{\lambda_1, ..., \lambda_k\}$ is a partition of n into k parts.

Similarly, suppose $\{\lambda_1, ..., \lambda_k\}$ is a partition of n into k parts. This corresponds to placing λ_i unlabeled balls into the ith urn. Since the $\lambda_i > 0$, no urn will be left empty. ■

It would be desirable to find a concise formula for $p(n, k)$. Unfortunately, we will have to be content with a recursive formula.

Theorem 4.3.4 *For all $n, k \in \mathbb{N}$, we have*

$$p(n,k) = p(n-1,k-1) + p(n-k,k).$$

Proof Note that $p(n, k)$ counts the number of partitions of n into k parts by definition. The right side also counts this by counting two disjoint, exhaustive sets:

(i) The set of partitions in which at least one of the parts is '1.' The remaining parts of the partition form a partition of $n-1$ into $k-1$ parts. There are $p(n-1, k-1)$ such partitions.
(ii) The set of partitions of n in which none of the parts are '1.' Since all of the parts of the partition are greater than one, we may subtract one from each part of the partition. The result is a valid partition of $n-k$ into k parts. There are $p(n-k, k)$ such partitions. ■

Using these partition numbers, we can also determine the number of distributions of n unlabeled balls into k unlabeled urns with no restriction.

Theorem 4.3.5 *The number of distributions of n unlabeled balls into k unlabeled urns is given by:*

$$\sum_{i=1}^{k} p(n,i).$$

Proof We count the set of all distributions of n unlabeled balls into k unlabeled urns in k disjoint, exhaustive sets. The ith set is the set of all distributions in which exactly i urns are non-empty. By Proposition 4.3.3, the number of distributions in this set is given by $p(n, i)$. ■

Also of interest is the partition function, denoted $p(n)$. This function counts the number of ways that n can be written as an unordered sum of positive integers. Thus,

$$p(n) = \sum_{k=1}^{n} p(n,k).$$

The values of $p(n)$ for small n are given in Table 4.4. The partition function is interesting in that it resembles the Fibonacci sequence for the first few values. However, the partition numbers diverge from the Fibonacci sequence beginning with $p(5)$.

A recursive formula for the values of $p(n)$ is

$$p(n) = p(n-1) + p(n-2) - p(n-5) - p(n-7) + p(n-12)$$

Table 4.4 The number of partitions of n

n	0	1	2	3	4	5	6	7	8	9	10	11	12	13
$p(n)$	1	1	2	3	5	7	11	15	22	30	42	56	77	101

$$+ p(n-15) - p(n-22) - \cdots$$

where the sum is taken over all $p(n-k)$ where $k = \frac{1}{2}i(3i-1)$ and $i = 1, -1, 2, -2, 3, -3, \ldots$ and the signs in the summation alternate $+$, $+$, $-$, $-$, $+$, $+$, $-$, $-$,... The proof of this result is beyond the scope of this text.

Example 4.3.6 Find the number of distributions of n unlabeled balls into k_1 unlabeled red urns and k_2 labeled white urns in which no urn is empty.

Solution We count the number of distributions by counting $n+1$ disjoint, exhaustive sets. The ith set, A_i, is the set of all distributions in which exactly i balls are placed in the red urns. There are $p(i, k_1)$ distributions of i unlabeled balls into k_1 unlabeled urns such that no urn is empty by Proposition 4.3.3. By Proposition 4.2.7 there are $\binom{n-i-1}{k_2-1}$ distributions of $n-i$ unlabeled balls into k_2 labeled white urns such that no urn is empty. By the Multiplication Principle, the number of distributions in this set is

$$|A_i| = p(i, k_1)\binom{n-i-1}{k_2-1}.$$

By the Addition Principle, the total number of distributions is given by

$$\sum_{i=0}^{n} |A_i| = \sum_{i=0}^{n} p(i, k_1)\binom{n-i-1}{k_2-1}.$$

□

By determining the number of ways to distribute n labeled balls into k unlabeled urns in such a way that no urn is empty, we can solve the remaining occupancy problems. As with the partition numbers, we must define new notation.

Definition 4.3.7 Let $S(n, k)$ denote the number of ways to distribute n labeled balls into k unlabeled urns in such a way that no urn is empty. These numbers are typically referred to as *Stirling numbers of the second kind.*

We note that $S(0, k) = 1$ and that $S(n, k) = 0$ for $n < k$. In Table 4.5, we list the values of $S(n, k)$ for small n and k. Later, we will show that the Stirling numbers of the second kind satisfy

$$S(n, k) = \frac{1}{k!}\sum_{i=0}^{k}(-1)^i\binom{k}{i}(k-i)^n.$$

For now, we will have to be satisfied by the following recurrence.

Table 4.5 Stirling numbers of the second kind, $S(n,k)$

$n\backslash k$	1	2	3	4	5	6	7	8	9	10
1	1									
2	1	1								
3	1	3	1							
4	1	7	6	1						
5	1	15	25	10	1					
6	1	31	90	65	15	1				
7	1	63	301	350	140	21	1			
8	1	127	966	1701	1050	266	28	1		
9	1	255	3025	7770	6951	2646	462	36	1	
10	1	311	9330	34105	42525	22827	5880	750	45	1

Theorem 4.3.8 *For all $n, k \in \mathbb{N}$, we have*

$$S(n,k) = kS(n-1,k) + S(n-1,k-1).$$

Proof Note that $S(n,k)$ counts the number of distributions of n labeled balls into k unlabeled urns in such a way that no urn is empty. The right side also counts this by counting two disjoint, exhaustive sets:

(i) The set of distributions in which the ball labeled '1' is in an urn with at least one other ball. Distribute the balls labeled $2, \ldots, n$ into the k unlabeled urns in such a way that no urn is empty. There are $S(n-1,k)$ ways to do this by definition. We then place the ball labeled '1' into one of the (now distinguishable) urns. There are k ways to do this. Hence, by the Multiplication Principle, there are $kS(n-1,k)$ such distributions.

(ii) The set of distributions in which the ball labeled '1' is in an urn by itself. We must now distribute the remaining $n-1$ labeled balls into the remaining $k-1$ unlabeled urns in such a way that no urn is empty. There are $S(n-1,k-1)$ ways to do this by definition. ■

Using the Stirling numbers of the second kind, we will be able to solve the remaining occupancy problems.

Theorem 4.3.9 *The number of ways to distribute n labeled balls into k unlabeled urns is given by*

$$\sum_{i=1}^{k} S(n,i).$$

Proof We count the set of distributions of n labeled balls into k unlabeled urns by counting k disjoint, exhaustive sets. The ith set is the set of all distributions in

Table 4.6 Summary of results for n balls into k urns

	No restriction	No empty urns	No multiple balls
Unlabeled balls Unlabeled urns	$\sum_{i=1}^{k} p(n,i)$	$p(n,k)$	If $n > k$, then 0 If $n \leq k$, then 1
Unlabeled balls Labeled urns	$\binom{n+k-1}{k-1}$	$\binom{n-1}{k-1}$	$\binom{k}{n}$
Labeled balls Unlabeled urns	$\sum_{i=1}^{k} S(n,i)$	$S(n,k)$	If $n > k$, then 0 If $n \leq k$, then 1
Labeled balls Labeled urns	k^n	$k!S(n,k)$	$P(k,n)$

which exactly i urns are non-empty. The number of elements in this set is $S(n,i)$ by definition. ■

Theorem 4.3.10 *The number of ways to distribute n labeled balls into k labeled urns such that no urn is empty is given by* $k!S(n,k)$.

Proof Begin by stripping the labels from all of the urns. There are $S(n,k)$ ways to distribute n labeled balls into k unlabeled urns such that no urn is empty. As the balls are labeled, the urns can now be distinguished by their contents. There are then $k!$ ways to assign labels to the urns. By the Multiplication Principle, there are $k!S(n,k)$ distributions of n labeled balls into k labeled urns such that no urn is left empty. ■

In Table 4.6, we give a summary of the results developed in this chapter.

Example 4.3.11 Find the number of distributions of n_1 labeled red balls and n_2 unlabeled white balls into k_1 unlabeled red urns and k_2 labeled white urns if each urn must contain a red ball and a white ball.

Solution Begin by distributing the labeled red balls. The number of distributions of labeled red balls can be counted in $n_1 + 1$ disjoint, exhaustive sets. The ith set counts the distributions in which i red balls are placed in the unlabeled red urns.

(i) Choose i of the n_1 labeled red balls to place in the red urns. There are $\binom{n_1}{i}$ ways to do this.
(ii) Distribute the chosen red balls. By definition, there are $S(i,k_1)$ distributions of i labeled red balls into k_1 unlabeled urns such that no urn can be empty.
(iii) Distribute the remaining red balls into the white urns. There are $k_2!S(n_1 - i, k_2)$ ways to distribute the remaining $n_1 - i$ labeled red balls unto the k_2 labeled white urns such that no urn is empty by Theorem 4.3.10.

Thus, by the Multiplication Principle, there are $\binom{n_1}{i}S(i,k_1)k_2!S(n_1 - i,,k_2)$ distributions in the ith set. Hence, by the Addition Principle, the total number of distributions of the red balls is given by

$$\sum_{i=0}^{n_1} \binom{n_1}{i} S(i,k_1)k_2!S(n_1 - i,,k_2).$$

Now, distribute the unlabeled white balls. The number of distributions of the white balls can be counted in $n_2 + 1$ disjoint, exhaustive sets. The jth set is the set of all distributions in which j white balls are placed in the unlabeled red urns.

(i) There are $p(j, k_1)$ distributions of j unlabeled balls into k_1 unlabeled red urns such that no urn is empty by Proposition 4.3.3.
(ii) There are $\binom{n_2-j-1}{k_2-1}$ ways to distribute the remaining $n_2 - j$ unlabeled white balls into k_2 labeled white urns such that no urn is empty by Theorem 4.2.7.

By the Multiplication Principle, there are $p(j, k_1)\binom{n_2-j-1}{k_2-1}$ distributions in the jth set. Then, the Addition Principle yields the total number of distributions of the white balls:

$$\sum_{j=0}^{n_2} p(j, k_1)\binom{n_2 - j - 1}{k_2 - 1}.$$

By the Multiplication Principle, the total number of distributions of n_1 labeled red balls and n_2 unlabeled white balls into k_1 unlabeled red urns and k_2 labeled urns if each urn must contain a red ball and a white ball is given by

$$\left(\sum_{i=0}^{n_1}\binom{n_1}{i} S(i, k_1) k_2! S(n_1 - i, , k_2)\right)\left(\sum_{j=0}^{n_2} p(j, k_1)\binom{n_2 - j - 1}{k_2 - 1}\right).$$

□

Example 4.3.12 Find the number of distributions of n labeled balls into k_1 unlabeled red urns and k_2 unlabeled white urns so that either the red urns or the white urns receive at least one ball.

Solution Let A be the set of all distributions in which at the red urns each receive at least one ball. Let B be the set of all distributions in which the white urns each receive at least one ball. Begin by computing the cardinality of A. This can be done by considering n disjoint, exhaustive sets. The ith set, A_i, is the set of all distributions in which exactly i balls have been placed in the red urns such that no urn is empty. The cardinality of A_i can be found by:

(i) Selecting i of the n labeled balls to place in the red urns. There are $\binom{n}{i}$ ways to do this.
(ii) Distribute the selected balls into the k_1 unlabeled red urns such that no urn is empty. There are $S(i, k_1)$ ways to do this.
(iii) Distribute the remaining $n - i$ labeled balls into the k_2 unlabeled white urns. There are $\sum_{j=1}^{k_2} S(n - i, j)$ ways to do this by Theorem 4.3.9.

By the Multiplication Principle, we have

$$|A_i| = \binom{n}{i} S(i, k_1)\left(\sum_{j=1}^{k_2} S(n - i, j)\right).$$

Hence, the Addition Principle yields

$$|A| = \sum_{i=1}^{n} |A_i| = \sum_{i=1}^{n} \binom{n}{i} S(i,k_1) \left(\sum_{j=1}^{k_2} S(n-i,j) \right).$$

By reversing the roles of the red and white urns yields the cardinality of B, namely

$$|B| = \sum_{i=1}^{n} \binom{n}{i} S(i,k_2) \left(\sum_{j=1}^{k_1} S(n-i,j) \right).$$

However, there are distributions in which the red and white urns are non-empty. In this case, we can ignore the colors and simply consider this as a distribution of n labeled balls into $k_1 + k_2$ unlabeled urns such that no urn is empty. The number of ways to do this is $|A \cap B| = S(n, k_1 + k_2)$.

Therefore, the number of appropriate distributions is

$$|A \cup B| = |A| + |B| - |A \cap B|$$

$$= \sum_{\ell=0}^{1} \left(\sum_{i=1}^{n} \binom{n}{i} S(i,k_{\ell+1}) \left(\sum_{j=1}^{k_{\ell+1+(-1)^\ell}} S(n-i,j) \right) \right) - S(n,k_1+k_2)$$

by the Principle of Inclusion and Exclusion. □

Exercise 4.3.13 Find the number of distributions of nine balls into sixteen urns if:

(i) Both the balls and the urns are unlabeled;
(ii) Both the balls and the urns are unlabeled and no urn is empty;
(iii) The balls are labeled and the urns are labeled;
(iv) The balls are labeled, the urns are unlabeled, and no urn is empty;
(v) The balls and urns are both labeled and no urn is empty.

Exercise 4.3.14 List all distributions of eleven unlabeled balls into five unlabeled urns such that no urn is left empty.

Exercise 4.3.15 Find the number of distributions of n_1 unlabeled red balls and n_2 labeled white balls into k_1 unlabeled red urns and k_2 labeled white urns.

Exercise 4.3.16 Find the number of distributions of n_1 unlabeled red balls and n_2 labeled white balls into k_1 unlabeled red urns and k_2 labeled white urns if each white urn must receive at least one white ball.

Exercise 4.3.17 Let $p_i(n,k)$ denote the number of partitions of n into k integers which are at least i.

(i) Prove combinatorially that

$$p_2(n,k) = p_2(n-2,k-1) + p_2(n-3,k-1) + p_2(n-2k,k);$$

(ii) Prove combinatorially that

$$p_i(n,k) = \sum_{j=1}^{2i-1} p_i(n-j,k-1) + p_i(n-ik,k);$$

(iii) Is it true that $p_i(n,k) = p(n-(i-1)k,k)$? Explain.

Exercise 4.3.18 Give a combinatorial proof that:

$$S(n+1,k) = \sum_{i=0}^{n} \binom{n}{i} S(i,k-1).$$

Exercise 4.3.19 Give a combinatorial proof that:

$$S(n,k) = \sum_{i=1}^{n} S(n-i,k-1)k^{i-1}.$$

Exercise 4.3.20 Give a combinatorial proof that

$$S(n,k) = \sum_{\substack{\lambda_1+\cdots+\lambda_k=n,\\ \lambda_1,\ldots,\lambda_n \geq 1}} \binom{n}{\lambda_1,\ldots,\lambda_k} \frac{1}{a_1!\ldots a_n!},$$

where a_i is the number of times i appears in the list $\lambda_1,\ldots,\lambda_k$ for $i=1,\ldots,n$.

Exercise 4.3.21 A *weak partition of n into k parts* is a collection of (not necessarily distinct) k non-negative integers that sum to n. Give a combinatorial proof that the number of weak partitions of n into k parts is given by:

$$\sum_{i=1}^{n} p(n,i).$$

Exercise 4.3.22 Find the number of distributions of n unlabeled balls into k_1 unlabeled red urns and k_2 unlabeled white urns so that either the red urns or the white urns receive at least one ball.

Exercise 4.3.23 Find the number of distributions of n labeled balls into k_1 labeled red urns and k_2 labeled white urns so that either the red urns or the white urns receive at least one ball.

4.4 The Twelvefold Way

The problem of distributing balls into urns seems innocent. However, it actually provides a framework for a more general, abstract problem. This framework is described as *the Twelvefold Way*. The Twelvefold Way is such an important part of advanced combinatorics that it is one of the first four sections in Richard Stanley's classic graduate text, *Enumerative Combinatorics* [40].

In the Twelvefold Way, the problem is to determine the number of functions from an n-set, N, and a k-set, K, according to certain restrictions. Each function, $f : N \to K$, corresponds to a specific distribution of n balls into k urns. Just as specific distributions of balls into urns satisfy certain restrictions, functions from N to K must satisfy certain restrictions. Typically, combinatorialists are interested in three possibilities for a function $f : N \to K$:

(i) *There are no restrictions on f.*
(ii) *We require that f must be an injective function.* Recall that if f is an injective function, then $f(x) = f(y)$ implies $x = y$. In other words, distinct elements of N must be mapped to distinct elements of K. Hence, there can be at most one element of N mapped to each element of K. This corresponds to distributions of n balls into k urns in which each urn receives at most one ball.
(iii) *We require that f must be a surjective function.* Recall that if f is a surjective function, then for each $y \in K$, there is an $x \in N$ such that $f(x) = y$. In other words, there is an element of N mapped to each element of K. This corresponds to distributions of n balls into k urns in which no urn is empty.

The astute reader will observe that there should be a *fourth possibility*. This fourth possibility is that: *We require f to be a bijective function.* Recall that bijections are both injective and surjective. As such, these functions correspond to distributions of n balls into k urns in which each urn receives exactly one ball. As we saw in Proposition 4.1.1, this case is nearly trivial. This may be why this case is traditionally omitted.

In the Twelvefold Way, there are also considerations as to which functions we consider to be identical. There are four such considerations:

(i) *All functions are considered distinct.* These functions correspond to distributions of n labeled balls into k labeled urns.
(ii) *Functions in which the elements of N are permuted are considered identical.* Since permutations on N are considered identical, we may consider the elements of N to be indistinguishable. As such, we are only concerned with the number of elements mapped to each element of K, not the specific elements involved. This corresponds to distributions of n unlabeled balls into k labeled urns.
(iii) *Functions in which the elements of K are permuted are considered identical.* Since permutations on K are considered identical, we may consider the

elements of K to be indistinguishable. This being the case, we are only concerned with which elements of N are grouped together. This corresponds to distributions of n labeled balls into k unlabeled urns.

(iv) *Functions in which the elements of N or K are permuted are considered identical.* Since permutations on N and K are considered identical, we may consider the elements of N to be indistinguishable. Similarly, the elements of K are considered indistinguishable. This corresponds to distributions of n unlabeled balls into k unlabeled urns.

Note that there are three (interesting) restrictions on f as well as four considerations on N and K. Hence, by the Multiplication Principle, there are $3 \times 4 = 12$ possibilities. Of course, there are 12 cells in Table 4.6. Ergo, the Twelvefold Way.

Exercise 4.4.1 Reconstruct Table 4.6 in terms of functions from an n-set to a k-set.

Chapter 5
Generating Functions

In this chapter, we will solve more difficult counting problems. For instance, consider the following problem: Find the number of unordered ten letter words from the alphabet {a,b,c,d} that satisfy the following:

(i) If 'a' is used, then it is used four times;
(ii) 'b' appears an even number of times;
(iii) 'c' appears at least three times;
(iv) 'd' appears no more than five times.

The techniques developed up to this point are insufficient to attack this problem. While brute force will (eventually) yield the solution, it will not help in the more general setting of finding all k letter words that satisfy the above.

To facilitate this solution, we give a brief review of certain underlying ideas. You will no doubt be familiar with several of these ideas such as factoring. However, we will be approaching these familiar ideas in a slightly unfamiliar way. Hence, it is imperative that you review these ideas before proceeding.

5.1 Review of Factoring and Partial Fractions

Let $n \in \mathbb{N}$. Recall that a *polynomial of degree n* is a function of the form

$$f(x) = a_n x^n + \cdots + a_1 x + a_0,$$

where $a_n \neq 0$. The a_i are assumed to be complex numbers called *coefficients*.

Fact 5.1.1 *Two polynomials are equal if and only if all of their coefficients are equal.*

One of the key problems in algebra is how to *factor* a polynomial. The polynomial $f(x)$ can be factored if there exists polynomials $g(x)$ and $h(x)$, both of which are of degree at least one, such that $f(x) = g(x)h(x)$. In this case, we say that $g(x)$ and $h(x)$ are *factors* of $f(x)$.

The problem of factoring a polynomial is equivalent to finding the *roots* of the polynomial. The complex number c is a root of $f(x)$ if $f(c) = 0$.

R. A. Beeler, *How to Count*, DOI 10.1007/978-3-319-13844-2_5

We now present two theorems regarding the roots of polynomials. The proofs of these theorems are omitted, but can be found in any algebra text.

Theorem 5.1.2 *Let $f(x)$ be a polynomial and let $c \in \mathbb{C}$. c is a root of $f(x)$ if and only if $1 - \frac{x}{c}$ is a factor of $f(x)$.*

Theorem 5.1.3 *(The Fundamental Theorem of Algebra) Let $f(x) = a_n x^n + \cdots + a_1 x + a_0$ be a polynomial with $a_i \in \mathbb{C}$. There exists (not necessarily distinct) complex numbers $c_1, ..., c_n, a$ such that:*

$$f(x) = a\left(1 - \frac{x}{c_1}\right)\cdots\left(1 - \frac{x}{c_n}\right).$$

In theory, the Fundamental Theorem of Algebra guarantees that a factorization exists for any polynomial. However, it may not be possible to factor a general polynomial. In fact, Galois Theory tells us that any polynomial of degree five or higher can not be factored using radicals. Even when a polynomial can be factored, often times the exact factors are too cumbersome to be of any practical use. While we will review factoring techniques that you may find useful, you may also want to use a computer algebra system, such as Maple, Sage, or MatLab.

You may have noticed that in the previous theorems, we have written the linear factors of the polynomial in an unusual way. In particular, if c is a root of the polynomial, we have written the corresponding linear factor as $1 - \frac{x}{c}$ or $1 - c^{-1}x$ as opposed to $x - c$. The reason for this choice will become apparent later.

Example 5.1.4 Factor $f(x) = 3x^2 - 9x + 6$.

Solution By the Fundamental Theorem of Algebra, we have:

$$\begin{aligned} f(x) = 6 - 9x + 3x^2 &= a(1 - c_1 x)(1 - c_2 x) \\ &= a - a(c_1 + c_2)x + a c_1 c_2 x^2. \end{aligned}$$

Comparing coefficients yields $a = 6$, $c_1 + c_2 = 3/2$, and $c_1 c_2 = 1/2$. Thus $c_2 = \frac{1}{2c_1}$. Substituting this into $c_1 + c_2 = 3/2$ yields:

$$c_1 + \frac{1}{2c_1} = \frac{3}{2}.$$

Multiplying by $2c_1$ gives:

$$\begin{aligned} & 2c_1^2 + 1 = 3c_1 \\ \Leftrightarrow\ & 2c_1^2 - 3c_1 + 1 = 0 \\ \Leftrightarrow\ & (2c_1 - 1)(c_1 - 1) = 0 \\ \Leftrightarrow\ & c_1 = \frac{1}{2} \quad \text{and} \quad c_1 = 1. \end{aligned}$$

If $c_1 = 1$, then $c_2 = 1/2$. Similarly, if $c_1 = 1/2$, then $c_2 = 1$. In either case, $f(x) = 6(1 - x)(1 - \frac{x}{2})$.

Alternately, if we can manipulate the "standard" factoring in algebra to achieve the desired form. Namely,

$$f(x) = 3x^2 - 9x + 6 = 3(x^2 - 3x + 2) = 3(x-2)(x-1)$$
$$= 3(-2)\left(1 - \frac{1}{2}x\right)(-1)(1-x) = 6\left(1 - \frac{1}{2}x\right)(1-x).$$

□

Few polynomials factor as easily as the one above. In general, the factorization may include irrational or even complex numbers. We now present two theorems that will aid in factoring such polynomials by hand.

Theorem 5.1.5 *(The Rational Root Theorem) Let $f(x) = a_n x^n + a_{n-1}x^{n-1} + \cdots + a_1 x + a_0$ be a polynomial with $a_i \in \mathbb{Z}$ for all i. Any rational root of $f(x)$ is of the form $x = \pm\frac{p}{q}$ where p divides a_0 and q divides a_n.*

The following is an interesting corollary of the Rational Root Theorem.

Corollary 5.1.6 *Let $n, m \in \mathbb{N}$. The number $\sqrt[m]{n}$ is either an integer or it is irrational.*

Proof We note that $\sqrt[m]{n}$ is a root of $f(x) = x^m - n$. The Rational Root Theorem implies that the only rational roots of $f(x)$ are factors of n. By definition, the factors of n are integers. Thus, it follows that $\sqrt[m]{n}$ is either an integer or it is irrational. ■

Another useful tool for factoring certain polynomials is the Quadratic Formula.

Theorem 5.1.7 *(The Quadratic Formula) Let $f(x) = ax^2 + bx + c$, where $a, b, c \in \mathbb{C}$. The roots of $f(x)$ are given by*

$$x = \frac{-b \pm \sqrt{b^2 - 4ac}}{2a}.$$

Proof To find the roots of $f(x)$, we set the polynomial equal to zero and solve for x.

$$\begin{aligned}
ax^2 + bx + c &= 0 \\
\Leftrightarrow x^2 + \frac{b}{a}x &= -\frac{c}{a} \\
\Leftrightarrow x^2 + \frac{b}{a}x + \left(\frac{b}{2a}\right)^2 &= -\frac{c}{a} + \left(\frac{b}{2a}\right)^2 \\
\Leftrightarrow \left(x + \frac{b}{2a}\right)^2 &= -\frac{4ac}{4a^2} + \frac{b^2}{4a^2} \\
\Leftrightarrow \left(x + \frac{b}{2a}\right)^2 &= \frac{b^2 - 4ac}{4a^2} \\
\Leftrightarrow x + \frac{b}{2a} &= \frac{\pm\sqrt{b^2 - 4ac}}{2a} \\
\Leftrightarrow x &= \frac{-b \pm \sqrt{b^2 - 4ac}}{2a}.
\end{aligned}$$

■

We now turn our attention to rational functions and their decomposition into partial fractions. A *rational function* is a function of the form $f(x) = p(x)/q(x)$, where $p(x)$ and $q(x)$ are polynomials such that $q(x) \neq 0$. If the degree of $p(x)$ is greater than or equal to the degree of $q(x)$, then we can use long division to reduce $f(x)$ to the form

$$f(x) = g(x) + \frac{p_1(x)}{q(x)},$$

where $g(x)$ and $p_1(x)$ are polynomials and the degree of $p_1(x)$ is strictly less than the degree of $q(x)$. Hence, without loss of generality, we restrict our attention to rational functions in which the degree of the numerator is strictly less than the degree of the denominator.

Assuming that the degree of $p(x)$ is strictly less that the degree of $q(x)$, we have two different possibilities for $q(x)$. In each case, we note that by the Fundamental Theorem of Algebra, we have:

$$q(x) = a(1 - c_1x) \cdots (1 - c_nx).$$

In the first case, we suppose that the c_i are distinct complex numbers. In other words, we can factor $q(x)$ into distinct linear factors. The *partial fraction decomposition* of $f(x)$ is of the form:

$$f(x) = \frac{A_1}{1 - c_1x} + \cdots + \frac{A_n}{1 - c_nx}$$

where the A_i are complex numbers.

Example 5.1.8 Find the partial fraction decomposition of

$$f(x) = \frac{2x + 1}{3x^2 - 9x + 6}.$$

Solution By above, $3x^2 - 9x + 6 = 6(1 - x)(1 - \frac{x}{2})$. Hence,

$$f(x) = \frac{2x + 1}{3x^2 - 9x + 6} = \frac{2x + 1}{6(1 - x)(1 - \frac{x}{2})}$$
$$= \frac{A_1}{1 - x} + \frac{A_2}{1 - \frac{x}{2}}.$$

Multiplying both sides by $6(1 - x)(1 - \frac{x}{2})$ yields

$$2x + 1 = 6A_1\left(1 - \frac{x}{2}\right) + 6A_2(1 - x).$$

Choosing two distinct values of x will yield a system of two equations with two unknowns, which we can then solve. We make the convenient choices of $x = 1$ and $x = 2$.

$$x = 1 \Rightarrow 3 = 3A_1 \Rightarrow A_1 = 1$$

$$x = 2 \Rightarrow 5 = -6A_2 \Rightarrow A_2 = \frac{-5}{6}.$$

So the required partial fraction decomposition is given by:

$$f(x) = \frac{2x+1}{3x^2 - 9x + 6} = \frac{1}{1-x} - \frac{5}{6(1-\frac{x}{2})}.$$

Adding the two fractions quickly confirms the result. □

In our second case, we assume that $q(x)$ has repeated linear factors. In other words, there is c_i such that $(1 - c_i x)^{m_i}$ appears in the factorization of $q(x)$. In this case, m_i is the *multiplicity* of the factor $(1 - c_i x)$. Thus,

$$q(x) = a(1 - c_1 x)^{m_1} \cdots (1 - c_n x)^{m_n}.$$

The *partial fraction decomposition* of $f(x)$ is of the form:

$$\begin{aligned} f(x) = \frac{p(x)}{q(x)} &= \frac{p(x)}{a(1 - c_1 x)^{m_1} \cdots (1 - c_n x)^{m_n}} \\ &= \frac{A_{1,1}}{1 - c_1 x} + \cdots + \frac{A_{1,m_1}}{(1 - c_1 x)^{m_1}} + \cdots + \frac{A_{n,1}}{1 - c_n x} + \cdots + \frac{A_{n,m_n}}{(1 - c_n x)^{m_n}}, \end{aligned}$$

where the $A_{i,j}$ are complex numbers.

Example 5.1.9 Find the partial fraction decomposition of

$$f(x) = \frac{4}{x^3 - 4x^2 + 5x - 2}.$$

Solution Note that $x^3 - 4x^2 + 5x - 2 = -2(1 - \frac{x}{2})(1 - x)^2$. Thus,

$$\begin{aligned} f(x) &= \frac{4}{x^3 - 4x^2 + 5x - 2} = \frac{-2}{(1 - \frac{x}{2})(1 - x)^2} \\ &= \frac{A_{1,1}}{1 - \frac{x}{2}} + \frac{A_{2,1}}{1 - x} + \frac{A_{2,2}}{(1 - x)^2}. \end{aligned}$$

Multiplying both sides by $(1 - \frac{x}{2})(1 - x)^2$ yields:

$$-2 = A_{1,1}(1 - x)^2 + A_{2,1}\left(1 - \frac{x}{2}\right)(1 - x) + A_{2,2}\left(1 - \frac{x}{2}\right).$$

Substituting in $x = 1$ and $x = 2$ yields:

$$x = 2 \Rightarrow -2 = A_{1,1} \quad \text{and}$$

$$x = 1 \Rightarrow -2 = A_{2,2}/2 \Rightarrow A_{2,2} = -4.$$

To solve for $A_{2,1}$, we can substitute a third value in for x (say $x = 0$). This yields $-2 = -2 + A_{2,1} - 4$. Hence $A_{2,1} = 4$.

Alternately, we can compare the coefficients on the same power of x on both sides of the equation. Comparing coefficients on x^2 yields $0x^2 = -2x^2 + \frac{A_{2,1}x^2}{2}$. Again, $A_{2,1} = 4$.

Thus the required decomposition is:

$$f(x) = \frac{4}{x^3 - 4x^2 + 5x - 2} = \frac{-2}{1 - \frac{x}{2}} + \frac{4}{1 - x} + \frac{-4}{(1 - x)^2}.$$

□

The astute reader may recall that when partial fractions were studied in calculus, an additional case may have been considered. Namely, there was the possibility that the factors of the denominator include *irreducible quadratics*, such as $x^2 + 1$. It is true that this polynomial cannot be written as a product of linear factors with *real* coefficients. However, the Fundamental Theorem of Algebra discusses factorization into linear factors with *complex* coefficients. In this case, we can write

$$1 + x^2 = (1 - ix)(1 + ix).$$

Thus putting it into the form we require.

Exercise 5.1.10 Factor the following polynomials into the form above.

(i) $f(x) = 4x^2 - 20x + 24$;
(ii) $f(x) = 2x^2 + 7x - 2$;
(iii) $f(x) = 2x^2 + 3x - 2$;
(iv) $f(x) = x^3 - 2x^2 - 5x + 6$.

Exercise 5.1.11 Find the partial fraction decomposition of each of the following rational functions.

(i)
$$f(x) = \frac{5x - 12}{x^2 - 5x + 6};$$

(ii)
$$f(x) = \frac{7}{2x^2 + 3x - 2};$$

(iii)
$$f(x) = \frac{1}{x^3 - 7x^2 + 16x - 12};$$

(iv)
$$f(x) = \frac{2}{x^4 - 1}.$$

Exercise 5.1.12 Prove the Rational Root Theorem.

5.2 Review of Power Series

Series are studied in calculus, however, their full potential is often not appreciated at that time. In combinatorics, we often take a different philosophy than that of calculus. However, some of the key ideas are worth revisiting. We will also examine the manipulation of these series in more depth than what is typically studied in calculus.

The *power series* of a function $f(x)$ centered around $x = a$ is a function of the form:

$$f(x) = \sum_{n=0}^{\infty} f_n (x - a)^n,$$

where the f_n are complex numbers. If the nth derivative of $f(x)$ exists at $x = a$, for all n, then $f_n = \frac{f^{(n)}(a)}{n!}$ where $f^{(n)}(x)$ denotes the nth derivative of $f(x)$. This is referred to as *Taylor's Theorem.*

As with polynomials of finite degree, two power series are equal if and only if their coefficients are all equal. In this book, we will be dealing with power series centered around $x = 0$ unless otherwise noted. These are called *MacLaurin Series* in elementary calculus.

One of the most useful and basic power series is known as a *geometric series*. A geometric series is of the form

$$f(x) = 1 + x + \cdots + x^n + \cdots = \sum_{n=0}^{\infty} x^n.$$

Theorem 5.2.1 *For $|x| < 1$, the geometric series*

$$f(x) = \sum_{n=0}^{\infty} x^n = \frac{1}{1-x}.$$

Proof Define the *partial sum* of the series as

$$S_k = \sum_{n=0}^{k} x^n = 1 + x + \cdots + x^k.$$

Note that

$$x S_k = \sum_{n=0}^{k} x^{n+1} = x + \cdots + x^k + x^{k+1}.$$

Hence $S_k - x S_k = 1 - x^{k+1}$. From this it follows that $S_k(1 - x) = 1 - x^{k+1}$. Since $x \neq 1$, we have that

$$S_k = \frac{1 - x^{k+1}}{1 - x}.$$

We note that for $|x| < 1$,

$$\lim_{k\to\infty} x^{k+1} = 0.$$

Thus

$$\sum_{n=0}^{\infty} x^n = \lim_{k\to\infty} S_k = \frac{1}{1-x}.$$

■

Note that the closed form of the geometric series is the same as the form studied in the previous section. This is the reason for the somewhat esoteric form that was used.

By manipulating the geometric series, we can obtain the power series for other functions as well. In a calculus or analysis class, such manipulations are usually preceded by discussions of convergence. In this book, we will omit these discussion. Instead, we assume that we are operating on some interval on which the series converge. There are two reasons for this convention. The first, is that since all of our power series are centered around zero, it is relatively easy to find a common interval in which all of the series converge. The second reason is that in combinatorics, we are more concerned with the *coefficients* of a power series, rather than the sum. This being the case, our arguments can typically be rewritten in terms of partial sums.

Example 5.2.2 Find the power series for $f(x) = \frac{1}{(1-x)^2}$.

Solution Note that:

$$\begin{aligned}\frac{d}{dx}\left(\frac{1}{1-x}\right) &= \frac{d}{dx}\left((1-x)^{-1}\right)\\ &= (-1)(1-x)^{-2}(-1) = \frac{1}{(1-x)^2}.\end{aligned}$$

Differentiating the power series for $\frac{1}{1-x}$ term by term yields:

$$\frac{d}{dx}\left(\sum_{n=0}^{\infty} x^n\right) = \sum_{n=0}^{\infty} nx^{n-1}.$$

Thus

$$\frac{1}{(1-x)^2} = \sum_{n=0}^{\infty} nx^{n-1}.$$

When manipulating series, it is usually necessary that the powers of x match. To do this, we *re-index* the summation. This is analogous to doing a u-substitution for definite integrals.

First note that $0x^{-1} = 0$. So, we can just as easily start the summation at $n = 1$. Hence,

$$\frac{1}{(1-x)^2} = \sum_{n=1}^{\infty} nx^{n-1}.$$

Let $m = n-1$. When we change variables in a definite integral, we must also change the bounds of the integral to match the new variable. Analogously, when the index of a summation is changes, we must change the bounds to match the new index. If $n = 1$, then $m = 0$. Similarly, if $n = \infty$ then $m = \infty$. Thus, the summation becomes

$$\frac{1}{(1-x)^2} = \sum_{m=0}^{\infty} (m+1)x^m.$$

Since the index of the summation is irrelevant, we have that

$$\frac{1}{(1-x)^2} = \sum_{n=0}^{\infty} (n+1)x^n.$$

□

Other rational functions can be expressed as the sum of geometric series as we will see in the next example.

Example 5.2.3 Find the power series for $f(x) = \frac{4}{15x^2-8x+1}$.

Solution By partial fraction decomposition, we have:

$$\begin{aligned} \frac{4}{15x^2-8x+1} &= \frac{4}{(1-3x)(1-5x)} \\ &= \frac{A_1}{1-3x} + \frac{A_2}{1-5x} \end{aligned}$$

$$\begin{aligned} &\Rightarrow 4 = A_1(1-5x) + A_2(1-3x) \\ x = \frac{1}{5} &\Rightarrow 4 = \frac{2A_2}{5} \Rightarrow A_2 = 10 \\ x = \frac{1}{3} &\Rightarrow 4 = \frac{-2A_1}{3} \Rightarrow A_1 = -6. \end{aligned}$$

Hence

$$\begin{aligned} f(x) &= \frac{-6}{1-3x} + \frac{10}{1-5x} \\ &= -6\sum_{n=0}^{\infty}(3x)^n + 10\sum_{n=0}^{\infty}(5x)^n \\ &= \sum_{n=0}^{\infty}(10 * 5^n - 6 * 3^n)x^n. \end{aligned}$$

■

We can also consider the multiplication of two series.

Theorem 5.2.4 *Let $f(x) = \sum_{n=0}^{\infty} f_n x^n$ and $g(x) = \sum_{n=0}^{\infty} g_n x^n$. It follows that*

$$f(x)g(x) = \sum_{n=0}^{\infty} \left(\sum_{k=0}^{n} f_k g_{n-k} \right) x^n.$$

Proof Note that

$$f(x)g(x) = (f_0 + f_1 x + \cdots + f_k x^k + \cdots)(g_0 + g_1 x + \cdots + g_{n-k} x^{n-k} + \cdots).$$

To achieve $c_n x^n$ in the power series of $f(x)g(x)$, we must choose a power of x from the series for $f(x)$. Without loss of generality, take $f_k x^k$ from $f(x)$. To achieve x^n, we must take a complimentary power in $g(x)$, namely $g_{n-k} x^{n-k}$. Summing up over all values of k yields the desired result. ■

Corollary 5.2.5 *Let $f(x) = \sum_{n=0}^{\infty} f_n x^n$.*

(i)
$$(f(x))^2 = \sum_{n=0}^{\infty} \left(\sum_{k=0}^{n} f_k f_{n-k} \right) x^n;$$

(ii)
$$(f(x))^k = \sum_{n=0}^{\infty} \left(\sum_{\substack{\lambda_1 + \cdots + \lambda_k = n, \\ \lambda_1, \ldots, \lambda_k \geq 0}} f_{\lambda_1} \cdots f_{\lambda_k} \right) x^n.$$

Proof

(i) Let $g(x) = f(x)$ and apply Theorem 5.2.4.
(ii) Left as an exercise to the reader ■

Note that in Corollary 5.2.5 (ii), the sum runs through all partitions of n into k non-negative integers.

Example 5.2.6 Let $f(x) = \sum_{n=0}^{\infty} f_n x^n$ with $f_0 > 0$. Find the first three coefficients in the power series for $\sqrt{f(x)}$.

Solution Let $\sqrt{f(x)} = \sum_{n=0}^{\infty} g_n x^n$. By Corollary 5.2.5, we have:

$$f(x) = \left(\sqrt{f(x)}\right)^2 = \sum_{n=0}^{\infty} \left(\sum_{k=0}^{n} g_k g_{n-k} \right) x^n.$$

Comparing coefficients yields:

$$f_0 = g_0^2 \Rightarrow g_0 = \sqrt{f_0};$$

$$f_1 = 2g_0 g_1 \Rightarrow g_1 = \frac{f_1}{2g_0} = \frac{f_1}{2\sqrt{f_0}};$$

$$f_2 = 2g_0g_2 + g_1^2 \Rightarrow g_2 = \frac{f_2 - g_1^2}{2g_0}$$
$$= \frac{f_2}{2g_0} - \frac{g_1^2}{2g_0} = \frac{f_2}{2\sqrt{f_0}} - \frac{f_1^2}{4f_0^{3/2}}.$$

□

Recall that one of our initial problems was to find the nth derivative of a composition function. By manipulating the power series for the functions involved, we can reduce this to a combinatorial problem.

Theorem 5.2.7 *(Faá di Bruno's Formula) Let $f(x) = \sum_{n=0}^{\infty} f_n x^n$ and $g(x) = \sum_{n=0}^{\infty} g_n x^n$. The power series for the composition is given by:*

$$f(g(x)) = \sum_{n=0}^{\infty} \left(\sum_{k=1}^{n} \sum_{\substack{\lambda_1+\cdots+\lambda_k=n,\\ \lambda_1,\ldots,\lambda_n \geq 0}} g_{\lambda_1} \cdots g_{\lambda_k} f_k \right) x^n.$$

Proof By Corollary 5.2.5, we have that

$$g(x)^k = \sum_{n=0}^{\infty} \left(\sum_{\substack{\lambda_1+\cdots+\lambda_k=n,\\ \lambda_1,\ldots,\lambda_k \geq 0}} g_{\lambda_1} \cdots g_{\lambda_k} \right) x^n.$$

Since $f(x) = \sum_{k=0}^{\infty} f_k x^k$, we have that:

$$f(g(x)) = \sum_{k=0}^{\infty} f_k g(x)^k.$$

We must compute the coefficient of x^n in this series. Note that the coefficient of x^n in $g(x)^k$ is:

$$\sum_{\substack{\lambda_1+\cdots+\lambda_k=n,\\ \lambda_1,\ldots,\lambda_k \geq 0}} g_{\lambda_1} \cdots g_{\lambda_k}.$$

Since $g(x)^k$ is multiplied by f_k, it follows that the coefficient of x^n in $f(g(x))$ is given by:

$$\sum_{k=1}^{n} \sum_{\substack{\lambda_1+\cdots+\lambda_k=n,\\ \lambda_1,\ldots,\lambda_k \geq 0}} g_{\lambda_1} \cdots g_{\lambda_k} f_k.$$

The result then follows. ■

From this, it is possible to extract the nth derivative of the composition function. This extraction is left as an exercise to the reader.

We end this section with a list of other common power series. The derivation of these series can be extracted using Taylor's Theorem and are found in any elementary calculus text.

(i)
$$e^x = \sum_{n=0}^{\infty} \frac{x^n}{n!};$$

(ii)
$$\sin(x) = \sum_{n=0}^{\infty} \frac{(-1)^n x^{2n+1}}{(2n+1)!};$$

(iii)
$$\cos(x) = \sum_{n=0}^{\infty} \frac{(-1)^n x^{2n}}{(2n)!};$$

(iv)
$$\tan^{-1}(x) = \sum_{n=0}^{\infty} \frac{(-1)^n x^{2n+1}}{2n+1};$$

(v)
$$\ln(1+x) = \sum_{n=0}^{\infty} \frac{(-1)^n x^{n+1}}{n+1}.$$

Exercise 5.2.8 Show that

$$f(x) = \frac{1}{(1-x)^3} = \sum_{n=0}^{\infty} \frac{(n+1)(n+2)}{2} x^n.$$

Exercise 5.2.9 Find the power series for $(f(x))^k$.

Exercise 5.2.10 Find the power series for

$$f(x) = \frac{3}{6x^2 - 5x + 1}.$$

Exercise 5.2.11 Let $f(x) = \sum_{n=0}^{\infty} f_n x^n$, with $f_0 \neq 0$. Find the first three coefficients in the power series for $\frac{1}{f(x)}$. Hint: Note that $f(x)\frac{1}{f(x)} = 1$.

Exercise 5.2.12 Assume that if $h(x) = \sum_{n=0}^{\infty} h_n x^n$, then $h_n = \frac{h^{(n)}(0)}{n!}$. Use this along with Faá di Bruno's formula to derive the nth derivative of $h(x) = f(g(x))$.

Exercise 5.2.13 Prove that for $-1 < x \leq 1$,

$$\tan^{-1}(x) = \sum_{n=0}^{\infty} \frac{(-1)^n x^{2n+1}}{2n+1}.$$

Exercise 5.2.14 Prove that for $-1 < x \leq 1$,

$$\ln(1+x) = \sum_{n=0}^{\infty} \frac{(-1)^n x^{n+1}}{n+1}.$$

Exercise 5.2.15 Define the *formal derivative* of $f(x) = \sum_{n=0}^{\infty} f_n x^n$ as

$$D[f] = \sum_{n=0}^{\infty} (n+1) f_{n+1} x^n.$$

Show that

$$D[fg] = D[f]g + fD[g].$$

5.3 Single Variable Generating Functions

In many combinatorial problems, we are interested in a sequence of values. For now, we are interested in sequences have a single index n. Examples of such sequences include Fibonacci numbers, partition numbers, the number of permutations on $[n]$, the number of derangements on $[n]$, and many others. If f_n is a sequence of values indexed by n, the *generating function* for f_n is an infinite series of the form

$$F(x) = \sum_{n=0}^{\infty} f_n x^n.$$

In his excellent text *generatingfunctionology*, Herbert Wilf describes a generating function as a "clothesline" on which to hang the terms of a sequence [45]. This is quite correct. The generating function effectively keeps track of the terms of the sequence, even as the corresponding series is being manipulated. For this reason, we do not concern ourselves with the convergence of such a series.

Fact 5.3.1 *Two sequences are equal if and only if they have the same generating function.*

As a motivating example, we consider the problem of determining the number of ways of making exact change for a dollar using pennies (cent pieces), nickels, dimes, quarters, and dollar bills (change for a dollar with a dollar involves simply handing them their dollar back).

First, we consider the problem of making change for an arbitrary amount using only pennies. Let n be an amount in pennies. The number of ways of making change for n using pennies is the coefficient of x^n in

$$(x^1)^0 + (x^1)^1 + \cdots + (x^1)^n + \cdots.$$

Normally, we would simplify this as

$$1 + x + \cdots x^n + \cdots.$$

However, in our initial exhibition, we prefer the expanded form. In this expanded form, the first exponent of one gives the value of penny and the second exponent of n gives the number of pennies used.

With this in mind, we can find the number of ways of making exact change for n using pennies and nickels. This is given by the coefficient of x^n in

$$\underbrace{\left((x^1)^0 + (x^1)^1 + \cdots + (x^1)^n + \cdots\right)}_{\text{pennies}} \underbrace{\left((x^5)^0 + (x^5)^1 + \cdots + (x^5)^n + \cdots\right)}_{\text{nickels}}$$

$$= \left(\frac{1}{1-x}\right)\left(\frac{1}{1-x^5}\right).$$

The reason that this gives the number of ways to make change is that each way of making change is represented by a term in the product. For instance, suppose we wanted to make change for a quarter using ten pennies and three nickels. This is obtained by selecting $(x^1)^{10}$ in the first term and $(x^5)^3$ in the second. When these terms are multiplied, they give a term of x^{25} in the product. When the full product is considered, all possible ways of writing 25 as a sum of 1's and 5's are accounted for. Each one of these possibilities contributes one x^{25} to the product. Hence, the coefficient of x^{25} in the product is the number of ways to make change for a quarter using pennies and nickels.

Continuing on in this matter, we can find the number of ways of making exact change for n using pennies, nickels, dimes, quarters, and dollar bills. This is given by the coefficient of x^n in

$$\underbrace{\left((x^1)^0 + (x^1)^1 + \cdots\right)}_{\text{pennies}} * \underbrace{\left((x^5)^0 + (x^5)^1 + \cdots\right)}_{\text{nickels}} *$$
$$\underbrace{\left((x^{10})^0 + (x^{10})^1 + \cdots\right)}_{\text{dimes}} * \underbrace{\left((x^{25})^0 + (x^{25})^1 + \cdots\right)}_{\text{quarters}} *$$
$$\underbrace{\left((x^{100})^0 + (x^{100})^1 + \cdots\right)}_{\text{dollars}}$$
$$= \left(\frac{1}{1-x}\right)\left(\frac{1}{1-x^5}\right)\left(\frac{1}{1-x^{10}}\right)\left(\frac{1}{1-x^{25}}\right)\left(\frac{1}{1-x^{100}}\right).$$

This can be justified as follows: Suppose that we consider one particular way of making change for a dollar. Namely, the way in which we use 15 pennies, five nickels, one dime, and two quarters. This corresponds to selecting the $(x^1)^{15}$ term in the first factor, the $(x^5)^5$ term in the second, the $(x^{10})^1$, the $(x^{25})^2$ term in the fourth factor, and the $(x^{100})^0$ term in the final factor. There is a unique way of making these selections. So, when the product is computed, the corresponding term $(x^1)^{15}(x^5)^5(x^{10})^1(x^{25})^2(x^{100})^0$ appears exactly once in the product. Thus, each way of making change for a dollar corresponds uniquely to a term in this product. For this reason, each way of making change for a dollar contributes exactly one to the coefficient of x^{100} in the product.

To determine this coefficient, we need only examine the individual sums up to the term involving x^{100}. Hence we can use a computer algebra system to multiply out the polynomial and determine the coefficient. Doing this, we determine that the coefficient of x^{100} to be 243.

As a second example, consider the problem of a store that sells shirts for \$5, pants for \$10, and shoes for \$25. We want to know how many ways there are to spend \$50 in the store. Similar to the coin changing example, the generating function has a term corresponding to shirts, namely

$$1 + x^5 + \cdots = \frac{1}{1 - x^5}.$$

Similarly, the second term

$$1 + x^{10} + \cdots = \frac{1}{1 - x^{10}}$$

will correspond to the pants. Finally, the third term

$$1 + x^{25} + \cdots = \frac{1}{1 - x^{25}}$$

will correspond to the shoes. Hence, the number of ways to spend exactly \$50 is given by the coefficient of x^{50} in the generating function

$$\left(\frac{1}{1 - x^5}\right)\left(\frac{1}{1 - x^{10}}\right)\left(\frac{1}{1 - x^{25}}\right).$$

Using a computer algebra system, we find this coefficient to be 10.

In reality, a person examining this problem would be more concerned with how many ways they can spend *at most* \$50 in the store. To find this, we first find the fiftieth degree Taylor polynomial for the function

$$f(x) = \left(\frac{1}{1 - x^5}\right)\left(\frac{1}{1 - x^{10}}\right)\left(\frac{1}{1 - x^{25}}\right).$$

This can be accomplished by truncating the polynomial

$$(1 + x^5 + \cdots + x^{50})(1 + x^{10} + \cdots + x^{50})(1 + x^{25} + x^{50}).$$

Namely, we only need to include terms in which the power of x is 50 or less. Thus the fiftieth degree Taylor polynomial is

$$T(x) = 10x^{50} + 8x^{45} + 7x^{40} + 6x^{35} + 5x^{30} + 4x^{25} + 3x^{20} + 2x^{15} + 2x^{10} + x^5 + 1.$$

The beauty of the Taylor polynomial is that we can evaluate it at a particular value of x, say $x = 1$. The effect of this is that it sums up all the coefficients of the powers of x up to x^{50}. Equivalently, this gives the number of ways we can spend at most 50 $ in the store. Using a computer algebra system, we find that $T(1) = 49$.

Example 5.3.2 Find the number of unordered ten letter words from the alphabet {a,b,c,d} that satisfy the following:

(i) If 'a' is used, then it is used four times;
(ii) 'b' appears an even number of times;
(iii) 'c' appears at least three times;
(iv) 'd' appears no more than five times.

Solution First we derive the generating function for the sequence that gives the number of unordered words that satisfy the above restrictions. The factor in this polynomial that determines whether 'a' is used is $(1 + x^4)$. Note that if 'a' is used four times, then we take the term 'x^4.' Similarly, if 'a' is not used, then we take the term '1.'

Since 'b' must be used an even number of times, the factor that determines the number of times 'b' is used is given by

$$1 + (x^2)^1 + \cdots = 1 + x^2 + \cdots = \frac{1}{1 - x^2}.$$

Because 'c' must appear at least three times, the exponent of x is at least three in its associated factor. Thus the corresponding factor in the generating function is

$$x^3 + x^4 + \cdots = x^3(1 + x + \cdots) = \frac{x^3}{1 - x}.$$

Finally, 'd' may appear no more than five times. Thus, the exponent of x is no more than five in its associated factor. Therefore, its associated factor is $1 + x + \cdots + x^5$.

From this it follows that the generating function is

$$(1 + x^4)\left(\frac{1}{1 - x^2}\right)\left(\frac{x^3}{1 - x}\right)(1 + x + \cdots + x^5).$$

Multiplying out the polynomial shows that the coefficient on x^{10} is 24. Hence there are 24 unordered words of length ten that satisfy the above requirements. □

We can also use generating functions to determine the number of integer solutions to a linear equation that is subject to certain constraints.

Example 5.3.3 Find the number of integer solutions to:

$$x_1 + x_2 + x_3 + x_4 = 15$$

Subject to:

$$0 \le x_1 \le 10;$$
$$0 \le x_2 \le 5;$$
$$4 \le x_3 \le 7;$$
$$2 \le x_4 \le 8.$$

Solution First we derive the generating function. The exponent of x in the ith factor gives the value of x_i in the sum. Thus the exponents of x_i must satisfy the above constraints. Moreover, the coefficient of x^{15} gives the number of integer solutions to the equation subject to the constraints. Thus the required generating function is

$$\underbrace{(1+x+\cdots+x^{10})}_{x_1} * \underbrace{(1+x+\cdots+x^5)}_{x_2} *$$
$$\underbrace{(x^4+x^5+x^6+x^7)}_{x_3} * \underbrace{(x^2+x^3+\cdots+x^8)}_{x_4}.$$

The coefficient of x^{15} in this polynomial is 134. Hence this is the number of integer solutions to the equation $x_1+x_2+x_3+x_4=15$ that satisfy the above requirements. □

Example 5.3.4 Find the number of integer solutions to:

$$x_1 + x_2 + 2x_3 + 3x_4 = 10$$

Subject to:

$$-1 \le x_1 \le 9;$$
$$-2 \le x_2 \le 4;$$
$$-3 \le x_3 \le 11;$$
$$0 \le x_4 \le 7.$$

Solution There are two possible ways of going about this.

In the first solution, we note that the constraints are equivalent to:

$$0 \le x_1 + 1 \le 10;$$
$$0 \le x_2 + 2 \le 6;$$
$$0 \le x_3 + 3 \le 14;$$
$$0 \le x_4 \le 7.$$

We now do a change of variables, namely $y_1 = x_1 + 1$, $y_2 = x_2 + 2$, $y_3 = x_3 + 3$, and $y_4 = x_4$. Thus our equation is equivalent to:

$$(x_1 + 1) + (x_2 + 2) + 2(x_3 + 3) + 3x_4 = 10 + 1 + 2 + 6 = 19$$
$$\Rightarrow y_1 + y_2 + 2y_3 + 3y_4 = 19$$

with constraints:

$$0 \le y_1 \le 10;$$
$$0 \le y_2 \le 6;$$
$$0 \le y_3 \le 14;$$
$$0 \le y_4 \le 7.$$

The associated generating function is:

$$\underbrace{(1 + y + \cdots + y^{10})}_{y_1} * \underbrace{(1 + y + \cdots + y^{6})}_{y_2} *$$
$$\underbrace{(1 + y^2 + \cdots + y^{28})}_{y_3} * \underbrace{(1 + y^3 + \cdots + y^{21})}_{y_4}.$$

The coefficient of y^{19} in this polynomial is 173. This gives the number of integer solutions that satisfy the above constraints.

As an alternative solution, we consider the possibility of using a "polynomial" in which some of the exponents on x are negative. In this case, the generating function is:

$$\underbrace{(x^{-1} + 1 + x + \cdots + x^{9})}_{x_1} * \underbrace{(x^{-2} + x^{-1} + 1 + \cdots + x^{4})}_{x_2} *$$
$$\underbrace{(x^{-6} + x^{-4} + \cdots + x^{22})}_{x_3} * \underbrace{(1 + x^3 + \cdots + x^{21})}_{x_4}.$$

The coefficient on x^{10} is again 173. Thus there are 173 integer solutions to the above equation subject to the constraints. □

A variation on this example is to consider the problem of distributing gumdrops to children.

Example 5.3.5 Suppose that we have 30 (indistinguishable) cherry gumdrops to be distributed among five children: Alice, Bob, Chad, Diane, and Edward. Alice must receive between three and eight gumdrops. Bob must receive between five and ten gumdrops. Chad must receive between six and fifteen gumdrops. Diane must receive an even number of gumdrops. Finally, Edward must receive exactly one more gumdrop than Alice. Find the number of distributions that satisfy all of the children's restrictions.

Solution With the exception of the last two restrictions, this problem is equivalent to the problem of determining the number of non-negative integer solutions to a linear equation subject to a set of constraints. We begin by finding the appropriate generating function.

Since the number of gumdrops that Edward receives is dependant on the number of gumdrops Alice receives, we treat them as a single unit. Note that the total number of gumdrops the duo receives is one of the numbers 7, 9,...,17. Thus the corresponding factor in the generating function is

$$(x^7 + x^9 + \cdots + x^{17}).$$

Similar to our previous examples, the factors corresponding to Bob and Chad's restrictions are

$$(x^5 + \cdots + x^{10}) \quad \text{and} \quad (x^6 + \cdots + x^{15}),$$

respectively.

Finally, since Diane must receive an even number of gumdrops, the powers of x in her corresponding factor are even numbers. Hence, the factor corresponding to her restriction is

$$(1 + x^2 + \cdots) = \frac{1}{1 - x^2}.$$

Thus the appropriate generating function is

$$(x^7 + x^9 + \cdots + x^{17}) * (x^5 + \cdots + x^{10}) *$$
$$(x^6 + \cdots + x^{15}) * \left(\frac{1}{1 - x^2}\right).$$

The number of distributions satisfying the constraints is the coefficient of x^{30} in the above generating function. Using a computer algebra system, we find this coefficient to be 104.

An alternate solution can be obtained by looking for the number of non-negative integer solutions to

$$a + b + c + 2d + (a + 1) = 30$$

subject to:

$$3 \leq a \leq 8;$$
$$5 \leq b \leq 10;$$
$$6 \leq c \leq 15.$$

Note that the coefficient of '2' in front of the d ensures that Diane's restriction is met. Further, the '$(a + 1)$' corresponds to Edward's restriction. So, we want to find the number of non-negative integer solutions to $2a + b + c + 2d = 29$ subject to the above constraints. Using similar methods to those described above, the corresponding generating function is

$$\underbrace{(x^6 + x^8 + \cdots + x^{16})}_{a}\underbrace{(x^5 + \cdots + x^{10})}_{b}\underbrace{(x^6 + \cdots + x^{15})}_{c}\underbrace{(1 + x^2 + x^4 + \cdots)}_{d}.$$

Again, using a computer algebra system, we find the coefficient of x^{29} to be 104. □

We now consider an example in which we only try to satisfy one child's restriction.

Example 5.3.6 Suppose that we have 20 indistinguishable cherry gumdrops to give to two children, Alice and Bob. Alice wants a multiple of three gumdrops. Bob wants a prime number of gumdrops. How many ways can we distribute the gumdrops such that at least one child's restriction is satisfied.

Solution Let A be the set of distributions in which Alice's restriction is met. Similarly, define B as the set of distributions in which Bob's restriction is satisfied. As usual, we define a factor in the generating function for each child's restriction.

We begin by determine $|A|$. Since Alice must have a multiple of three gumdrops, the factor corresponding to her restriction is

$$1 + x^3 + x^6 + \cdots = \frac{1}{1 - x^3}.$$

Because Bob can receive any number of gumdrops, the factor corresponding to the number of gumdrops is

$$1 + x + x^2 + \cdots = \frac{1}{1 - x}.$$

The cardinality of A is given by the coefficient of x^{20} in

$$\left(\frac{1}{1 - x^3}\right)\left(\frac{1}{1 - x}\right).$$

Using a computer algebra system, we find that $|A| = 7$.

We compute $|B|$ in a similar manner. Bob must have a prime number of gumdrops. Thus, the factor corresponding to his restriction is

$$x^2 + x^3 + x^5 + x^7 + x^{11} + x^{13} + x^{17} + x^{19}.$$

Since we are ignoring Alice's restriction, she may have any number of gumdrops. Hence, the factor corresponding to the number of gumdrops she receives is $\frac{1}{1-x}$. To find $|B|$, we simply find the coefficient of x^{20} in

$$\left(\frac{1}{1 - x}\right)\left(x^2 + x^3 + x^5 + x^7 + x^{11} + x^{13} + x^{17} + x^{19}\right).$$

Using a computer algebra system, we find that $|B| = 8$.

Finally, we determine $|A \cap B|$. Using a similar argument as above, this is given by the coefficient of x^{20} in

$$\left(\frac{1}{1-x^3}\right)(x^2+x^3+x^5+x^7+x^{11}+x^{13}+x^{17}+x^{19}).$$

Using a computer algebra system, we find that $|A \cap B| = 4$.

Applying the Principle of Inclusion and Exclusion, the number of ways that we can satisfy at least one child's restriction is given by

$$\begin{aligned}|A \cup B| &= |A| + |B| - |A \cap B| \\ &= 7 + 8 - 4 = 11.\end{aligned}$$

□

Recall that the partition function of n, $p(n)$, gives the number of ways that n can be written as an unordered sum of positive integers. While a concise formula for $p(n)$ is unknown, the generating formula for $p(n)$ can be illuminating.

Theorem 5.3.7 *The generating function associated with $p(n)$ is given by*

$$P(x) = \frac{1}{(1-x)(1-x^2)\cdots} = \prod_{i=1}^{\infty}\left(\frac{1}{1-x^i}\right).$$

Proof Suppose that we have a partition of n in which the number i is used j times. The corresponding term in the generating function is $(x^i)^j$. Thus the factor in the generating function corresponding to sums including i is

$$(x^i)^0 + (x^i)^1 + \cdots + (x^i)^j + \cdots = \frac{1}{1-x^i}.$$

Multiplying over all possible values of i yields the required generating function,

$$P(x) = \frac{1}{(1-x)(1-x^2)\cdots} = \prod_{i=1}^{\infty}\left(\frac{1}{1-x^i}\right).$$

■

Exercise 5.3.8 Confirm the entries in Table 2.2.

Exercise 5.3.9 How many ways are there to make change for a hundred dollar bill using one, five, ten, twenty, fifty, and hundred dollar bills?

Exercise 5.3.10 Suppose that a restaurant sells drinks for one dollar, sides for three dollars, and entrees for five dollars. How many ways are there to spend at most fifteen dollars?

Exercise 5.3.11 Realistically, an individual will only carry a certain number of coins with them. Suppose that a person will carry at most 15 pennies, seven nickels, five dimes, and three quarters. How many ways can this person make exact change for a dollar?

Exercise 5.3.12 Suppose that we allow people to make change for a dollar using 50-cent pieces as well as pennies, nickels, dimes, and quarters. How many ways can they make exact change for a dollar?

Exercise 5.3.13 In the fictional country of Combinatoria, the denizens use cent pieces, 3 cent pieces, 12 cent pieces, and 30 cent pieces. Find the number of ways of making exact change for their dollar bill worth 100 cents.

Exercise 5.3.14 Find the number of unordered 12 letter words from the alphabet {a,b,c,d} that satisfy the following:

(i) If 'a' is used, then it is used five times;
(ii) The number of times 'b' appears is a multiple of three;
(iii) 'c' appears at least four times;
(iv) 'd' appears no more than six times.

Exercise 5.3.15 Find the number of integer solutions to $x_1 + x_2 + 2x_3 + 3x_4 = 17$ that satisfy:

$$0 \le x_1 \le 7;$$
$$4 \le x_2 \le 10;$$
$$2 \le x_3 \le 7;$$
$$0 \le x_4 \le 5.$$

Exercise 5.3.16 Find the number of integer solutions to $x_1 + x_2 + 2x_3 + 2x_4 = 14$ that satisfy:

$$-2 \le x_1 \le 10;$$
$$-3 \le x_2 \le 7;$$
$$-5 \le x_3 \le 10;$$
$$0 \le x_4 \le 8.$$

Exercise 5.3.17 Suppose that we have 40 (indistinguishable) cherry gumdrops to be distributed among six children: Alice, Bob, Chad, Diane, Edward, and Fran. Alice must receive between 5 and 12 gumdrops. Bob must receive between 7 and 11 gumdrops. Chad must receive between 4 and 16 gumdrops. The number of gumdrops that Diane receives must be a multiple of three. Edward must receive exactly three more gumdrops than Alice. Finally, Fran must receive exactly one less gumdrop than Diane. Find the number of distributions that satisfy all of the children's restrictions.

Exercise 5.3.18 Suppose that we have 25 gumdrops to distribute to two children, Alice and Bob. Alice wants to receive a multiple of four gumdrops. Bob wants to receive a multiple of five gumdrops. How many ways are there to distribute the gumdrops such that at least one child's restriction is satisfied?

Exercise 5.3.19 A *distinct partition of* n is a way of writing n as the sum of distinct positive integers. Give the generating function for $d(n)$, the number of distinct partitions of n.

Exercise 5.3.20 An *odd partition of* n is a way of writing n as the sum of odd positive integers. Give the generating function for $o(n)$, the number of odd partitions of n.

Exercise 5.3.21 Show that $d(n) = o(n)$ for all n.

5.4 Generating Functions with Two or More Variables

In some cases, a generating function with a single variable is not sufficient. As a motivating example, we consider the problem of finding the number of non-negative integer solutions to:

$$x_1 + x_2 + x_3 + x_4 = 10;$$
$$x_1 + 2x_2 + 3x_3 + 4x_4 = 20.$$

In this case, we use a generating function with two variables. The exponent on the first variable, x, will give the contribution of each of the x_i to the first equation. The exponent on the second variable, y, will give the contribution of each of the x_i to the second equation.

The factor corresponding to x_1 in the generating function is of the form $(1+xy+x^2y^2+\cdots)$. This is because x_1 will contribute equal amounts to both equations.

Similarly, the factor corresponding to x_2 in the generating function is of the form $(1+xy^2+x^2y^4+\cdots)$. Notice that the exponent on y is always twice the exponent of x. This is because x_2 contributes twice as much in the second equation as the first.

By the same logic, the factors corresponding to x_3 and x_4 in the generating function are of the form $(1+xy^3+x^2y^6+\cdots)$ and $(1+xy^4+x^2y^8+\cdots)$, respectively. Thus the generating function is:

$$\underbrace{(1+xy+x^2y^2+\cdots)}_{x_1} * \underbrace{(1+xy^2+x^2y^4+\cdots)}_{x_2} *$$
$$\underbrace{(1+xy^3+x^2y^6+\cdots)}_{x_3} * \underbrace{(1+xy^4+x^2y^8+\cdots)}_{x_4}$$
$$= \frac{1}{(1-xy)(1-xy^2)(1-xy^3)(1-xy^4)}.$$

Using our computer algebra system to multiply out the polynomial, we find that the coefficient of $x^{10}y^{20}$ is 14. Hence, there are 14 non-negative integer solutions to the above equations.

The above example can be generalized to give the number of non-negative integer solutions to a system of inequalities. Such systems appear often in integer programming problems.

Example 5.4.1 Find the number of non-negative integer solutions to:

$$x_1 + 3x_2 + 2x_3 + x_4 \leq 11;$$
$$x_1 + 2x_2 + 5x_3 + 7x_4 \leq 23.$$

Solution As before, we will use a two variable generating function. Each factor of the generating function will correspond to one of the variables in the system of inequalities. The variable x in the generating function will represent the contribution of x_i to the first inequality. Similarly, the variable y in the generating function will represent the contribution of x_i to the second inequality.

Since x_1 contributes the same to each inequality, its factor is

$$1 + xy + x^2y^2 + \cdots = \frac{1}{1 - xy}.$$

The variable x_2 contributes three to the first inequality and two to the second inequality. Thus, the corresponding factor in the generating function is

$$1 + x^3y^2 + x^6y^4 + \cdots = \frac{1}{1 - x^3y^2}.$$

Similarly, x_3 contributes two to the first inequality and five to the second inequality. Its corresponding factor in the generating function is

$$1 + x^2y^5 + x^4y^{10} + \cdots = \frac{1}{1 - x^2y^5}.$$

Finally, x_4 contributes seven times as much to the second inequality as the first. Hence, the corresponding factor in the generating function is

$$1 + xy^7 + x^2y^{14} + \cdots = \frac{1}{1 - xy^7}.$$

Ergo, the generating function for this problem is

$$\left(\frac{1}{1 - xy}\right)\left(\frac{1}{1 - x^3y^2}\right)\left(\frac{1}{1 - x^2y^5}\right)\left(\frac{1}{1 - xy^7}\right).$$

To solve this problem, we first find all terms in this generating function such that the exponent on x is less than or equal to 11 and the exponent on y is less than or equal to 23. Using a computer algebra system, we find these terms to be:

$$f(x, y) = 3x^{11}y^{23} + 3x^{11}y^{22} + x^{10}y^{23} + 2x^{11}y^{21}$$
$$+3x^{10}y^{22} + x^9y^{23} + 3x^{11}y^{20} + 2x^{10}y^{21} + 2x^8y^{23} + 2x^{11}y^{19} + x^{10}y^{20} + 3x^9y^{21} + x^8y^{22}$$
$$+x^{11}y^{18} + 2x^{10}y^{19} + 2x^9y^{20} + 2x^7y^{22} + x^6y^{23} + 2x^{11}y^{17} + 2x^{10}y^{18} + x^9y^{19} + 3x^8y^{20}$$
$$+x^7y^{21} + x^5y^{23} + 2x^{11}y^{16} + x^{10}y^{17} + 2x^9y^{18} + 2x^8y^{19} + x^6y^{21} + 2x^{11}y^{15} + 2x^{10}y^{16}$$
$$+2x^9y^{17} + x^8y^{18} + 2x^7y^{19} + x^4y^{22} + 2x^{11}y^{14} + 2x^{10}y^{15} + x^9y^{16} + 2x^8y^{17} + x^7y^{18}$$
$$+x^5y^{20} + x^{11}y^{13} + 2x^{10}y^{14} + 2x^9y^{15} + x^8y^{16} + 2x^6y^{18} + x^3y^{21} + x^{11}y^{12} + 2x^{10}y^{13}$$
$$+2x^9y^{14} + 2x^7y^{16} + x^6y^{17} + x^4y^{19} + 2x^{11}y^{11} + x^{10}y^{12} + x^9y^{13} + 2x^8y^{14} + x^7y^{15}$$
$$+2x^5y^{17} + x^{11}y^{10} + x^{10}y^{11} + x^9y^{12} + 2x^8y^{13} + 2x^6y^{15} + x^5y^{16} + x^{11}y^9 + x^{10}y^{10}$$
$$+x^9y^{11} + x^8y^{12} + 2x^7y^{13} + x^6y^{14} + x^4y^{16} + x^{11}y^8 + x^{10}y^9 + x^9y^{10} + x^8y^{11}$$
$$+2x^7y^{12} + x^5y^{14} + x^{10}y^8 + x^9y^9 + x^8y^{10} + x^7y^{11} + 2x^6y^{12} + x^3y^{15} + x^{10}y^7$$
$$+x^9y^8 + x^8y^9 + x^7y^{10} + x^6y^{11} + x^4y^{13} + x^9y^7 + x^8y^8 + x^7y^9 + 2x^5y^{11}$$
$$+x^2y^{14} + x^9y^6 + x^8y^7 + x^6y^9 + x^5y^{10} + x^3y^{12} + x^8y^6 + x^7y^7 + x^6y^8$$
$$+2x^4y^{10} + x^7y^6 + x^5y^8 + x^4y^9 + x^7y^5 + x^6y^6 + x^5y^7 + x^3y^9 + x^6y^5$$
$$+x^4y^7 + x^6y^4 + x^5y^5 + x^2y^8 + x^5y^4 + x^3y^6 + x^4y^4 + xy^7 + x^4y^3$$
$$+x^2y^5 + x^3y^3 + x^3y^2 + x^2y^2 + xy + 1.$$

Evaluating the resulting polynomial at $x = 1$ and $y = 1$ yields the number of non-negative integer solutions to the above system of inequalities. We find this to be 174. □

The method above can be extended to systems of more than two linear equations. In the next example, we consider a system of three linear equations.

Example 5.4.2 Find the generating function, $F(x, y, z)$ in which the coefficient on $x^k y^m z^n$ is the number of non-negative integer solutions to:

$$x_1 + x_2 + 3x_3 + 2x_4 = k;$$
$$x_1 + 2x_2 + 5x_3 + x_4 = m;$$
$$2x_2 + 3x_2 + x_3 + 3x_4 = n.$$

Solution Again, we will have a factor in the generating function for each of the x_i and a variable for each of the three equations.

Since x_1 contributes one to each of the first two equations and two to the third equation, the terms in its corresponding factor will be powers of xyz^2.

Similarly, as x_2 contributes one to the first equation, two to the second, and three to the third, the terms in its corresponding factor will be powers of xy^2z^3.

Moreover, x_3 will contribute three to the first equation, five to the second, and one to the third. So the terms in its corresponding factor will be powers of x^3y^5z.

Finally, x_4 will contribute two to the first equation, one to the second, and three to the third. Thus the terms in its corresponding factor will be powers of x^2yz^3.

Thus the required generating function is given by:

$$F(x,y,z) = \underbrace{(1 + xyz^2 + x^2y^2z^4 + \cdots)}_{x_1} * \underbrace{(1 + xy^2z^3 + x^2y^4z^6 + \cdots)}_{x_2} *$$

$$\underbrace{(1 + x^3y^5z + x^6y^{10}z^2 + \cdots)}_{x_3} * \underbrace{(1 + x^2yz^3 + x^4y^2z^6 + \cdots)}_{x_4}$$

$$= \frac{1}{(1 - xyz^2)(1 - xy^2z^3)(1 - x^3y^5z)(1 - x^2yz^3)}.$$

□

We now consider a more complicated example which incorporates different kinds of gumdrops.

Example 5.4.3 Suppose that we have 20 cherry gumdrops and 30 lime gumdrops to distribute among three children: Alice, Bob, and Chad. Alice must receive between five and ten of each type of gumdrop. Bob must receive between 6 and 12 cherry gumdrops and between 7 and 14 lime gumdrops. Chad must receive between 7 and 11 cherry gumdrops and between 8 and 13 lime gumdrops.

Further, each child has restrictions as to how many *total* gumdrops they receive. Alice must receive between 12 and 19 total gumdrops. Bob must receive between 15 and 20 total gumdrops. Chad must receive between 17 and 20 total gumdrops.

Find the number of distributions which satisfy each child's restrictions.

Solution In this problem, we use two variables. The first variable x will represent the cherry gumdrops. The second variable y will represent the lime gumdrops.

To facilitate our solution, we think of this as three separate problems. The first problem will deal with the number of gumdrops that Alice will receive. The generating function corresponding to her restriction will have two factors. The first factor will correspond to how many cherry gumdrops she will receive. The second factor will correspond to the number of lime gumdrops she will receive. Thus the generating function corresponding to the number of gumdrops she will receive is:

$$\left(x^5 + x^6 + \cdots + x^{10}\right)\left(y^5 + y^6 + \cdots + y^{10}\right) = x^{10}y^{10} + x^{10}y^9 + x^9y^{10} + x^{10}y^8 + x^9y^9$$
$$+x^8y^{10} + x^{10}y^7 + x^9y^8 + x^8y^9 + x^7y^{10} + x^{10}y^6 + x^9y^7 + x^8y^8 + x^7y^9 + x^6y^{10} + x^{10}y^5$$
$$+x^9y^6 + x^8y^7 + x^7y^8 + x^6y^9 + x^5y^{10} + x^9y^5 + x^8y^6 + x^7y^7 + x^6y^8 + x^5y^9 + x^8y^5$$
$$+x^7y^6 + x^6y^7 + x^5y^8 + x^7y^5 + x^6y^6 + x^5y^7 + x^6y^5 + x^5y^6 + x^5y^5.$$

We now apply her restriction on the total number of gumdrops she will receive. This involves removing the terms in which the power of x^ny^m is less than 12 or more than 19. This leaves us with the factor

$$A(x,y) = x^{10}y^9 + x^9y^{10} + x^{10}y^8 + x^9y^9 + x^8y^{10} + x^{10}y^7 + x^9y^8 + x^8y^9 + x^7y^{10} + x^{10}y^6$$
$$+x^9y^7 + x^8y^8 + x^7y^9 + x^6y^{10} + x^{10}y^5 + x^9y^6 + x^8y^7 + x^7y^8 + x^6y^9 + x^5y^{10} + x^9y^5$$
$$+x^8y^6 + x^7y^7 + x^6y^8 + x^5y^9 + x^8y^5 + x^7y^6 + x^6y^7 + x^5y^8 + x^7y^5 + x^6y^6 + x^5y^7.$$

Similarly, the generating function corresponding to the number of gumdrops Bob will receive is

$$\left(x^6+x^7+\cdots+x^{12}\right)\left(y^7+y^8+\cdots+y^{14}\right)$$

Again, we truncate the generating function to account for the restriction on the total number of gumdrops Bob can receive. This leaves us with the factor

$$B(x,y)=x^{12}y^8+x^{11}y^9+x^{10}y^{10}+x^9y^{11}+x^8y^{12}+x^7y^{13}+x^6y^{14}+x^{12}y^7$$
$$+x^{11}y^8+x^{10}y^9+x^9y^{10}+x^8y^{11}+x^7y^{12}+x^6y^{13}+x^{11}y^7+x^{10}y^8+x^9y^9+x^8y^{10}$$
$$+x^7y^{11}+x^6y^{12}+x^{10}y^7+x^9y^8+x^8y^9+x^7y^{10}+x^6y^{11}+x^9y^7+x^8y^8+x^7y^9+x^6y^{10}$$
$$+x^8y^7+x^7y^8+x^6y^9.$$

Finally, the generating function corresponding to the number of gumdrops Chad will receive is

$$\left(x^7+x^8+\cdots+x^{11}\right)\left(y^8+y^9+\cdots+y^{13}\right).$$

Truncating this function to account for Chad's restriction on the total number of gumdrops he receives yields

$$C(x,y)=x^{11}y^9+x^{10}y^{10}+x^9y^{11}+x^8y^{12}+x^7y^{13}+x^{11}y^8+x^{10}y^9+x^9y^{10}+x^8y^{11}$$
$$+x^7y^{12}+x^{10}y^8+x^9y^9+x^8y^{10}+x^7y^{11}+x^9y^8+x^8y^9+x^7y^{10}.$$

The total number of distributions that satisfy the children's restrictions is given by the coefficient of $x^{20}y^{30}$ in $A(x,y)B(x,y)C(x,y)$. Using a computer algebra system, we find this coefficient to be 101. □

The above example can be extended to any number of gumdrop types and any number of children. However, for purposes of exhibition three children and two types of gumdrops will suffice.

As a final example in this section, we return to the problem of determining the number of partitions of n into k parts, namely $p(n,k)$.

Theorem 5.4.4 *The generating function associated with $p(n,k)$, the number of partitions of n into k positive integers, is given by*

$$P(x,y)=\frac{1}{(1-xy)(1-x^2y)\cdots}=\prod_{i=1}^{\infty}\left(\frac{1}{1-x^iy}\right).$$

Proof The generating function will have a factor associated with each positive integer i. If i is used j times, then $(x^iy)^j$ is associated with this term. Hence the terms in the factor corresponding to the positive integer i are powers of x^iy. Thus the generating function is of the form:

$$P(x,y)=(1+xy+\cdots)(1+x^2y+\cdots)(1+x^3y+\cdots)\cdots$$
$$=\frac{1}{(1-xy)(1-x^2y)\cdots}=\prod_{i=1}^{\infty}\left(\frac{1}{1-x^iy}\right).$$

■

Exercise 5.4.5 Find the number of non-negative integer solutions to:

$$x_1 + x_2 + x_3 + 2x_4 = 7;$$
$$2x_1 + x_2 + 4x_3 + 5x_4 = 22.$$

Exercise 5.4.6 Find the number of non-negative integer solutions such that

$$x_1 + 3x_2 + 5x_2 + 7x_2 = 15$$

or

$$3x_1 + 4x_2 + 6x_3 + 2x_4 = 17$$

is satisfied.

Exercise 5.4.7 Find the number of non-negative integer solutions to:

$$x_1 + 5x_2 + 4x_3 + 2x_4 = 13;$$
$$2x_1 + 3x_2 + 7x_3 + x_4 \leq 25.$$

Exercise 5.4.8 Find the number of non-negative integer solutions to:

$$x_1 + 7x_2 + 9x_3 + 5x_4 \leq 16;$$
$$3x_1 + x_2 + 2x_3 + 7x_4 \leq 17.$$

Exercise 5.4.9 Find the number of non-negative integer solutions to:

$$3 \leq x_1 + 4x_2 + 5x_3 + 6x_4 \leq 11;$$
$$5 \leq 2x_1 + 3x_2 + 5x_3 + 7x_4 \leq 13.$$

Exercise 5.4.10 Find the generating function, $F(x, y, z)$ in which the coefficient on $x^k y^m z^n$ is the number of non-negative integer solutions to:

$$x_1 + 2x_2 + 3x_3 + x_4 = k;$$
$$2x_1 + 3x_2 + 7x_3 + 5x_4 = m;$$
$$3x_2 + x_2 + x_3 + 2x_4 = n.$$

Exercise 5.4.11 Confirm the entries in Table 4.3.

Exercise 5.4.12 Suppose that we have 25 cherry gumdrops and 35 lime gumdrops to distribute among three children: Alice, Bob, and Chad. Alice must receive between 7 and 12 of each type of gumdrop. Bob must receive between 5 and 13 cherry gumdrops and between 8 and 19 lime gumdrops. Chad must receive between 4 and 20 cherry gumdrops and between 9 and 19 lime gumdrops.

Further, each child has restrictions as to how many *total* gumdrops they receive. Alice must receive between 15 and 20 total gumdrops. Bob must receive between 15 and 21 total gumdrops. Chad must receive between 15 and 30 total gumdrops.

Find the number of distribution which satisfy each child's restrictions.

Exercise 5.4.13 Suppose that we have 20 cherry gumdrops, 30 lime gumdrops, and 25 grape gumdrops to distribute among four children: Alice, Bob, Chad, and Diane. Alice must receive between 6 and 11 of each type of gumdrop. Bob must receive between 6 and 13 cherry gumdrops, between 8 and 11 lime gumdrops, and less than 5 grape gumdrops. Chad must receive between 5 and 12 cherry gumdrops, less than 7 lime gumdrops, and between 10 and 15 grape gumdrops. Diane must receive less than eight cherry gumdrops, between 4 and 18 lime gumdrops, and between 7 and 10 grape gumdrops.

Further, each child has restrictions as to how many *total* gumdrops they receive. Alice must receive between 20 and 30 total gumdrops. Bob must receive between 15 and 20 total gumdrops. Chad must receive between 20 and 30 total gumdrops. Diane must receive between 12 and 28 total gumdrops.

Find the number of distribution which satisfy each child's restrictions.

5.5 Ordered Words with a Given Set of Restrictions

Previously, we used generating functions to find the number of unordered ten letter words from the alphabet {a,b,c,d} that satisfy the following:

(i) If 'a' is used, then it is used four times;
(ii) 'b' appears an even number of times;
(iii) 'c' appears at least three times;
(iv) 'd' appears no more than five times.

In general, the number of unordered words of length n that satisfy the above restrictions is given by the coefficient of x^n in the generating function,

$$(1+x^4)\left(\frac{1}{1-x^2}\right)\left(\frac{x^3}{1-x}\right)(1+x+\cdots+x^5).$$

However, this does not answer the more interesting problem of how many *ordered* words satisfy the above restrictions. As a motivating example, we consider *one* of these unordered words of length ten, namely

$$aaaabbcccd.$$

Note that this word has four a's, two b's, three c's, and one d. Hence the number of ordered words that can be obtained from these letters is given by

$$\binom{10}{4,2,3,1}=\frac{10!}{4!2!3!1!}.$$

In the generating function, the word "aaaabbcccd" corresponds to the x^4 term in the first factor, x^2 in the second factor, x^3 in the third factor, and x in the last factor.

Suppose that in each factor of the generating function, we divide the term of x^k by $k!$. For example, x^4 is divided by $4!$, x^2 is divided by $2!$, x^3 is divided by $3!$, and x is divided by $1!$. Hence the associated generating function is

$$f(x) = \left(1 + \frac{x^4}{4!}\right)\left(1 + \frac{x^2}{2!} + \frac{x^4}{4!} + \cdots\right)\left(\frac{x^3}{3!} + \frac{x^4}{4!} + \cdots\right)\left(1 + \frac{x}{1!} + \cdots + \frac{x^5}{5!}\right).$$

Now, the product of the associated terms in the above generating functions yields

$$\frac{x^{10}}{4!2!3!1!}.$$

Notice that the denominator in this term is precisely the denominator of the multinomial coefficient used to compute the number of anagrams of the word "aaaabbcccd."

Hence, to find the number of ordered words of length n that satisfy the above restrictions, we compute the coefficient of x^n in the generating function, $f(x)$, and then multiply it by $n!$. From this it follows that the number of ten letter ordered words that satisfy the above restrictions is 44933.

In general, we consider all possibilities assuming that all objects are distinct. For instance, in the ten letter words, we consider all the letters to be distinct (whether they are or not). Thus there are 10! (or in general, $n!$) ways to arrange ten distinct letters. For each object of the same type, we divide by $k!$, where k is the number of objects of that type. Thus, since there are four a's, we divide by 4!.

In our second example, we consider a distribution of gumdrops to children. In our previous examples, we had assumed that the order in which the gumdrops were distributed was irrelevant. In this example, we consider a variation in which the order in which the children are served matters.

Example 5.5.1 Suppose that we are to distribute 20 gumdrops to three children, Alice, Bob, and Chad. Alice must receive at least five gumdrops. Bob must receive an even number of gumdrops. Chad must receive no more that seven gumdrops. How many ways are there to distribute the gumdrops of the order in which the children are served matters.

Solution In the generating function, the term corresponding to Alice's restriction is:

$$\frac{x^5}{5!} + \frac{x^6}{6!} + \cdots.$$

Similarly, the term corresponding to Bob's restriction is:

$$1 + \frac{x^2}{2!} + \frac{x^4}{4!} + \cdots$$

and the term corresponding to Chad's restriction is:

$$1 + \frac{x}{1!} + \frac{x^2}{2!} + \cdots + \frac{x^7}{7!}.$$

So the generating function is

$$f(x) = \left(\frac{x^5}{5!} + \frac{x^6}{6!} + \cdots\right) *$$

$$\left(1 + \frac{x^2}{2!} + \frac{x^4}{4!} + \cdots\right) * \left(1 + \frac{x}{1!} + \frac{x^2}{2!} + \cdots + \frac{x^7}{7!}\right).$$

Again, we use a computer algebra system to find the coefficient of x^{20} in $f(x)$. Multiplying by 20! gives the number of distributions at 1069540003.

□

The above example can be generalized to any number of children and any number of gumdrops. Incorporating multiple variables will allow this method to work for the case where we are distributing several kinds of gudrops as well

Exercise 5.5.2 Find the number of ordered twelve letter words from the alphabet {a,b,c,d} that satisfy the following:

(i) If 'a' is used, then it is used five times;
(ii) The number of times 'b' appears is a multiple of three;
(iii) 'c' appears at least four times;
(iv) 'd' appears no more than six times.

Exercise 5.5.3 Find the number of ordered twelve letter words from the alphabet {a,b,c,d} that satisfy the following:

(i) If 'a' is used, then it is used five times;
(ii) The number of times 'b' appears is a multiple of three;
(iii) 'c' appears at least four times;
(iv) 'd' appears no more than six times.

Exercise 5.5.4 Find the number of ordered 15 letter words from the alphabet {a,b,c,d,e} that satisfy the following:

(i) If 'a' is used, then it is used seven times;
(ii) The number of times 'b' appears is a multiple of four;
(iii) 'c' appears at least two times;
(iv) 'd' appears no more than five times;
(v) 'e' appears a prime number of times.

Exercise 5.5.5 Suppose that we are to distribute 15 gumdrops to three children, Alice, Bob, and Chad. Alice must receive at least three gumdrops. The number of gumdrops that Bob must receive is a multiple of three. Chad must receive no more that five gumdrops. How many ways are there to distribute the gumdrops of the order in which the children are served matters.

Exercise 5.5.6 Suppose that we are to distribute 18 gumdrops to four children, Alice, Bob, Chad, and Diane. Alice must receive at least seven gumdrops. The number of gumdrops that Bob must receive is a multiple of four. Chad must receive no more that five gumdrops. Diane must receive one more gumdrop than Chad. How many ways are there to distribute the gumdrops of the order in which the children are served matters.

Chapter 6
Recurrence Relations

In this chapter, we examine the problem of determining concise formulas for the recurrence relations found in earlier sections. You will find that the terminology and techniques involved are quite similar to the terminology and techniques discussed in an elementary differential equations course. This is because recurrence relations are the discrete analog to differential equations. In fact, recurrence relations are often referred to as *difference equations*. Just as entire books have been written on the subject of differential equations, entire books can (and have) been written about solving recurrence relations.

6.1 Finding Recurrence Relations

Throughout this text, we have discussed several sequences that correspond to counting problems. In some cases, we were able to give a closed form for these sequences. In other cases, we were only able to give a *recurrence relation*. A recurrence relation for the sequence $\{R_n\}$ is a function that defines the current value of the sequence, R_n, in terms of the previous values of the sequence. For example, the *Fibonacci sequence* is often defined with the recurrence $F_n = F_{n-1} + F_{n-2}$, where $F_0 = 0$ and $F_1 = 1$. The values, $F_0 = 0$ and $F_1 = 1$ are called the *initial values* of the recurrence. In other words, given these initial values and the recurrence $F_n = F_{n-1} + F_{n-2}$, we could compute the entire Fibonacci sequence.

Unfortunately, if we wished to compute the millionth Fibonnaci number, we might not want to compute the nine-hundred ninety-nine thousand nine-hundred ninety-nine previous Fibonacci numbers. To do this, we would require a *closed form* for the recurrence. The exact definition of a "closed form" is not always clear in mathematics. However, for our purposes we will assume that a closed form for a recurrence R_n is a function of n that:

(i) Requires finitely many steps to compute. In other words, we will not allow infinite series.
(ii) Makes no reference to values of the sequence, other than perhaps the initial values.

R. A. Beeler, *How to Count*, DOI 10.1007/978-3-319-13844-2_6

This closed form is often referred to as the *solution* to the recurrence.

Remark 6.1.1 Often there is some ambiguity as to what is allowed in a "closed form." For instance, some people would argue that factorials and exponentials are not allowed as part of a "closed form." One might also argue that radicals should not be allowed in closed forms, because in general they require infinitely many steps to compute exactly. In this text, we will always permit factorials, exponentials, and radicals in closed forms.

The goal of this chapter will be to find closed forms for various recurrence relations. To that end, we will be defining the relevant terminology in this section. We will also use this opportunity to show other ways of discovering recurrence relations.

If $\{R_n\}$ is a sequence defined recursively, then the recurrence relation can be rewritten in the form $f(R_{n+k}, R_{n+k-1}, \ldots, R_n) = g(n)$, where f is a function of the previous values in the sequence and g is a function of n. If $g(n) = 0$, then we say that the recurrence is *homogenous*. If $g(n) \neq 0$, then we say that the recurrence is *non-homogenous*.

We say that $f(R_{n+k}, R_{n+k-1}, \ldots, R_n)$ is a kth order recurrence relation. If f can be written in the form

$$f(R_{n+k}, R_{n+k-1}, \ldots, R_n) = f_k(n)R_{n+k} + f_{k-1}(n)R_{n+k-1} + \cdots + f_0(n)R_n,$$

where the $f_i(n)$ are functions of n that do not depend on the values of R_n, then f is a kth order *linear* recurrence relation. If $f_i(n) = c_i \in \mathbb{C}$ for all i, then the recurrence relation has *constant coefficients*.

Example 6.1.2 Identify the following recurrences as completely as possible:

(i) $F_n = F_{n-1} + F_{n-2}$;
(ii) $R_n = 2R_{n-1} + 5R_{n-2} + (-1)^n$;
(iii) $S_n = S_{n-1} + 5S_{n-2} + (-1)^n S_{n-3}$;
(iv) $J_n = nJ_{n-1} + (-1)^n J_{n-2} + 5J_{n-3} + 2$;
(v) $K_n = K_{n-1}K_{n-2}$.

Solution

(i) We can rewrite this recurrence as $F_n - F_{n-1} - F_{n-2} = 0$. Adding two to each index yields $F_{n+2} - F_{n+1} - F_n = 0$. This is a homogenous second-order linear recurrence with constant coefficients.
(ii) This recurrence can be rewritten as $R_{n+2} - 2R_{n+1} - 5R_n = (-1)^n$. This is a non-homogeneous second-order linear recurrence with constant coefficients.
(iii) Rewrite this recurrence as $S_{n+3} - S_{n+2} - 5S_{n+1} - (-1)^n S_n = 0$. This is a homogenous third-order linear recurrence. However, it does not have constant coefficients.
(iv) The recurrence can be rewritten as $J_{n+3} - nJ_{n+2} - (-1)^n J_{n+1} - 5J_n = 2$. This is a non-homogenous third-order linear recurrence. However, it does not have constant coefficients.
(v) This recurrence can be rewritten as $K_{n+2} - K_{n+1}K_n = 0$. This is a homogenous second-order recurrence with constant coefficients. However, it is not linear. ■

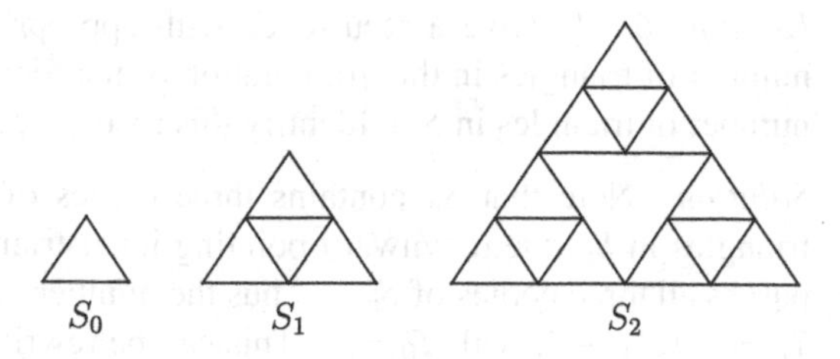

Fig. 6.1 Sierpiński graphs

New recurrences can often be obtained from other recurrences. We now give an example of this.

Example 6.1.3 Let $\{F_n\}$ be the Fibonacci sequence, in other words, $F_n = F_{n-1} + F_{n-2}$, where $F_0 = 0$ and $F_1 = 1$. Find a recurrence with appropriate initial conditions for S_n, where

$$S_n = \sum_{k=0}^{n} F_k.$$

Identify this recurrence as completely as possible.

Solution Note that $F_{n+2} - F_{n+1} - F_n = 0$. Further note that

$$S_{n+1} = \sum_{k=0}^{n+1} F_k = F_{n+1} + \sum_{k=0}^{n} F_k = F_{n+1} + S_n$$

$$\Leftrightarrow S_{n+1} - S_n = F_{n+1}.$$

Similarly, $S_{n+2} - S_{n+1} = F_{n+2}$ and $S_n - S_{n-1} = F_n$. From this it follows that

$$\begin{aligned} 0 &= F_{n+2} - F_{n+1} - F_n \\ &= (S_{n+2} - S_{n+1}) - (S_{n+1} - S_n) - (S_n - S_{n-1}) \\ &= S_{n+2} - 2S_{n+1} + S_{n-1}. \end{aligned}$$

Rewriting this recurrence yields $S_{n+3} - 2S_{n+2} + S_n = 0$. This is a homogenous third-order linear recurrence with constant coefficients. The appropriate initial values for this recurrence are $S_0 = F_0 = 0$, $S_1 = F_0 + F_1 = 1$, and $S_2 = F_2 + S_1 = 2$. □

As a second example, consider the problem of determining the number of triangles in the *Sierpiński graph*. The Sierpiński graph is constructed iteratively as follows: Begin with a triangle, we call this graph S_0. To obtain S_1, we connect the midpoints on each edge of the triangle (see Fig. 6.1). S_n can be obtained by replacing each of the upward pointing triangles in S_1 with a copy of S_{n-1}.

Example 6.1.4 Give a recurrence with appropriate initial conditions for T_n, the number of triangles in the nth iteration of the Sierpiński graph (in other words, the number of triangles in S_n). Identify this recurrence as completely as possible.

Solution Note that S_n contains three copies of S_{n-1}. There are two additional triangles in S_n, the downward pointing inner triangle and the triangle that encompasses all three copies of S_{n-1}. Thus the number of triangles satisfies the recurrence $T_n = 3T_{n-1} + 2$, with $T_0 = 1$. This can be rewritten as $T_{n+1} - 3T_n = 2$. This is a non-homogeneous first-order linear recurrence with constant coefficients. □

Recall that a *derangement* is a permutation on $[n]$ in which there are no fixed points. For instance, 43251 is a derangement on $[5]$. Let D_n be the number of derangements on $[n]$. We now give two recurrences for D_n.

Theorem 6.1.5 *The number of derangements on* $[n]$, D_n, *satisfies the recurrence*

$$D_n = (n-1)(D_{n-1} + D_{n-2}),$$

for $n \geq 2$ *with* $D_0 = 1$ *and* $D_1 = 0$.

Proof There is one derangement on $[0]$, the empty derangement. There are no derangements on $[1]$, as this element would necessarily be mapped to itself. Thus, we need only show that both sides of the recurrence count the same set.

By definition, D_n counts the number of derangements on $[n]$. The right side of the equation also counts this by first choosing an element k to be placed in the first position. Since k is to be placed in the first position, there are $n-1$ choices for k, namely $2, 3, \ldots, n$. The number of ways to arrange the remaining elements can be counted using two disjoint, exhaustive sets:

(i) The set of all derangements in which the element '1' is not placed in the kth position. Thus, any permutation in this set must be a derangement on the elements $2, 3, \ldots, k-1, 1, k+1, \ldots, n$. Thus there are D_{n-1} elements in this set.
(ii) The set of all derangements in which the element '1' is placed in the kth position. Since 1 and k have been exchanged, the placement of the remaining $n-2$ elements must be a derangement. Thus the number of elements in this set is given by D_{n-2}.

By the Multiplication Principle and the Addition Principle, the number of derangements on $[n]$ is given by $(n-1)(D_{n-1} + D_{n-2})$. ■

We note that the recurrence obtained in Theorem 6.1.5 is a homogenous second-order linear recurrence. However, it does not have constant coefficients. Our next result will be to obtain a non-homogeneous first-order linear recurrence. This second result will make it easier to find a closed form for the recurrence.

Corollary 6.1.6 *The number of derangements on* $[n]$, D_n, *satisfies the recurrence* $D_n = nD_{n-1} + (-1)^n$, *where* $D_0 = 1$.

Proof We proceed by induction on n. For $n = 1$, we have $0 = D_1$ and $1\ D_0 + (-1) = 1(1) - 1 = 0$. Assume that the result holds for some $n \geq 1$. We need only confirm that the result holds for $n + 1$.

Note that $(-1)^{n+1} + (-1)^n = 0$ for all $n \in \mathbb{N}$. By Theorem 6.1.5, D_{n+1} satisfies the recurrence

$$\begin{aligned} D_{n+1} &= n(D_n + D_{n-1}) \\ &= nD_n + nD_{n-1} + (-1)^{n+1} + (-1)^n \\ &= nD_n + (-1)^{n+1} + (nD_{n-1} + (-1)^n). \end{aligned}$$

By the inductive hypothesis, we have $D_n = nD_{n-1} + (-1)^n$. Thus,

$$\begin{aligned} D_{n+1} &= nD_n + (-1)^{n+1} + (nD_{n-1} + (-1)^n) \\ &= nD_n + (-1)^{n+1} + D_n = (n+1)D_n + (-1)^{n+1}. \end{aligned}$$

Hence the result holds by the Principle of Mathematical Induction. ∎

Suppose that we want to make a necklace with n beads. Each bead may be one of m different colors and we have unlimited beads of each color. Normally, two necklaces are considered the same if one can be obtained from the other by rotating or reflecting the other. However, this problem will wait until Chap. 8. For the next example, we will consider the necklaces to be fixed (in other words, rotations and reflections are different).

Theorem 6.1.7 *Suppose that we want to make a fixed necklace with n beads, where $n \geq 3$. Each bead may be one of m different colors and we have unlimited beads of each color. However, we do not want two adjacent beads to have the same color. Let $c_m(n)$ be the number of ways to do this. Here, $c_m(3) = m(m-1)(m-2)$ and for $n \geq 4$,*

$$c_m(n) = m(m-1)^{n-1} - c_m(n-1).$$

Proof For $n = 3$, the first bead of the necklace may be any of m colors. The second bead may be any of the $m - 1$ colors not used. Similarly, the third bead may be any of the $m - 2$ not already used. Thus $c_m(3) = m(m-1)(m-2)$ by the Multiplication Principle.

For $n \geq 4$, we prove the equivalent statement that

$$m(m-1)^{n-1} = c_m(n) + c_m(n-1). \tag{6.1}$$

Recall that $m(m-1)^{n-1}$ counts the number of ways to make a fixed string with n beads, where the beads may be any of m colors and no two adjacent beads may be the same color (see Example 2.1.3). The right side of Eq. (6.1) also counts this by counting two disjoint, exhaustive sets:

(i) The set of all acceptable strings in which the first and last bead are different colors. In this case, we can add string to connect the first and last beads. This forms a fixed necklace on n beads, where the beads may be any of m colors and no two adjacent beads may be the same color. Thus, the number of elements in this set is counted by $c_m(n)$ by definition.

(ii) The set of all acceptable strings in which the first and last bead are the same color. It follows that the $(n-1)$st bead is a different color than the first bead. Thus, we remove the last bead from the string. Now, the first bead and the $(n-1)$st bead can be connected by a string. This gives us a necklace with $n-1$ beads, where the beads may be any of m colors and no two adjacent beads may be the same color. The number of elements in this set is counted by $c_m(n-1)$ by definition.

Thus, Eq. (6.1) holds by the Addition Principle. ■

The recurrence in Theorem 6.1.7 is a non-homogeneous first-order linear recurrence with constant coefficients.

When studying definite integrals in calculus, *power sums* appear when computing the Riemann sums. A *power sum* is a sum of the form

$$\sum_{i=1}^{n} i^k,$$

where $k \in \mathbb{N}$. Because of their utility in computing certain Riemann sums, it is desirable to obtain a recurrence. Later, we will find closed forms for these power sums.

Example 6.1.8 Find a recurrence relation for the power sum $S_n = \sum_{i=0}^{n} i^2$. Classify this recurrence as completely as possible.

Solution Note that

$$S_{n+1} = \sum_{i=1}^{n+1} i^2 = \sum_{i=1}^{n} i^2 + (n+1)^2 = S_n + (n+1)^2.$$

We can rewrite this recurrence as $S_{n+1} - S_n = (n+1)^2$. Hence, this is a non-homogeneous first-order linear recurrence with constant coefficients. □

As a final example, we consider a problem from linear algebra. A common method for computing the *determinant* of an $n \times n$ matrix is the technique of *cofactor expansion*. In cofactor expansion, the determinant is found by selecting either rows or columns, then expanding on the cofactors of that row or column (see [5]). For instance, suppose that we want to compute the determinant of

$$\begin{bmatrix} a_{1,1} & a_{1,2} & a_{1,3} \\ a_{2,1} & a_{2,2} & a_{2,3} \\ a_{3,1} & a_{3,2} & a_{3,3} \end{bmatrix}.$$

Begin by selecting the first row. The associated cofactor expansion by the first row is:

$$\begin{vmatrix} a_{1,1} & a_{1,2} & a_{1,3} \\ a_{2,1} & a_{2,2} & a_{2,3} \\ a_{3,1} & a_{3,2} & a_{3,3} \end{vmatrix} = a_{1,1} \begin{vmatrix} a_{2,2} & a_{2,3} \\ a_{3,2} & a_{3,3} \end{vmatrix} - a_{1,2} \begin{vmatrix} a_{2,1} & a_{2,3} \\ a_{3,1} & a_{3,3} \end{vmatrix} + a_{1,3} \begin{vmatrix} a_{2,1} & a_{2,2} \\ a_{3,1} & a_{3,2} \end{vmatrix}.$$

The determinants of the smaller matrices can likewise be computed using cofactor expansion. Thus, a combinatorial question is to determine the number of ways to compute a determinant using cofactor expansion (see [7]).

Theorem 6.1.9 *Let A_n be the number of ways to compute the determinant of an $n \times n$ matrix using cofactor expansion. A_n satisfies the recurrence $A_n = 2nA_{n-1}^n$ with $A_1 = 1$.*

Proof By definition, the determinant of a 1×1 matrix is the lone element in that matrix. Hence $A_1 = 1$. For all other values of A_n:

(i) Choose to expand along either a row or a column. There are two ways to do this.
(ii) Choose the particular row or column to expand on. There are n ways to do this.
(iii) For each of the n $(n-1) \times (n-1)$ submatrices, there are A_{n-1} ways to compute the determinant by cofactor expansion by definition. Hence there are A_{n-1}^n ways to compute these determinants by Corollary 2.1.4.

By the Multiplication Principle, $A_n = 2nA_{n-1}^n$. ■

Note that the recurrence found in Theorem 6.1.9 is a first-order homogeneous equation. It is not linear, nor does it have constant coefficients.

Exercise 6.1.10 Let $R_n = 5R_{n-1} - 2R_{n-2}$, with $R_0 = 6$ and $R_1 = 2$. Identify this recurrence as completely as possible. Give the first ten values of the sequence R_n.

Exercise 6.1.11 Identify the following recurrences as completely as possible:

(i) $F_n = 3F_{n-1} + 5F_{n-2} - 7F_{n-3}$;
(ii) $R_n = 3R_{n-1} + 2R_{n-2} + n$;
(iii) $S_n = 6S_{n-1} - 8S_{n-2} + nS_{n-3} + S_{n-4} + 4$;
(iv) $J_n = 6J_{n-1} + 4J_{n-2} + 2^n J_{n-3}$;
(v) $K_n = (-1)^n K_{n-1} K_{n-3}$.

Exercise 6.1.12 Let $R_n = 5R_{n-1} - 2R_{n-2}$, with $R_0 = 6$ and $R_1 = 2$. Find a recurrence for S_n, where

$$S_n = \sum_{k=0}^{n} R_k.$$

Identify the recurrence for S_n as completely as possible. Give initial values for S_n.

Exercise 6.1.13 Let T_n be the number of triangles in the nth iteration of the Sierpiński Graph (see Example 6.1.4 and Fig. 6.1). Find a recurrence for R_n, where

$$R_n = \sum_{k=0}^{n} T_k.$$

Identify this recurrence as completely as possible.

Exercise 6.1.14 Give a recurrence with appropriate initial conditions for the number of points in which two or more lines meet in the nth iteration of the Sierpiński graph (see Example 6.1.4 and Fig. 6.1). Identify this recurrence as completely as possible.

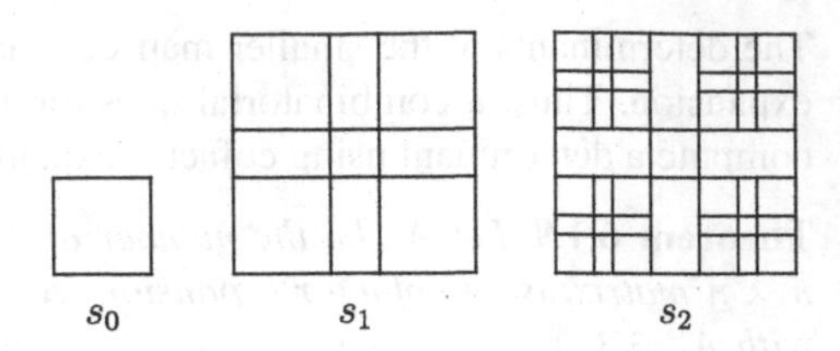

Fig. 6.2 Iterated squares

Exercise 6.1.15 Give a recurrence for the power sum $S_n = \sum_{i=0}^{n} i^3$. Classify this recurrence as closely as possible.

Exercise 6.1.16 Give a recurrence for the power sum $S_n = \sum_{i=0}^{n} i^4$. Classify this recurrence as closely as possible.

Exercise 6.1.17 Give a recurrence for the power sum $S_n = \sum_{i=0}^{n} i(i-1)$. Classify this recurrence as closely as possible.

Exercise 6.1.18 Let d_n be the number of divisions in the plane created by n lines which satisfy:

(i) No two lines are parallel;
(ii) No three lines meet at the same point.

Give a recurrence relation and appropriate initial conditions for d_n. Identify this recurrence as completely as possible.

Exercise 6.1.19 A *triangulation* of a regular n-gon is a way of placing $n-3$ non-intersecting diagonals in the n-gon. Give a recurrence with appropriate initial conditions for t_n, the number of triangulations on the regular n-gon. Identify the recurrence as completely as possible.

Exercise 6.1.20 The *Iterated square graph* can be constructed as follows: Begin with a square, call this graph s_0. To obtain s_1, we place lines going across the square at two-fifths and three-fifths the length of each side (see Fig. 6.2). The graph s_n can be obtained by replacing each of the larger squares in s_1 with a copy of s_{n-1}. Give a recurrence with appropriate initial conditions for the number of squares in the nth iterated square graph. Identify this recurrence as completely as possible.

Exercise 6.1.21 Give a recurrence with appropriate initial conditions for the number of points in which two or more lines meet in the nth iterated square graph (see Exercise 6.1.20 and Fig. 6.2). Identify this recurrence as completely as possible.

Exercise 6.1.22 Give a combinatorial proof for the recurrence

$$n! = (n-1)((n-1)! + (n-2)!).$$

Exercise 6.1.23 Find the number of ways to compute the determinant of a $n \times n$ matrix using cofactor expansion for $n = 1, 2, 3, 4, 5$.

6.2 The Method of Generating Functions

We now examine the problem of finding solutions, or closed forms, for various recurrences. To achieve this, we utilize the generating functions developed in the last chapter. We begin by finding a closed form for the Fibonacci recurrence.

Example 6.2.1 Find a closed form for the Fibonacci sequence defined by $F_0 = 0$, $F_1 = 1$, and $F_n = F_{n-1} + F_{n-2}$ for $n \geq 2$.

Solution Note that the generating function for the Fibonacci sequence is given by

$$F(x) = \sum_{n=0}^{\infty} F_n x^n.$$

We now algebraically manipulate this series to take advantage of the Fibonacci recurrence.

$$\begin{aligned}
F(x) &= \sum_{n=0}^{\infty} F_n x^n \\
&= F_0 + F_1 x + \sum_{n=2}^{\infty} F_n x^n \\
&= F_0 + F_1 x + \sum_{n=2}^{\infty} (F_{n-1} + F_{n-2}) x^n \\
&= F_0 + F_1 x + \sum_{n=2}^{\infty} F_{n-1} x^n + \sum_{n=2}^{\infty} F_{n-2} x^n \\
&= F_0 + F_1 x + x \sum_{n=2}^{\infty} F_{n-1} x^{n-1} + x^2 \sum_{n=2}^{\infty} F_{n-2} x^{n-2} \\
&= F_0 + F_1 x - F_0 x + x \sum_{n=1}^{\infty} F_{n-1} x^{n-1} + x^2 \sum_{n=2}^{\infty} F_{n-2} x^{n-2}.
\end{aligned}$$

We now re-index the first summation using the change of variables $i = n - 1$ and the second summation using the change of variables $j = n - 2$. Thus,

$$F(x) = F_0 + F_1 x - F_0 x + x \sum_{i=0}^{\infty} F_i x^i + x^2 \sum_{j=0}^{\infty} F_j x^j.$$

Since the variable used in the summation is irrelevant, we have that

$$F(x) = F_0 + F_1 x - F_0 x + x F(x) + x^2 F(x).$$

This gives an algebraic equation that we can solve for $F(x)$:

$$F(x) - x F(x) - x^2 F(x) = F_0 + F_1 x - F_0 x$$

$$\Rightarrow F(x)(1-x-x^2) = F_0 + F_1x - F_0x$$
$$\Rightarrow F(x) = \frac{F_0 + F_1x - F_0x}{1-x-x^2}.$$

Substituting the initial values, $F_0 = 0$ and $F_1 = 1$, yields:

$$F(x) = \frac{x}{1-x-x^2}.$$

Unfortunately, we have only found a closed form for the generating function associated with the Fibonacci sequence. We have yet to find a closed form for the recurrence itself.

Using partial fraction decomposition, we have

$$F(x) = \frac{x}{1-x-x^2} = \frac{A}{1-ax} + \frac{B}{1-bx}.$$

To properly find the partial fraction decomposition, we must solve for A, B, a, and b. Note that

$$1-x-x^2 = (1-ax)(1-bx) = 1-(a+b)x+abx^2$$
$$\Rightarrow a+b=1 \quad \text{and} \quad ab=-1$$
$$\Rightarrow b = \frac{-1}{a} \Rightarrow a - \frac{1}{a} = 1$$
$$\Rightarrow a^2 - a - 1 = 0.$$

Using the Quadratic Formula yields

$$a = \frac{1 \pm \sqrt{5}}{2}.$$

Taking a to be the "positive" solution, we can then back substitute to find b

$$b = 1 - a = 1 - \left(\frac{1+\sqrt{5}}{2}\right) = \frac{1-\sqrt{5}}{2}.$$

To find A and B, we note that

$$\frac{x}{1-x-x^2} = \frac{A}{1-ax} + \frac{B}{1-bx}$$

$$\Rightarrow x = A(1-bx) + B(1-ax).$$

Letting $x = 1/b$, yields:

$$x = \frac{2}{1-\sqrt{5}} \Rightarrow \frac{2}{1-\sqrt{5}} = B\left(1 - \left(\frac{1+\sqrt{5}}{2}\right)\left(\frac{2}{1-\sqrt{5}}\right)\right)$$

$$\Rightarrow \frac{-2-2\sqrt{5}}{4} = B\left(\frac{5+\sqrt{5}}{2}\right)$$

$$\Rightarrow B = \frac{-1}{\sqrt{5}}.$$

Similarly, substituting $x = \frac{1}{a}$ will yield A:

$$x = \frac{2}{1+\sqrt{5}} \Rightarrow \frac{2}{1+\sqrt{5}} = A\left(1-\left(\frac{1-\sqrt{5}}{2}\right)\left(\frac{2}{1+\sqrt{5}}\right)\right)$$

$$\Rightarrow \frac{2}{1+\sqrt{5}} = A\left(\frac{5-\sqrt{5}}{2}\right)$$

$$\Rightarrow A = \frac{1}{\sqrt{5}}.$$

Thus the generating function associated with the Fibonacci sequence can be rewritten as

$$F(x) = \frac{x}{1-x-x^2} = \frac{1}{\sqrt{5}}\left(\frac{1}{1-\left(\frac{1+\sqrt{5}}{2}\right)x} - \frac{1}{1-\left(\frac{1-\sqrt{5}}{2}\right)x}\right).$$

This in turn can be rewritten in terms of geometric series as follows:

$$F(x) = \frac{1}{\sqrt{5}}\left(\sum_{n=0}^{\infty}\left(\frac{1+\sqrt{5}}{2}\right)^n x^n - \sum_{n=0}^{\infty}\left(\frac{1-\sqrt{5}}{2}\right)^n x^n\right)$$

$$= \frac{1}{\sqrt{5}}\sum_{n=0}^{\infty}\left(\left(\frac{1+\sqrt{5}}{2}\right)^n - \left(\frac{1-\sqrt{5}}{2}\right)^n\right)x^n.$$

However, by definition

$$F(x) = \sum_{n=0}^{\infty} F_n x^n = \frac{1}{\sqrt{5}}\sum_{n=0}^{\infty}\left(\left(\frac{1+\sqrt{5}}{2}\right)^n - \left(\frac{1-\sqrt{5}}{2}\right)^n\right)x^n.$$

Comparing the coefficients on x^n on each side of the equation yields

$$F_n = \frac{1}{\sqrt{5}}\left[\left(\frac{1+\sqrt{5}}{2}\right)^n - \left(\frac{1-\sqrt{5}}{2}\right)^n\right].$$

□

We can use this same technique to obtained closed forms for other recurrences as well.

Example 6.2.2 Find a closed form for T_n, the number of triangles in the nth iteration of the Sierpiński graph (see Example 6.1.4 and Fig. 6.1).

Solution In Example 6.1.4, we found that T_n satisfies the recurrence $T_n = 3T_{n-1}+2$, with $T_0 = 1$. The generating function associated with this sequence is given by

$$T(x) = \sum_{n=0}^{\infty} T_n x^n.$$

Again, we will manipulate the summation in a way that will allow us to apply the recurrence.

$$\begin{aligned} T(x) &= T_0 + \sum_{n=1}^{\infty} T_n x^n \\ &= T_0 + \sum_{n=1}^{\infty} (3T_{n-1} + 2)x^n \\ &= T_0 + 3\sum_{n=1}^{\infty} T_{n-1}x^n + 2\sum_{n=1}^{\infty} x^n \\ &= T_0 + 3x\sum_{n=1}^{\infty} T_{n-1}x^{n-1} + 2x\sum_{n=1}^{\infty} x^{n-1}. \end{aligned}$$

We now use the change of variables $i = n - 1$ to re-index both summations. This yields

$$T(x) = T_0 + 3x\sum_{i=0}^{\infty} T_i x^i + 2x\sum_{i=0}^{\infty} x^i.$$

Since the variable used to index the summation is irrelevant, we have that

$$T(x) = T_0 + 3xT(x) + \frac{2x}{1-x}.$$

Inputting the initial value $T_0 = 1$ gives us an algebraic equation that we can solve for $T(x)$

$$\begin{aligned} T(x) - 3xT(x) &= 1 + \frac{2x}{1-x} \\ \Rightarrow T(x)(1-3x) &= \frac{1-x}{1-x} + \frac{2x}{1-x} = \frac{1+x}{1-x} \\ \Rightarrow T(x) &= \frac{1+x}{(1-3x)(1-x)} = \frac{A}{1-3x} + \frac{B}{1-x}. \end{aligned}$$

We now must find the coefficients A and B. To do this note that

$$1 + x = A(1-x) + B(1-3x).$$

Substituting $x = 1$ and $x = \frac{1}{3}$ into this equation yields

$$x = 1 \Rightarrow 2 = B(-2) \Rightarrow B = -1;$$

$$x = \frac{1}{3} \Rightarrow \frac{4}{3} = A\left(\frac{2}{3}\right) \Rightarrow A = 2.$$

Thus,

$$T(x) = \frac{2}{1-3x} - \frac{1}{1-x}.$$

Rewriting in terms of geometric series yields

$$T(x) = 2\sum_{n=0}^{\infty} 3^n x^n - \sum_{n=0}^{\infty} x^n$$

$$= \sum_{n=0}^{\infty} (2 * 3^n - 1)x^n.$$

By definition,

$$T(x) = \sum_{n=0}^{\infty} T_n x^n = \sum_{n=0}^{\infty} (2 * 3^n - 1)x^n.$$

Comparing the coefficient on x^n in both sides of this equation yields $T_n = 2 * 3^n - 1$. □

In Proposition 1.2.1, we gave an inductive proof for the fact that $\sum_{i=0}^{n} i = \frac{n(n+1)}{2}$. We now derive and prove this result simultaneously using generating functions.

Proposition 6.2.3 *For* $n \in \mathbb{N}$, $\sum_{i=0}^{n} i = \frac{n(n+1)}{2}$.

Proof Let $S_n = \sum_{i=0}^{n} i$. Using a similar method as Example 6.1.8, we can show that $S_0 = 0$ and $S_n = S_{n-1} + n$ for $n \geq 1$. Let

$$S(x) = \sum_{n=0}^{\infty} S_n x^n.$$

We now manipulate this series to obtain an algebraic expression for $S(x)$. So,

$$S(x) = \sum_{n=0}^{\infty} S_n x^n = \sum_{n=1}^{\infty} S_n x^n$$

$$= \sum_{n=1}^{\infty} (S_{n-1} + n)x^n = x\sum_{n=1}^{\infty} S_{n-1}x^{n-1} + x\sum_{n=1}^{\infty} nx^{n-1}.$$

We now change the index of the summation for both terms. Let $m = n - 1$ and note that $n = 1$ implies $m = 0$ and $n = \infty$ implies $m = \infty$. This gives us the equivalent expression

$$S(x) = x\sum_{m=0}^{\infty} S_m x^m + x\sum_{m=0}^{\infty}(m+1)x^m.$$

The first summation is $xS(x)$. Using Example 5.2.2 yields

$$\sum_{m=0}^{\infty}(m+1)x^m = \frac{1}{(1-x)^2}.$$

Thus,

$$\begin{aligned}
S(x) &= xS(x) + \frac{x}{(1-x)^2} \\
\Rightarrow\ S(x) - xS(x) &= \frac{x}{(1-x)^2} \\
\Rightarrow\ S(x)(1-x) &= \frac{x}{(1-x)^2} \\
\Rightarrow\ S(x) &= \frac{x}{(1-x)^3}.
\end{aligned}$$

From Exercise 5.2.8, we have that

$$\frac{1}{(1-x)^3} = \sum_{n=0}^{\infty}(n+1)(n+2)x^n.$$

Hence,

$$S(x) = \sum_{n=0}^{\infty}\frac{(n+1)(n+2)}{2}x^{n+1}.$$

We now apply a change of variable. Namely, we let $\ell = n + 1$. Thus, when $n = 0$, $\ell = 1$ and when $n = \infty$, $\ell = \infty$. This yields

$$S(x) = \sum_{\ell=1}^{\infty}\frac{\ell(\ell+1)}{2}x^{\ell} = \sum_{\ell=0}^{\infty}\frac{\ell(\ell+1)}{2}x^{\ell}.$$

Since the index of summation is irrelevant, this gives us

$$S(x) = \sum_{n=0}^{\infty} S_n x^n = \sum_{n=0}^{\infty}\frac{n(n+1)}{2}x^n.$$

Comparing the coefficient of x^n in this expression yields

$$S_n = \frac{n(n+1)}{2}.$$

■

We now use the method of generating functions to find a closed form for the number of derangements on $[n]$.

Theorem 6.2.4 *For all $n \in \mathbb{N}$, the number of derangements on $[n]$, D_n, satisfies*

$$D_n = n! \sum_{k=0}^{n} \frac{(-1)^k}{k!}.$$

Proof We instead find a closed form for $D_n/n!$. Let

$$D(x) = \sum_{n=0}^{\infty} \frac{D_n}{n!} x^n.$$

By Corollary 6.1.6, D_n satisfies the recurrence $D_n = nD_{n-1} + (-1)^n$. We now manipulate the series to take advantage of this recurrence.

$$\begin{aligned}
D(x) &= \sum_{n=0}^{\infty} \frac{D_n}{n!} x^n \\
&= 1 + \sum_{n=1}^{\infty} \frac{D_n}{n!} x^n \\
&= 1 + \sum_{n=1}^{\infty} \frac{nD_{n-1} + (-1)^n}{n!} x^n \\
&= 1 + \sum_{n=1}^{\infty} \frac{nD_{n-1}}{n!} x^n + \sum_{n=1}^{\infty} \frac{(-1)^n}{n!} x^n \\
&= 1 + x \sum_{n=1}^{\infty} \frac{D_{n-1}}{(n-1)!} x^{n-1} + \sum_{n=0}^{\infty} \frac{(-1)^n}{n!} x^n - \frac{1}{0!} x^0. \\
&= x \sum_{n=1}^{\infty} \frac{D_{n-1}}{(n-1)!} x^{n-1} + \sum_{n=0}^{\infty} \frac{(-1)^n}{n!} x^n.
\end{aligned}$$

Using the change of variables $i = n - 1$ in the first summation yields

$$D(x) = x \sum_{i=0}^{\infty} \frac{D_i}{i!} x^i + \sum_{n=0}^{\infty} \frac{(-1)^n}{n!} x^n.$$

Since the index of the summation is irrelevant, the first summation is simply $D(x)$. The second summation is equivalent to e^{-x}, as shown in an elementary calculus course. With these observations in mind, we have

$$\begin{aligned} D(x) &= xD(x) + e^{-x} \\ \Rightarrow D(x) - xD(x) &= e^{-x} \\ \Rightarrow D(x)(1-x) &= e^{-x} \\ \Rightarrow D(x) &= \frac{e^{-x}}{1-x}. \end{aligned}$$

To derive a closed form for $\frac{D_n}{n!}$, we must rewrite the generating function in such a way that we can read the coefficient on x^n. Note that

$$\begin{aligned} D(x) &= e^{-x}\left(\frac{1}{1-x}\right) \\ &= \left(\sum_{n=0}^{\infty} \frac{(-1)^n}{n!} x^n\right)\left(\sum_{n=0}^{\infty} x^n\right) \\ &= \sum_{n=0}^{\infty}\left(\sum_{k=0}^{n} \frac{(-1)^k}{k!}\right) x^n \text{ by Theorem 5.2.4.} \end{aligned}$$

By definition, we have

$$D(x) = \sum_{n=0}^{\infty} \frac{D_n}{n!} x^n = \sum_{n=0}^{\infty}\left(\sum_{k=0}^{n} \frac{(-1)^k}{k!}\right) x^n.$$

Comparing the coefficient on x^n in both sides of the equation yields

$$\begin{aligned} \frac{D_n}{n!} &= \sum_{k=0}^{n} \frac{(-1)^k}{k!} \\ \Rightarrow D_n &= n! \sum_{k=0}^{n} \frac{(-1)^k}{k!}. \end{aligned}$$

■

The generating function used to find a closed form for the number of derangements is more properly called an *exponential generating function*. An *exponential generating function* for a sequence $\{A_n\}$ is a function of the form

$$A(x) = \sum_{n=0}^{\infty} A_n \frac{x^n}{n!}.$$

Exponential generating functions (as well as other types of generating functions) are often studied in more advanced texts.

Exercise 6.2.5 Find a closed form for the recurrence $R_n = R_{n-1} + 6R_{n-2}$, where $R_0 = 0$ and $R_1 = 1$.

Exercise 6.2.6 Find a closed form for the recurrence $R_n = 2R_{n-1} + 35R_{n-2}$, where $R_0 = 0$ and $R_1 = 1$.

Exercise 6.2.7 Find a closed form for the recurrence $R_n = 3R_{n-1} - R_{n-2}$, where $R_0 = 0$ and $R_1 = 1$.

Exercise 6.2.8 Find a closed form for the recurrence $R_n = 5R_{n-1} - 29R_{n-2} + 105R_{n-3}$, where $R_0 = 0$, $R_1 = 1$, and $R_2 = 1$.

Exercise 6.2.9 Find a closed form for the recurrence $R_n = 4R_{n-1} + 6$, where $R_0 = 1$.

Exercise 6.2.10 Find a closed form for the recurrence $R_n = 4R_{n-1} + (-1)^n$, where $R_0 = 1$.

Exercise 6.2.11 Find a closed form for the recurrence $R_n = nR_{n-1} + 2$, where $R_0 = 1$.

6.3 The Method of Characteristic Polynomials

In solving recurrences, you may find that the method of generating functions to be a bit complicated. The series manipulations alone can be intimidating. In this section, we describe a method that will allow us to "shortcut" many of the steps involved when using generating functions.

Suppose that $\{R_n\}$ is a homogeneous kth-order linear recurrence with constant coefficients. In other words, $\{R_n\}$ satisfies

$$c_k R_{n+k} + \cdots + c_1 R_{n+1} + c_0 R_n = 0,$$

where $c_0, c_1, \ldots, c_k \in \mathbb{C}$ are constants. The *characteristic polynomial* associated with $\{R_n\}$ is

$$c_k x^k + \cdots + c_1 x + c_0.$$

This polynomial has k (not necessarily distinct) roots, $\lambda_1, \ldots, \lambda_k \in \mathbb{C}$. These are the *characteristic roots* of the polynomial.

To illustrate the connection between these characteristic roots and the closed form of the recurrence, we begin with a motivating example.

Example 6.3.1 Find a closed form for $\{R_n\}$, where $R_0 = 1$, $R_1 = 4$, and for all $n \geq 2$, we have $R_n = 5R_{n-1} - 6R_{n-2}$.

Solution Note that the recurrence can be rewritten as $R_{n+2} - 5R_{n+1} + 6R_n = 0$. The characteristic polynomial associated with this recurrence is $x^2 - 5x + 6$. The characteristic roots can be found using basic algebra.

$$x^2 - 5x + 6 = 0 \Rightarrow (x-2)(x-3) = 0.$$

Hence the characteristic roots are $\lambda_1 = 2$ and $\lambda_2 = 3$.

With these characteristic roots in mind, we now apply the method of generating functions. Define

$$R(x) = \sum_{n=0}^{\infty} R_n x^n.$$

As usual, manipulate the series in a way that will take advantage of the recurrence.

$$\begin{aligned} R(x) &= R_0 + R_1 x + \sum_{n=2}^{\infty} R_n x^n \\ &= 1 + 4x + \sum_{n=2}^{\infty} (5R_{n-1} - 6R_{n-2}) x^n \\ &= 1 + 4x + 5x \sum_{n=2}^{\infty} R_{n-1} x^{n-1} - 6x^2 \sum_{n=2}^{\infty} R_{n-2} x^{n-2} \\ &= 1 + 4x - 5x + 5x \sum_{n=1}^{\infty} R_{n-1} x^{n-1} - 6x^2 \sum_{n=2}^{\infty} R_{n-2} x^{n-2}. \end{aligned}$$

We use the change of variables $i = n - 1$ in the first summation and $j = n - 2$ in the second summation. This yields

$$R(x) = 1 - x + 5x \sum_{i=0}^{\infty} R_i x^i - 6x^2 \sum_{j=0}^{\infty} R_j x^j.$$

Since the variable used to index the summation is irrelevant, we have that

$$R(x) = 1 - x + 5x R(x) - 6x^2 R(x).$$

Solving this equation for $R(x)$ gives us

$$\begin{aligned} & R(x) - 5x R(x) + 6x^2 R(x) = 1 - x \\ \Rightarrow\ & R(x)(1 - 5x + 6x^2) = 1 - x \\ \Rightarrow\ & R(x) = \frac{1 - x}{1 - 5x + 6x^2} \\ \Rightarrow\ & R(x) = \frac{A}{1 - ax} + \frac{B}{1 - bx}. \end{aligned}$$

We now must solve for $A, B, a,$ and b in the partial fraction decomposition. Note that:

$$1 - 5x + 6x^2 = (1 - ax)(1 - bx) = 1 - (a + b)x + abx^2$$

$$\Rightarrow a + b = 5 \quad \text{and} \quad ab = 6$$
$$\Rightarrow b = 5 - a \Rightarrow a(5 - a) = 6$$
$$\Rightarrow -a^2 + 5a = 6 \Rightarrow a^2 - 5a + 6 = 0.$$

The polynomial $a^2 - 5a + 6$ should look familiar. In fact, this is the characteristic polynomial using the variable a instead of x. As we saw earlier, the roots of this polynomial are $a = \lambda_1 = 2$ and $b = \lambda_2 = 3$. Thus,

$$R(x) = \frac{1 - x}{1 - 5x + 6x^2} = \frac{A}{1 - 2x} + \frac{B}{1 - 3x}.$$

Using geometric series, we find that

$$R(x) = A \sum_{n=0}^{\infty} 2^n x^n + B \sum_{n=0}^{\infty} 3^n x^n$$
$$= \sum_{n=0}^{\infty} (A * 2^n + B * 3^n) x^n = \sum_{n=0}^{\infty} R_n x^n.$$

Comparing coefficients on x^n in both sides of this equation yields $R_n = A{*}2^n{+}B{*}3^n$. Notice that the numbers being exponentiated are the characteristic roots associated with the recurrence.

To find A and B, we could return to the partial fraction decomposition, as we did in the last section. Instead, we apply the initial conditions $R_0 = 1$ and $R_1 = 4$. Substituting $n = 0$ and $n = 1$ into the equation $R_n = A * 2^n + B * 3^n$ yields:

$$n = 0 \Rightarrow R_0 = A + B = 1 \quad \text{and}$$
$$n = 1 \Rightarrow R_1 = 2A + 3B = 4.$$

Note that $B = 1 - A$ implies that $2A + 3(1 - A) = 4$. Solving this equation for A gives us $A = -1$. Substituting this value into $B = 1 - A$ gives us $B = 2$.

Thus the closed form of the recurrence is $R_n = 2 * 3^n - 2^n$. □

Notice that in the last example, we found that the recurrence satisfied $R_n = A_1\lambda_1^n + A_2\lambda_2^2$, where λ_1 and λ_2 were the characteristic roots of the polynomial associated with the recurrence relation. If this can be generalized for all homogeneous linear recurrences with constant coefficients, then we could solve these recurrences in a much more compact manner. For instance, in the last example this observation would have compacted the solution by nearly two-thirds. We will state this result, however we will delay proving it until the end of the section.

Theorem 6.3.2 *Suppose that $\{R_n\}$ is a sequence defined recursively by:*

$$R_n = -(c_{k-1}R_{n-1} + c_{k-2}R_{n-2} + \cdots + c_1 R_{n-k+1} + c_0 R_{n-k}),$$

where $c_{k-1}, \ldots, c_0 \in \mathbb{C}$. Let $\lambda_1, \ldots, \lambda_\ell$ be the characteristic roots associated with this recursion, where λ_i is a root of multiplicity m_i. The closed form of R_n satisfies

$$R_n = A_{1,0}\lambda_1^n + A_{1,1}n\lambda_1^n + \cdots + A_{1,m_1-1}n^{m_1-1}\lambda_1^n +$$
$$\cdots + A_{\ell,0}\lambda_\ell^n + A_{\ell,1}n\lambda_\ell^n + \cdots + A_{\ell,m_\ell-1}n^{m_\ell-1}\lambda_\ell^n.$$

We can use Theorem 6.3.2 to find the closed form of recurrences.

Example 6.3.3 Find a closed form for $R_n = 7R_{n-1} - 16R_{n-2} + 12R_{n-3}$, where $R_0 = 7$, $R_1 = 16$, and $R_2 = 40$.

Solution Note that

$$R_{n+3} - 7R_{n+2} + 16R_{n+1} - 12R_n = 0.$$

Hence the characteristic polynomial is

$$x^3 - 7x^2 + 16x - 12 = (x-2)^2(x-3).$$

By Theorem 6.3.2, the closed form of R_n satisfies:

$$R_n = A2^n + Bn2^n + C3^n.$$

Substituting in $n = 0, 1, 2$ yields

$$n = 0 \Rightarrow 7 = R_0 = A + C;$$
$$n = 1 \Rightarrow 16 = R_1 = 2A + 2B + 3C;$$
$$n = 2 \Rightarrow 40 = R_2 = 4A + 8B + 9C.$$

Using a computer algebra system to solve for these coefficients yields $A = 3$, $B = -1$, and $C = 4$. Thus,

$$R_n = 3 * 2^n - n2^n + 4 * 3^n.$$

□

Example 6.3.4 Find a closed form for the recurrence $S_n = 2S_{n-1} - S_{n-3}$, where $S_0 = 0$, $S_1 = 1$, and $S_2 = 2$.

Solution Rewriting the recurrence as $S_{n+3} - 2S_{n+2} + S_n = 0$ yields the characteristic polynomial:

$$x^3 - 2x^2 + 1 = (x-1)(x^2 - x - 1).$$

From the Quadratic Formula, the characteristic roots of the polynomial are $\lambda_1 = 1$, $\lambda_2 = \frac{1+\sqrt{5}}{2}$, and $\lambda_3 = \frac{1-\sqrt{5}}{2}$. Thus by Theorem 6.3.2, the recurrence satisfies the closed form

$$S_n = A + B\left(\frac{1+\sqrt{5}}{2}\right)^n + C\left(\frac{1-\sqrt{5}}{2}\right)^n.$$

Substituting in the initial values yields the system:

$$n = 0 \Rightarrow 0 = A + B + C;$$

$$n = 1 \Rightarrow 1 = A + B\left(\frac{1+\sqrt{5}}{2}\right) + C\left(\frac{1-\sqrt{5}}{2}\right);$$

$$n = 2 \Rightarrow 2 = A + B\left(\frac{1+\sqrt{5}}{2}\right)^2 + C\left(\frac{1-\sqrt{5}}{2}\right)^2.$$

Using a computer algebra system to solve for the coefficients yields $A = -1$, $B = \frac{5+3\sqrt{5}}{10}$, and $C = \frac{5-3\sqrt{5}}{10}$. Thus,

$$S_n = -1 + \frac{5+3\sqrt{5}}{10}\left(\frac{1+\sqrt{5}}{2}\right)^n + \frac{5-3\sqrt{5}}{10}\left(\frac{1-\sqrt{5}}{2}\right)^n.$$

□

Sometimes, we may only want the *form* of the recurrence relation, not the closed form itself. In these cases, we do not apply the initial conditions, if they are indeed available. For instance, suppose that we choose two arbitrary numbers and apply the Fibonacci recurrence, $F_n = F_{n-1} + F_{n-2}$. As we showed previously (and can easily confirm using the method in this section), F_n satisfies the form

$$F_n = A\left(\frac{1+\sqrt{5}}{2}\right)^n + B\left(\frac{1-\sqrt{5}}{2}\right)^n.$$

The form of a recurrence when the initial conditions have not been applied is called the *general solution* to the recurrence. The *particular solution* is obtained by applying the initial conditions and solving for the coefficients involved. Thus the solution to any linear recurrence with constant coefficients can be found by first finding the general solution, then finding the particular solution.

Example 6.3.5 Find the general solution to a recurrence satisfying $R_n = 4R_{n-1} - 5R_{n-2} + 4R_{n-3} - 4R_{n-4}$.

Solution We rewrite the recurrence as

$$R_{n+4} - 4R_{n+3} + 5R_{n+2} - 4R_{n+1} + 4R_n = 0.$$

This gives us the characteristic polynomial

$$x^4 - 4x^3 + 5x^2 - 4x + 4.$$

Using the Rational Root Theorem or a computer algebra system to aid in factoring, we see that

$$x^4 - 4x^3 + 5x^2 - 4x + 4 = (x - i)(x + i)(x - 2)^2.$$

Thus, the recurrence satisfies

$$R_n = Ai^n + B(-i)^n + C2^n + Dn2^n.$$

□

One interesting application of the general solution is that it allows us to find the *growth rate* of the sequence. The growth rate of the sequence $\{R_n\}$ is defined as

$$\lim_{n\to\infty} \frac{R_n}{R_{n-1}}.$$

This growth rate is one way of comparing two distinct sequences. Suppose that $\{R_n\}$ and $\{S_n\}$ are two sequences such that $R_n > 0$ and $S_n > 0$ for sufficiently large n. If $\{R_n\}$ and $\{S_n\}$ have the same growth rate, then we say that they are *asymptotic* to each other. Two sequences, $\{R_n\}$ and $\{S_n\}$ are asymptotic to each another if and only if

$$\lim_{n\to\infty} \frac{R_n}{S_n} = c,$$

where c is a non-zero constant. Asymptotics are one of the key tools in combinatorial analysis. The interested reader is referred to the more advanced texts on this subject.

The growth rate allows us to do an interesting party trick. Have everyone in the room pick two positive numbers. Have them apply the Fibonacci recurrence, $F_n = F_{n-1} + F_{n-2}$, to these numbers fifteen to twenty times. Then have them divide the last two numbers they generated. Because the growth rate is independent of the initial values, this ratio should be very close to $\frac{1+\sqrt{5}}{2} \approx 1.61803$.

We end this section with a formal proof of Theorem 6.3.2. We first prove two lemmas.

Lemma 6.3.6 *For $\lambda \in \mathbb{C}$ and $k, n \in \mathbb{N}$, the coefficient on x^n in $\frac{1}{(1-\lambda x)^k}$ is*

$$\lambda^n(A_0 + A_1 n + \cdots A_{k-1}n^{k-1}),$$

where $A_0, \ldots, A_{k-1} \in \mathbb{C}$ are constants.

Proof Recall that

$$\frac{1}{1-\lambda x} = \sum_{n=0}^{\infty} \lambda^n x^n.$$

Taking the $(k-1)$st derivative of the left side yields:

$$\frac{d^{k-1}}{dx^{k-1}}\left(\frac{1}{1-\lambda x}\right) = \frac{\lambda^{k-1}(k-1)!}{(1-\lambda x)^k} = \frac{c'}{(1-\lambda x)^k},$$

where c' is a constant.

We can use term-by-term differentiation to obtain the $(k-1)st$ derivative of the right side. Namely,

$$\frac{d^{k-1}}{dx^{k-1}}\left(\sum_{n=0}^{\infty}\lambda^n x^n\right) = \sum_{n=k-1}^{\infty} n(n-1)\cdots(n-k+2)\lambda^n x^{n-k+1}$$

$$= \lambda^{k-1}\sum_{n=k-1}^{\infty} n(n-1)\cdots(n-k+2)\lambda^{n-k+1}x^{n-k+1}.$$

We now use the change of variables $i = n-k+1$. This yields the equivalent summation

$$\lambda^{k-1}\sum_{i=0}^{\infty}(i+k-1)(i+k-2)\ldots(i+1)\lambda^i x^i.$$

Because the index of summation is irrelevant, this is equivalent to

$$\lambda^{k-1}\sum_{n=0}^{\infty}(n+k-1)(n+k-2)\ldots(n+1)\lambda^n x^n.$$

From this it follows that

$$\frac{1}{(1-\lambda x)^k} = \frac{\lambda^{k-1}}{c'}\sum_{n=0}^{\infty}(n+k-1)(n+k-2)\cdots(n+1)\lambda^n x^n.$$

We must determine the coefficient of x^n in the right side of the equation. To facilitate this, we note that

$$\frac{\lambda^{k-1}}{c'}(n+k-1)(n+k-2)\ldots(n+1) = A_0 + A_1 n + \cdots + A_{k-1}n^{k-1},$$

where the $A_0, A_1, \ldots, A_{k-1} \in \mathbb{C}$ are constants. Thus,

$$\frac{1}{(1-\lambda x)^k} = \sum_{n=0}^{\infty}\lambda^n(A_0 + A_1 n + \cdots + A_{k-1}n^{k-1})x^n.$$

Thus the coefficient of x^n in $\frac{1}{(1-\lambda x)^k}$ is

$$\lambda^n(A_0 + A_1 n + \cdots A_{k-1}n^{k-1}),$$

where $A_0, \ldots, A_{k-1} \in \mathbb{C}$ are constants. ∎

If we actually *wanted* the power series of $\frac{1}{(1-\lambda x)^k}$ for its own sake, then we could determine the value of the constants $A_0, A_1, \ldots, A_{k-1}$ more precisely. For our current purposes, we solve for both these coefficients and the coefficients involved in the partial fraction decomposition simultaneously by substituting in the initial values of the recurrence. This being the case, we leave the precise determination of the A_i as an exercise to the reader.

Lemma 6.3.7 *Suppose that* $\lambda_1, \ldots, \lambda_n, c_0, c_1, \ldots, c_n \in \mathbb{C}$ *are constants. We have*

$$c_n(1-\lambda_1 x)\cdots(1-\lambda_n x) = c_n + c_{n-1}x + \cdots + c_1 x^{n-1} + c_0 x^n$$

if and only if

$$c_n(x-\lambda_1)\cdots(x-\lambda_n) = c_n x^n + c_{n-1}x^{n-1} + \cdots + c_1 x + c_0.$$

Proof Without loss of generality, we may assume that $c_n = 1$. We proceed by induction on n.

Let $n = 1$. If $1-\lambda_1 x = 1+(-\lambda)x$, then $c_0 = -\lambda_1$. Similarly $x - \lambda_1 = x + (-\lambda)$ implies $c_0 = -\lambda$. Thus the result holds for $n = 1$.

Suppose that for some $n = k$, we have that

$$(1-\lambda_1 x)\cdots(1-\lambda_k x) = 1 + c_{k-1}x + \cdots + c_1 x^{k-1} + c_0 x^k$$

if and only if

$$(x-\lambda_1)\cdots(x-\lambda_k) = x^k + c_{k-1}x^{k-1} + \cdots + c_1 x + c_0.$$

We now consider the case for $n = k+1$. Suppose that

$$(1-\lambda_1 x)\cdots(1-\lambda_k x)(1-\lambda_{k+1}x) = 1 + c_k' x + \cdots + c_1' x^k + c_0' x^{k+1}.$$

Note that

$$(1-\lambda_1 x)\cdots(1-\lambda_k x) = 1 + c_{k-1}x + \cdots + c_1 x^{k-1} + c_0 x^k.$$

This implies that

$$(1-\lambda_1 x)\cdots(1-\lambda_k x)(1-\lambda_{k+1}x) = (1 + c_{k-1}x + \cdots + c_1 x^{k-1} + c_0 x^k)(1-\lambda_{k+1}x)$$

$$= (1 + c_{k-1}x + \cdots + c_1 x^{k-1} + c_0 x^k) - (\lambda_{k+1}x + c_{k-1}\lambda_{k+1}x^2 + \cdots + c_0\lambda_{k+1}x^{k+1})$$

$$= 1 + (c_{k-1} - \lambda_{k+1})x + (c_{k-2} - c_{k-1}\lambda_{k+1})x^2 +$$

$$\cdots + (c_0 - c_1\lambda_{k+1})x^k - c_0\lambda_{k+1}x^{k+1}.$$

Comparing coefficients yields

$$c_{n-1} - \lambda_{k+1} = c_n',\ c_{n-2} - c_{n-1}\lambda_{k+1} = c_{n-1}', \ldots, c_0 - c_1\lambda_{k+1} = c_1', \text{and} - c_0\lambda_{k+1} = c_0'.$$

Now consider the polynomial

$$(x-\lambda_1)\cdots(x-\lambda_k)(x-\lambda_{k+1}).$$

By inductive hypothesis, this is equivalent to:

$$\begin{aligned}&(x^k+c_{k-1}x^{k-1}+\cdots+c_1x+c_0)(x-\lambda_{k+1})\\&=(x^{k+1}+c_{k-1}x^k+\cdots+c_1x^2+c_0x)\\&\quad-(\lambda_{k+1}x^k+c_{k-1}\lambda_{k+1}x^{k-1}+\cdots+c_1\lambda_{k+1}x+c_0\lambda_{k+1})\\&=x^{k+1}+(c_{k-1}-\lambda_{k+1})x^k+(c_{k-2}-c_{k-1}\lambda_{k+1})x^{k-1}+\cdots+\\&\quad(c_0-c_1\lambda_{k+1})x-c_0\lambda_{n+1}.\end{aligned}$$

Applying the observation that

$$c_{n-1}-\lambda_{k+1}=c'_n, c_{n-2}-c_{n-1}\lambda_{k+1}=c'_{n-1},\ldots,c_0-c_1\lambda_{k+1}=c'_1, \text{and} -c_0\lambda_{k+1}=c'_0$$

yields

$$(x-\lambda_1)\cdots(x-\lambda_k)(x-\lambda_{k+1})=c'_{k+1}x^{k+1}+c'_kx^k+\cdots+c'_1x+c'_0.$$

The reverse implication follows in a similar manner. ■

We now proceed with a proof of Theorem 6.3.2.

Proof Note that the recursion has characteristic polynomial

$$x^k+c_{k-1}x^{k-1}+\cdots+c_1x+c_0=0.$$

Since λ_i is a root of multiplicity m_i for $i=1,\ldots,\ell$, it follows that:

$$x^k+c_{k-1}x^{k-1}+\cdots+c_1x+c_0=(x-\lambda_1)^{m_1}\cdots(x-\lambda_\ell)^{m_\ell}.$$

Suppose that $R(x)$ is the generating function associated with the sequence R_n. In other words,

$$R(x)=\sum_{n=0}^{\infty}R_nx^n.$$

We now manipulate the series to take advantage of the recurrence.

$$R(x)=R_0+R_1x+\cdots+R_{k-1}x^{k-1}+\sum_{n=k}^{\infty}R_n.$$

Define $p_{k-1}(x)=R_0+R_1x+\cdots+R_{k-1}x^{k-1}$. Further, for $n\geq k$ we have that $R_n=c_{k-1}R_{n-1}+c_{k-2}R_{n-2}+\cdots+c_1R_{n-k+1}+c_0R_{n-k}$. Thus,

$$R(x) = p_{k-1}(x) - \sum_{n=k}^{\infty}(c_{k-1}R_{n-1} + c_{k-2}R_{n-2} + \cdots + c_1R_{n-k+1} + c_0R_{n-k})x^n$$

$$= p_{k-1}(x) - \left(c_{k-1}\sum_{n=k}^{\infty} R_{n-1}x^n + c_{k-2}\sum_{n=k}^{\infty} R_{n-2}x^n + \cdots + c_1\sum_{n=k}^{\infty} R_{n-k+1}x^n + c_0\sum_{n=k}^{\infty} R_{n-k}x^n\right)$$

$$= p_{k-1}(x) - \left(c_{k-1}x\sum_{n=k}^{\infty} R_{n-1}x^{n-1} + c_{k-2}x^2\sum_{n=k}^{\infty} R_{n-2}x^{n-2} + \cdots + c_1x^{k-1}\sum_{n=k}^{\infty} R_{n-k+1}x^{n-k+1} + c_0x^k\sum_{n=k}^{\infty} R_{n-k}x^{n-k}\right)$$

$$= p_{k-1}(x) + c_{k-1}x(R_0 + R_1x + \cdots + R_{k-2}x^{k-2}) - c_{k-1}x\sum_{n=1}^{\infty} R_{n-1}x^{n-1} + c_{k-2}x^2(R_0 + R_1x + \cdots + R_{k-3}x^{k-3}) - c_{k-2}x^2\sum_{n=2}^{\infty} R_{n-2}x^{n-2} - \cdots - c_1x^{k-1}R_0 - c_1x^{k-1}\sum_{n=k-1}^{\infty} R_{n-k+1}x^{n-k+1} - c_0x^k\sum_{n=k}^{\infty} R_{n-k}x^{n-k}$$

$$= q_{k-1}(x) - c_{k-1}x\sum_{n=1}^{\infty} R_{n-1}x^{n-1} - c_{k-2}x^2\sum_{n=2}^{\infty} R_{n-2}x^{n-2} - \cdots - c_1x^{k-1}\sum_{n=k-1}^{\infty} R_{n-k+1}x^{n-k+1} - c_0x^k\sum_{n=k}^{\infty} R_{n-k}x^{n-k},$$

where $p_{k-1}(x)$ and $q_{k-1}(x)$ are polynomials of degree at most $k-1$. We now re-index the summations using the change of variables $n_i = n - i$. This gives us

$$R(x) = q_{k-1}(x) - c_{k-1}x\sum_{n_1=0}^{\infty} R_{n_1}x^{n_1} - c_{k-2}x^2\sum_{n_2=0}^{\infty} R_{n_2}x^{n_2} - \cdots - c_1x^{k-1}\sum_{n_{k-1}=0}^{\infty} R_{n_{k-1}}x^{n_{k-1}} - c_0x^k\sum_{n_k=0}^{\infty} R_{n_k}x^{n_k}.$$

Since the variable used to index the summation is irrelevant, we have that

$$R(x) = q_{k-1}(x) - c_{k-1}xR(x) - c_{k-2}x^2R(x) - \cdots - c_1x^{k-1}R(x) - c_0x^kR(x)$$

$$\Rightarrow R(x) + c_{k-1}xR(x) + \cdots + c_1x^{k-1}R(x) + c_0x^kR(x) = q_{k-1}(x)$$

$$\Rightarrow R(x)(1 + c_{k-1}x + \cdots + c_1x^{k-1} + c_0x^k) = q_{k-1}(x)$$

$$\Rightarrow R(x) = \frac{q_{k-1}(x)}{1 + c_{k-1}x + \cdots + c_1x^{k-1} + c_0x^k}.$$

The characteristic polynomial factors as

$$x^k + c_{k-1}x^{k-1} + \cdots + c_1x + c_0 = (x - \lambda_1)^{m_1} \cdots (x - \lambda_\ell)^{m_\ell}.$$

It follows from Lemma 6.3.7 that

$$1 + c_{k-1}x + \cdots + c_1x^{k-1} + c_0x^k = (1 - \lambda_1x)^{m_1} \cdots (1 - \lambda_\ell)^{m_\ell}.$$

Since the degree of q_{k-1} is strictly less than k, we can perform a partial fraction decomposition on the generating function. In other words,

$$\begin{aligned} R(x) &= \frac{q_{k-1}(x)}{(1 - \lambda_1x)^{m_1} \cdots (1 - \lambda_\ell)^{m_\ell}} \\ &= \frac{a_{1,1}}{1 - \lambda_1x} + \cdots + \frac{a_{1,m_1}}{(1 - \lambda_1x)^{m_1}} \\ &+ \cdots + \frac{a_{\ell,1}}{1 - \lambda_\ell x} + \cdots + \frac{a_{\ell,m_\ell}}{(1 - \lambda_\ell x)^{m_\ell}}. \end{aligned}$$

By Lemma 6.3.6, the coefficient on x^n in $\frac{a_{i,j}}{(1-\lambda_ix)^j}$ is given by

$$a_{i,j}\lambda_i^n(A_0 + A_1n + \cdots + A_{j-1}n^{j-1}).$$

From this it follows that the coefficient of x^n in

$$\frac{a_{i,1}}{1 - \lambda_ix} + \cdots + \frac{a_{i,m_i}}{(1 - \lambda_ix)^{m_i}}$$

is

$$\lambda_i^n(A_{i,0} + A_{i,1}n + \cdots + A_{i,m_i-1}n^{m_i-1}).$$

By definition, the coefficient of x^n in $R(x)$ is R_n. Hence:

$$R_n = A_{1,0}\lambda_1 + A_{1,1}n\lambda_1^n + \cdots + A_{1,m_1-1}n^{m_1-1}\lambda_1^n +$$

$$\cdots + A_{\ell,0}\lambda_\ell^n + A_{\ell,1}n\lambda_\ell^n + \cdots + A_{\ell,m_\ell-1}n^{m_\ell-1}\lambda_\ell^n.$$

■

Exercise 6.3.8 Determine the values of $A_0, \ldots, A_{k-1}$ in Lemma 6.3.6. In other words, give a formula for A_i.

Exercise 6.3.9 Show that if a sequence satisfies $F_n = F_{n-1} + F_{n-2}$, then

$$\lim_{n\to\infty} \frac{F_n}{F_{n-1}} = \frac{1 + \sqrt{5}}{2}.$$

Exercise 6.3.10 Find a closed form for the recurrence $R_n = R_{n-1} + 6R_{n-2}$, where $R_0 = 0$ and $R_1 = 1$.

Exercise 6.3.11 Find a closed form for the recurrence $R_n = 3R_{n-1} - R_{n-2}$, where $R_0 = 0$ and $R_1 = 1$.

Exercise 6.3.12 Find a closed form for the recurrence $R_n = 5R_{n-1} + 29R_{n-2} - 105R_{n-3}$, where $R_0 = 0$, $R_1 = 1$, and $R_2 = 1$.

Exercise 6.3.13 Find a closed form for the recurrence $R_n = 7R_{n-1} + 5R_{n-2} - 75R_{n-3}$, where $R_0 = R_1 = 0$ and $R_2 = 1$.

Exercise 6.3.14 Find a closed form for the recurrence $R_n = 8R_{n-1} - 21R_{n-2} + 18R_{n-3}$, where $R_0 = R_1 = 0$ and $R_2 = 1$.

Exercise 6.3.15 Find the general solution for the recurrence

$$R_{n+5} - 9R_{n+4} + 31R_{n+3} - 63R_{n+2} + 108R_{n+1} - 108R_n.$$

Exercise 6.3.16 Prove that two sequences, $\{R_n\}$ and $\{S_n\}$, are asymptotic to each other if and only if

$$\lim_{n\to\infty} \frac{R_n}{S_n} = c,$$

where c is a constant.

6.4 The Method of Symbolic Differentiation

In general, a recurrence relation is of the form,

$$f(R_{n+k}, R_{n+k-1}, \dots, R_n) = g(n).$$

Until now, we have concentrated our efforts on those recurrences in which $g(n) = 0$, in other words, a *homogenous* recurrence relation. Now we consider the case of non-homogeneous linear recurrences with constant coefficients. As a reminder, these recurrences will be of the form

$$a_k R_{n+k} + a_{k-1} R_{n+k-1} + \cdots + a_1 R_{n+1} + a_0 R_n = g(n),$$

where the a_i are constants.

In this section, we employ *method of symbolic differentiation* to solve these recurrences. The basic idea of this method is to manipulate the non-homogeneous recurrence to obtain a homogeneous recurrence.

Recall that the number of triangles in the nth iteration of the Sierpiński graph is given by the non-homogeneous recurrence $T_n = 3T_{n-1} + 2$, where $T_0 = 1$.

Previously, we found the closed form of this recurrence to be $T_n = 2 * 3^n - 1$. We will use this recurrence to motivate the method of symbolic differentiation.

Example 6.4.1 Find a closed form for $T_n = 3T_{n-1} + 2$, where $T_0 = 1$.

Solution We rewrite the recursion as $T_n - 3T_{n-1} = 2$. Since this relation holds true for all $n \geq 1$, it is also true that $T_{n+1} - 3T_n = 2$ (this is the symbolic differentiation). From this it follows that:

$$T_n - 3T_{n-1} = 2 = T_{n+1} - 3T_n$$

$$\Rightarrow T_{n+1} - 4T_n + 3T_{n-1} = 0.$$

We now have a homogeneous linear recurrence with constant coefficients. So, it can be solved with the method described in the previous section. Note that the new recurrence can be rewritten as $T_{n+2} - 4T_{n+1} + 3T_n = 0$. The associated characteristic polynomial associated with this recurrence is

$$x^2 - 4x + 3 = (x - 1)(x - 3).$$

Thus the characteristic roots of this polynomial are $\lambda_1 = 1$ and $\lambda_2 = 3$. Hence, all possible solutions to the recurrence are of the form

$$T_n = A + B * 3^n. \tag{6.2}$$

To find the coefficients, we must find another value of the sequence. In particular, $T_1 = 3(1) + 2 = 5$. Substituting $n = 0$ and $n = 1$ into Eq. (6.2) yields

$$n = 0 \Rightarrow T_0 = A + B = 1;$$

$$n = 1 \Rightarrow T_1 = A + 3B = 5.$$

We now solve the above system.

$$A = 1 - B \Rightarrow 1 - B + 3B = 5$$

$$\Rightarrow 2B = 4 \Rightarrow B = 2 \Rightarrow A = -1.$$

Thus $T_n = 2 * 3^n - 1$, as we found with the previous method. ∎

We can also use this method to solve other non-homogeneous recurrences.

Example 6.4.2 Find a closed form for the recurrence $R_n = 7R_{n-1} - 12R_{n-2} + n^2$, where $R_0 = R_1 = 1$.

Solution Begin by rewriting the recurrence as

$$R_n - 7R_{n-1} + 12R_{n-2} = n^2. \tag{6.3}$$

Symbolically differentiating Eq. (6.3) yields

$$R_{n+1} - 7R_n + 12R_{n-1} = (n+1)^2 = n^2 + 2n + 1$$

$$\Rightarrow R_{n+1} - 7R_n + 12R_{n-1} - 2n - 1 = n^2.$$

Equating this with Eq. (6.3) gives the equation

$$R_{n+1} - 7R_n + 12R_{n-1} - 2n - 1 = R_n - 7R_{n-1} + 12R_{n-2}$$

$$\Rightarrow R_{n+1} - 8R_n + 19R_{n-1} - 12R_{n-2} = 2n + 1. \tag{6.4}$$

Unfortunately, this equation is not homogeneous. Hence, we symbolically differentiate again to obtain

$$R_{n+2} - 8R_{n+1} + 19R_n - 12R_{n-1} = 2(n+1) + 1 = 2n + 3$$

$$\Rightarrow R_{n+2} - 8R_{n+1} + 19R_n - 12R_{n-1} - 2 = 2n + 1.$$

Equating this with Eq. (6.4) yields

$$\begin{aligned} R_{n+2} - 8R_{n+1} + 19R_n - 12R_{n-1} - 2 \\ = R_{n+1} - 8R_n + 19R_{n-1} - 12R_{n-2} \\ \Rightarrow R_{n+2} - 9R_{n+1} + 27R_n - 31R_{n-1} + 12R_{n-2} = 2 \end{aligned} \tag{6.5}$$

Symbolically differentiating Eq. (6.5), yields

$$R_{n+3} - 9R_{n+2} + 27R_{n+1} - 31R_n + 12R_{n-1} = 2.$$

Equating this with Eq. (6.5) yields

$$\begin{aligned} R_{n+2} - 9R_{n+1} + 27R_n - 31R_{n-1} + 12R_{n-2} \\ = R_{n+3} - 9R_{n+2} + 27R_{n+1} - 31R_n + 12R_{n-1} \end{aligned}$$

$$\Rightarrow R_{n+3} - 10R_{n+2} + 36R_{n+1} - 58R_n + 43R_{n-1} - 12R_{n-2} = 0.$$

Hence, we now have a homogenous linear recurrence with constant coefficients. This recurrence can be rewritten as

$$R_{n+5} - 10R_{n+4} + 36R_{n+3} - 58R_{n+2} + 43R_{n+1} - 12R_n = 0.$$

The associated characteristic polynomial is

$$x^5 - 10x^4 + 36x^3 - 58x^2 + 43x - 12.$$

Using a combination of the Rational Root Theorem and long division (or a computer algebra system), we find that the characteristic roots of the polynomial are $\lambda_1 = \lambda_2 = \lambda_3 = 1$, $\lambda_4 = 3$, and $\lambda_5 = 4$. Hence, all possible solutions to the recurrence are of the form

$$R_n = A + Bn + Cn^2 + D * 3^n + E * 4^n. \tag{6.6}$$

To determine these coefficients, we must determine an additional three values of the sequence. In particular,

$$\begin{aligned}
R_2 &= 7R_1 - 12R_0 + 2^2 = 7 - 12 + 4 = -1; \\
R_3 &= 7(-1) - 12(1) + 3^2 = -10; \\
R_4 &= 7(-10) - 12(-1) + 4^2 = -42.
\end{aligned}$$

Substituting the values $n = 0, 1, 2, 3, 4$ into Eq. 6.6 yields the system

$$\begin{aligned}
n = 0 \Rightarrow R_0 &= A + D + E = 1; \\
n = 1 \Rightarrow R_1 &= A + B + C + 3D + 4E = 1; \\
n = 2 \Rightarrow R_2 &= A + 2B + 4C + 9D + 16E = -1; \\
n = 3 \Rightarrow R_3 &= A + 3B + 9C + 27D + 64E = -10; \\
n = 4 \Rightarrow R_4 &= A + 4B + 16C + 81D + 256E = -42.
\end{aligned}$$

Using a computer algebra system, we find the coefficients to be

$$A = \frac{83}{54}, B = \frac{17}{18}, C = \frac{1}{6}, D = \frac{-1}{2}, \quad \text{and} \quad E = \frac{-1}{27}.$$

Hence, the closed form of the recurrence is

$$R_n = \frac{83}{54} + \frac{17n}{18} + \frac{n^2}{6} - \frac{3^n}{2} - \frac{4^n}{27}.$$

□

Note that in the previous example, each time we employed symbolical differentiation, the degree of the polynomial on the right side of the equation decreased by one. Recall that differentiation in the traditional sense decreases the power of a polynomial by one as well. Hence the terminology "symbolic differentiation" is very appropriate for this process. This observation is generalized in the next section.

Example 6.4.3 Find a closed form for the recurrence $R_n = 9R_{n-1} - 14R_{n-2} + 5^n$, where $R_0 = R_1 = 1$.

Solution Rewrite the recurrence as

$$R_n - 9R_{n-1} + 14R_{n-2} = 5^n. \tag{6.7}$$

Symbolically differentiate Eq. (6.7) to obtain

$$R_{n+1} - 9R_n + 14R_{n-1} = 5^{n+1}.$$

Multiplying Eq. (6.7) by 5 yields

$$5R_n - 45R_{n-1} + 70R_{n-2} = 5^{n+1}.$$

This yields the equation

$$R_{n+1} - 9R_n + 14R_{n-1} = 5^{n+1} = 5R_n - 45R_{n-1} + 70R_{n-2}.$$

Thus the corresponding homogeneous equation is

$$R_{n+1} - 14R_n + 59R_{n-1} - 70R_{n-2} = 0.$$

This can be rewritten as

$$R_{n+3} - 14R_{n+2} + 59R_{n+1} - 70R_n = 0.$$

The characteristic polynomial is

$$x^3 - 14x^2 + 59x - 70 = (x-2)(x-5)(x-7).$$

The characteristic roots of this polynomial are $\lambda_1 = 2$, $\lambda_2 = 5$, and $\lambda_3 = 7$. Hence all possible solutions to the recurrence are of the form

$$R_n = A2^n + B5^n + C7^n. \tag{6.8}$$

To obtain the coefficients, we expect to need one additional value of the sequence. In particular,

$$R_2 = 9R_1 - 14R_0 + 5^2 = 9(1) - 14(1) + 5^2 = 20.$$

Inputting $n = 0, 1, 2$ into Eq. (6.8) yields the system:

$$\begin{aligned} n = 0 &\Rightarrow R_0 = A + B + C = 1; \\ n = 1 &\Rightarrow R_1 = 2A + 5B + 7C = 1; \\ n = 2 &\Rightarrow R_2 = 4A + 25B + 49C = 20. \end{aligned}$$

Using a computer algebra system, we find the coefficients to be

$$A = \frac{43}{15}, B = \frac{-25}{6}, \quad \text{and} \quad C = \frac{23}{10}.$$

Thus the closed form of the recurrence is

$$R_n = \frac{43}{15} * 2^n - \frac{25}{6} * 5^n + \frac{23}{10} * 7^n.$$

□

Notice that in the previous example, $g(n) = 5^n$ and 5 was a characteristic root of the resulting homogeneous equation. This will be generalized in the next section.

Example 6.4.4 Find all possible closed forms for the recurrence $R_n = 9R_{n-1} - 18R_{n-2} + 4^n - 3^n$.

Solution We rewrite the recurrence as

$$R_n - 9R_{n-1} + 18R_{n-2} = 4^n - 3^n. \tag{6.9}$$

Symbolically differentiating Eq. (6.9) yields

$$R_{n+1} - 9R_n + 18R_{n-1} = 4^{n+1} - 3^{n+1}$$

$$\Rightarrow R_{n+1} - 9R_n + 18R_{n-1} + 3^{n+1} = 4^{n+1}. \tag{6.10}$$

Multiplying Eq. (6.9) by 4 yields

$$4R_n - 36R_{n-1} + 72R_{n-2} = 4^{n+1} - 4 * 3^n$$
$$\Rightarrow 4R_n - 36R_{n-1} + 72R_{n-2} + 4 * 3^n = 4^{n+1}.$$

Setting this equation equal to Eq. (6.10) yields:

$$R_{n+1} - 9R_n + 18R_{n-1} + 3^{n+1} = 4^{n+1}$$
$$= 4R_n - 36R_{n-1} + 72R_{n-2} + 4 * 3^n$$
$$\Rightarrow R_{n+1} - 13R_n + 54R_{n-1} - 72R_{n-2} + 3^{n+1} - 4 * 3^n = 0$$
$$\Rightarrow R_{n+1} - 13R_n + 54R_{n-1} - 72R_{n-2} + 3^n(3 - 4) = 0$$
$$R_{n+1} - 13R_n + 54R_{n-1} - 72R_{n-2} = 3^n. \tag{6.11}$$

Symbolically differentiating Eq. (6.11) yields

$$R_{n+2} - 13R_{n+1} + 54R_n - 72R_{n-1} = 3^{n+1}. \tag{6.12}$$

Multiplying Eq. (6.11) by 3 yields

$$3R_{n+1} - 39R_n + 162R_{n-1} - 216R_{n-2} = 3^{n+1}.$$

By setting this equation equal to Eq. (6.12), we obtain

$$R_{n+2} - 13R_{n+1} + 54R_n - 72R_{n-1} = 3^{n+1}$$
$$= 3R_{n+1} - 39R_n + 162R_{n-1} - 216R_{n-2}$$
$$\Rightarrow R_{n+2} - 16R_{n+1} + 93R_n - 234R_{n-1} + 216R_{n-2} = 0.$$

This equation can be rewritten as

$$R_{n+4} - 16R_{n+3} + 93R_{n+2} - 234R_{n+1} + 216R_n = 0.$$

The associated characteristic polynomial is

$$x^4 - 16x^3 + 93x^2 - 234x + 216 = (x-3)^2(x-4)(x-6).$$

The characteristic roots associated with the recurrence are $\lambda_1 = \lambda_2 = 3$, $\lambda_3 = 4$, and $\lambda_4 = 6$. From this it follows that all solutions to the recurrence are of the form

$$R_n = A3^n + Bn3^n + C4^n + D6^n.$$

□

Notice that $g(n) = 4^n - 3^n$ and that 3 and 4 appear as characteristic roots of the resulting homogeneous equation.

Example 6.4.5 Find all possible solutions to the recurrence $R_n = 5R_{n-1} + n + 3^n$.

Solution As usual, we begin be rewriting the equation as

$$R_n - 5R_{n-1} - 3^n = n. \tag{6.13}$$

Using symbolic differentiation, we obtain

$$\begin{aligned} R_{n+1} - 5R_n - 3^{n+1} &= n + 1 \\ \Rightarrow R_{n+1} - 5R_n - 3^{n+1} - 1 &= n. \end{aligned}$$

Setting this equal to Eq. (6.13) yields

$$\begin{aligned} R_n - 5R_{n-1} - 3^n = n &= R_{n+1} - 5R_n - 3^{n+1} - 1 \\ \Rightarrow R_{n+1} - 6R_n + 5R_{n-1} - 3^{n+1} + 3^n &= 1 \\ \Rightarrow R_{n+1} - 6R_n + 5R_{n-1} - 3^n(-3+1) &= 1 \\ \Rightarrow R_{n+1} - 6R_n + 5R_{n-1} - 2 * 3^n &= 1. \end{aligned} \tag{6.14}$$

Symbolically differentiating Eq. (6.14) yields

$$R_{n+2} - 6R_{n+1} + 5R_n - 2 * 3^{n+1} = 1.$$

Setting this equal to Eq. (6.14) gives us

$$\begin{aligned} R_{n+2} - 6R_{n+1} + 5R_n - 2 * 3^{n+1} &= 1 = R_{n+1} - 6R_n + 5R_{n-1} - 2 * 3^n \\ \Rightarrow R_{n+2} - 7R_{n+1} + 11R_n - 5R_{n-1} &= 2 * 3^{n+1} - 2 * 3^n = 2 * 3^n(3-1) \\ R_{n+2} - 7R_{n+1} + 11R_n - 5R_{n-1} &= 4 * 3^n. \end{aligned} \tag{6.15}$$

Symbolically differentiating Eq. (6.15) a final time yields

$$R_{n+3} - 7R_{n+2} + 11R_{n+1} - 5R_n = 4 * 3^{n+1}. \tag{6.16}$$

Multiplying Eq. (6.15) by 3 gives us

$$3R_{n+2} - 21R_{n+1} + 33R_n - 15R_{n-1} = 4 * 3^{n+1}.$$

By setting this equal to Eq. (6.16), we obtain

$$3R_{n+2} - 21R_{n+1} + 33R_n - 15R_{n-1} = 4 * 3^{n+1} = R_{n+3} - 7R_{n+2} + 11R_{n+1} - 5R_n$$
$$\Rightarrow R_{n+3} - 10R_{n+2} + 32R_{n+1} - 38R_n + 15R_{n-1} = 0.$$

This recurrence can be rewritten as

$$R_{n+4} - 10R_{n+3} + 32R_{n+2} - 38R_{n+1} + 15R_n = 0.$$

The characteristic polynomial associated with this recurrence is

$$x^4 - 10x^3 + 32x^2 - 38x + 15 = (x-1)^2(x-3)(x-5).$$

The characteristic roots are $\lambda_1 = \lambda_2 = 1$, $\lambda_3 = 3$, and $\lambda_4 = 5$. From this it follows that all possible solutions to the recurrence are of the form

$$R_n = A + Bn + C3^n + D5^n.$$

□

Like the differentiation studied in calculus, symbolic differentiation is a *linear operator*. An operator is linear if it preserves addition and scalar multiplication. We will prove this result in the next section.

While trigonometric functions are not usual in combinatorial recurrences, they do sometimes occur. For instance, the function $\cos\left(\frac{n\pi}{2}\right)$ will generate the sequence $\{1, 0, -1, 0, 1, \ldots\}$. The next example will illustrate this further.

Example 6.4.6 Find all possible solutions to the recurrence $R_n = 7R_{n-1} + \sin\left(\frac{n\pi}{2}\right)$.

Solution We rewrite the recurrence as

$$R_n - 7R_{n-1} = \sin\left(\frac{n\pi}{2}\right). \tag{6.17}$$

Symbolically differentiating Eq. (6.17) yields

$$R_{n+1} - 7R_n = \sin\left(\frac{\pi(n+1)}{2}\right)$$
$$= \sin\left(\frac{n\pi}{2}\right)\cos\left(\frac{\pi}{2}\right) + \cos\left(\frac{n\pi}{2}\right)\sin\left(\frac{\pi}{2}\right)$$
$$= \cos\left(\frac{n\pi}{2}\right).$$

Symbolically differentiating this equation yields

$$R_{n+2} - 7R_{n+1} = \cos\left(\frac{\pi(n+1)}{2}\right)$$

$$= \cos\left(\frac{n\pi}{2}\right)\cos\left(\frac{\pi}{2}\right) - \sin\left(\frac{n\pi}{2}\right)\sin\left(\frac{\pi}{2}\right)$$
$$= -\sin\left(\frac{n\pi}{2}\right).$$

In other words,

$$-R_{n+2} + 7R_{n+1} = \sin\left(\frac{n\pi}{2}\right).$$

Setting this equal to Eq. (6.17) yields

$$R_n - 7R_{n-1} = \sin\left(\frac{n\pi}{2}\right) = -R_{n+2} + 7R_{n+1}$$
$$\Rightarrow R_{n+2} - 7R_{n+1} + R_n - 7R_{n-1} = 0$$
$$\Rightarrow R_{n+3} - 7R_{n+2} + R_{n+1} - 7R_n = 0.$$

The associated characteristic polynomial is

$$x^3 - 7x^2 + x - 7 = x^2(x-7) + (x-7)$$
$$= (x^2+1)(x-7) = (x-i)(x+i)(x-7).$$

Thus the characteristic roots are $\lambda_1 = i$, $\lambda_2 = -i$, and $\lambda_3 = 7$. From this it follows that all solutions to the recurrence are of the form

$$R_n = Ai^n + B(-i)^n + C7^n.$$

□

Exercise 6.4.7 Find a closed form for the recurrence $R_n = 4R_{n-1} + 6$, where $R_0 = 1$.

Exercise 6.4.8 Find a closed form for the recurrence $R_n = 3R_{n-1} + n^2 + n + 1$, where $R_0 = 1$.

Exercise 6.4.9 Find a closed form for the recurrence $R_n = 5R_{n-1} + 4^n$, where $R_0 = 1$.

Exercise 6.4.10 Find a closed form for the recurrence $R_n = 7R_{n-1} + 5^n - n^2$, where $R_0 = 1$.

Exercise 6.4.11 Find a closed form for the recurrence $R_n = 2R_{n-1} + 2^n$, where $R_0 = 1$.

Exercise 6.4.12 Find a closed form for the recurrence $R_n = 8R_{n-1} + \sin(n\pi)$, where $R_0 = 1$.

Exercise 6.4.13 Find a closed form for the recurrence $R_n = 7R_{n-1} - 12R_{n-2} + 5n^2 - 3n + 7$, where $R_0 = R_1 = 1$.

Exercise 6.4.14 Find a closed form for the recurrence $R_n = 10R_{n-1} - 33R_{n-2} + 36R_{n-3} + 4^n - n^2 + 2n + 1$, where $R_0 = R_1 = R_2 = 1$.

Exercise 6.4.15 Find a closed form for the recurrence $R_n = R_{n-1}+R_{n-2}+\cos(n\pi)$, where $R_0 = R_1 = 1$.

Exercise 6.4.16 Find a closed form for the recurrence $R_n = 8R_{n-1} - 16R_{n-2}+4^n$, where $R_0 = R_1 = 1$.

Exercise 6.4.17 Find a closed form for the recurrence $R_n = 8R_{n-1} - 20R_{n-2} + 32R_{n-3} - 64R_{n-4} + (-1)^n + 4^n + n^2 - \sin(n\pi)$, where $R_0 = R_1 = R_2 = R_3 = 1$.

6.5 The Method of Undetermined Coefficients

We now continue with our discussion of finding solutions of non-homogenous linear recurrences. In this section, we employ *the method of undetermined coefficients* to find the closed form of these recurrences. To find a solution to the recurrence

$$a_k R_{n+k} + \cdots + a_1 R_{n+1} + a_0 R_n = g(n),$$

we first find $h(n)$ which is a solution to

$$a_k R_{n+k} + \cdots + a_1 R_{n+1} + a_0 R_n = 0.$$

The idea behind the method of undetermined coefficients is to find $p(n)$, which is an intelligent "guess" based on $g(n)$. The recurrence will then be of the form $R_n = h(n) + p(n)$. To determine this "guess," we use the method of symbolic differentiation on several common families of functions.

Proposition 6.5.1 *Suppose that $\{R_n\}$ is a sequence satisfying the recurrence*

$$a_k R_{n+k} + \cdots + a_1 R_{n+1} + a_0 R_n = a^n.$$

Let $p(x)$ be the characteristic polynomial associated with

$$a_k R_{n+k} + \cdots + a_1 R_{n+1} + a_0 R_n.$$

If a is root of multiplicity $m \in \mathbb{N}$ in $p(x)$, then the characteristic polynomial associated with the homogenous equation has a as a characteristic root of multiplicity $m + 1$.

Proof We begin by symbolically differentiating

$$a_k R_{n+k} + \cdots + a_1 R_{n+1} + a_0 R_n = a^n. \tag{6.18}$$

This yields

$$a_k R_{n+k+1} + \cdots + a_1 R_{n+2} + a_0 R_{n+1} = a^{n+1}. \tag{6.19}$$

Multiplying Eq. (6.18) by a gives us

$$a\,(a_k R_{n+k} + \cdots + a_1 R_{n+1} + a_0 R_n) = a^{n+1}.$$

Equating this with Eq. (6.19) gives the homogeneous equation

$$a_k R_{n+k+1} + \cdots + a_1 R_{n+2} + a_0 R_{n+1} = a^{n+1} = a\left(a_k R_{n+k} + \cdots + a_1 R_{n+1} + a_0 R_n\right)$$
$$\Rightarrow a_k R_{n+k+1} + \cdots + a_1 R_{n+2} + a_0 R_{n+1} - a\left(a_k R_{n+k} + \cdots + a_1 R_{n+1} + a_0 R_n\right) = 0.$$

The corresponding characteristic polynomial is

$$\begin{aligned} & a_k x^{k+1} + \cdots + a_1 x^2 + a_0 x - ap(x) \\ &= x\left(a_k x^k + \cdots + a_1 x + a_0\right) - ap(x) \\ &= xp(x) - ap(x) = (x-a)p(x). \end{aligned}$$

Note that $p(x)$ has a as a root of multiplicity m. It follows that the characteristic polynomial associated with the homogenous equation has a as a root of multiplicity $m+1$. ■

Note that Proposition 6.5.1 implies that if $g(n) = a^n$, then an intelligent guess for $p(n)$ would be $p(n) = An^{m+1}a^n$.

Proposition 6.5.2 *Suppose that $\{R_n\}$ is a sequence satisfying the recurrence*

$$a_k R_{n+k} + \cdots + a_1 R_{n+1} + a_0 R_n = g_d(n),$$

where $g_d(n)$ is a polynomial in n of degree d. Let $p(x)$ be the characteristic polynomial associated with

$$a_k R_{n+k} + \cdots + a_1 R_{n+1} + a_0 R_n.$$

If 1 is root of multiplicity $m \in \mathbb{N}$ in $p(x)$, then the characteristic polynomial associated with the homogenous recurrence has 1 as a characteristic root of multiplicity $m+d$.

Proof We proceed by induction on d. If $d = 0$, then the result follows from Lemma 6.5.1. Assume that the result holds for some $d \geq 0$. Consider the case where

$$a_k R_{n+k} + \cdots + a_1 R_{n+1} + a_0 R_n = g_{d+1}(n). \tag{6.20}$$

Here, we assume that $g_{d+1}(n)$ can be written as

$$g_{d+1}(n) = c_{d+1} n^{d+1} + g'_d(n),$$

where $g'_d(n)$ is a polynomial of degree at most d. Symbolically differentiating Eq. (6.20) yields

$$a_k R_{n+k+1} + \cdots + a_1 R_{n+2} + a_0 R_{n+1} = g_{d+1}(n+1). \tag{6.21}$$

Again, we write

$$g_{d+1}(n+1) = c_{d+1} n^{d+1} + g''_d(n),$$

where $g''_d(n)$ is a polynomial of degree at most d. Thus Eqs. (6.20) and (6.21) imply that

$$a_k R_{n+k} + \cdots + a_1 R_{n+1} + a_0 R_n - g'_d(n) = a_k R_{n+k+1} + \cdots + a_1 R_{n+2} + a_0 R_{n+1} - g''_d(n)$$

$$\Rightarrow a_k R_{n+k+1} + \cdots + a_1 R_{n+2} + a_0 R_{n+1} - (a_k R_{n+k} + \cdots + a_1 R_{n+1} + a_0 R_n) = g''_d(n) - g'_d(n).$$

We note that $g''_d(n) - g'_d(n)$ is a polynomial of degree at most d. Further note that the characteristic polynomial associated with

$$a_k R_{n+k+1} + \cdots + a_1 R_{n+2} + a_0 R_{n+1} - (a_k R_{n+k} + \cdots + a_1 R_{n+1} + a_0 R_n)$$

is

$$xp(x) - p(x) = (x - 1)p(x).$$

Hence if 1 is characteristic root of multiplicity m in the original recurrence, then it is a characteristic root of multiplicity $m + 1$. Therefore, the claim follows by the Principle of Mathematical Induction. ■

Proposition 6.5.2 implies that if $g(n)$ is a polynomial of degree d, then an intelligent guess for $p(n)$ would be

$$p(n) = n^{m+1}(A_d n^d + \cdots + A_1 n + A_0).$$

Proposition 6.5.3 *Suppose that $\{R_n\}$ is a sequence satisfying the recurrence*

$$a_k R_{n+k} + \cdots + a_1 R_{n+1} + a_0 R_n = \cos\left(n\frac{\pi}{2}\right).$$

Let $p(x)$ be the characteristic polynomial associated with

$$a_k R_{n+k} + \cdots + a_1 R_{n+1} + a_0 R_n.$$

If i is root of multiplicity $m \in \mathbb{N}$ in $p(x)$, then the characteristic polynomial associated with the homogenous recurrence has i and $-i$ as characteristic roots of multiplicity $m + 1$.

Proof We begin by symbolically differentiating

$$a_k R_{n+k} + \cdots + a_1 R_{n+1} + a_0 R_n = \cos\left(n\frac{\pi}{2}\right). \tag{6.22}$$

This yields

$$a_k R_{n+k+1} + \cdots + a_1 R_{n+2} + a_0 R_{n+1} = \cos\left((n+1)\frac{\pi}{2}\right)$$

$$= \cos\left(n\frac{\pi}{2}\right)\cos\left(\frac{\pi}{2}\right) - \sin\left(n\frac{\pi}{2}\right)\sin\left(\frac{\pi}{2}\right)$$

$$a_k R_{n+k+1} + \cdots + a_1 R_{n+2} + a_0 R_{n+1} = -\sin\left(n\frac{\pi}{2}\right). \tag{6.23}$$

Symbolically differentiating Eq. (6.23) yields

$$a_k R_{n+k+2} + \cdots + a_1 R_{n+3} + a_0 R_{n+2} = -\sin\left((n+1)\frac{\pi}{2}\right)$$

$$= -\left(\sin\left(n\frac{\pi}{2}\right)\cos\left(\frac{\pi}{2}\right) + \cos\left(n\frac{\pi}{2}\right)\sin\left(\frac{\pi}{2}\right)\right)$$

$$a_k R_{n+k+2} + \cdots + a_1 R_{n+3} + a_0 R_{n+2} = -\cos\left(n\frac{\pi}{2}\right). \tag{6.24}$$

Multiplying Eq. (6.22) by -1 and equating it with Eq. (6.24) yields

$$a_k R_{n+k+2} + \cdots + a_1 R_{n+3} + a_0 R_{n+2} = -(a_k R_{n+k} + \cdots + a_1 R_{n+1} + a_0 R_n) \Rightarrow$$

$$(a_k R_{n+k+2} + \cdots + a_1 R_{n+3} + a_0 R_{n+2}) + (a_k R_{n+k} + \cdots + a_1 R_{n+1} + a_0 R_n) = 0.$$

The associated characteristic polynomial is

$$x^2 p(x) + p(x) = p(x)(x^2+1) = p(x)(x+i)(x-i).$$

Thus if i is a root of multiplicity m in $p(x)$, it follows that i is a root of multiplicity $m+1$ in the new recurrence. Since complex roots come in conjugate pairs, the same is true for $-i$. ∎

The case for $g(n) = \sin\left(n\frac{\pi}{2}\right)$ follows similarly to Proposition 6.5.3. In fact, the intelligent guess for both functions is

$$p(n) = n^m(Ai^n + B(-i)^n).$$

While we are only concerned with recurrences involving integer sequences, one could also consider recurrence involving real number or even complex numbers. In these cases, we would consider $g(n) = a\cos(nk) + b\sin(kn)$. The intelligent guess for this function would be $p(n) = n^m(Ai^n + B(-i)^n)$. However, the guess of $p(n) = n^m(A\sin(kn) + B\cos(kn))$ is more common in practice. The proof of this follows in a similar manner to the above.

Table 6.1 Good guesses for $p(n)$

Right side	Guess
a^n	$An^d a^n$
n^p	$n^d(A_p n^p + \cdots + A_1 n + A_0)$
$n^p a^n$	$a^n n^d(A_p n^p + \cdots + A_1 n + A_0)$
$a \sin(kn) + b \cos(kn)$	$n^d(A \sin(kn) + B \cos(kn))$
$a^n(b \sin(kn) + c \cos(kn))$	$a^n n^d(A \sin(kn) + B \cos(kn))$
$n^p(b \sin(kn) + c \cos(kn))$	$n^d \sin(kn)(A_p n^p + \cdots + A_1 n + A_0)$ $+n^d \cos(kn)(B_p n^p + B_{p-1} n^{p-1} + \cdots + B_1 n + B_0)$

Fact 6.5.4 *Suppose that $h(n)$ is a solution to the homogeneous linear recurrence*

$$a_k R_{n+k} + \cdots + a_1 R_{n+1} + a_0 R_n = 0.$$

All solutions to

$$a_k R_{n+k} + \cdots + a_1 R_{n+1} + a_0 R_n = g(n)$$

are of the form $h(n) + p(n)$, where $p(n)$ is a function determined by $g(n)$.

In Table 6.1, we provide several good guesses for $p(n)$, based on various $g(n)$. We note that n^d is the smallest power of n so that our guess is distinct from $h(n)$. Several of these proofs are included in this section. Others are more tedious and rely on similar methods to the above propositions. This being the case, we omit these proofs.

We now give several examples illustrating this method. Recall that the number of triangles in the nth iteration of the Sierpiński graph satisfies the recurrence $T_n = 3T_{n-1} + 2$, where $T_0 = 1$. Previously, we used the method of generating functions and symbolic differentiation to find that $T_n = 2 * 3^n - 1$ is the closed form of this recurrence. We now present an alternative solution with the method discussed in this section.

Example 6.5.5 Find a closed form for the recurrence $T_n = 3T_{n-1} + 2$, with $T_0 = 1$.

Solution Note that the homogeneous recurrence corresponding to this problem is $T_{n+1} - 3T_n = 0$. Thus the characteristic polynomial associated with this recurrence is $x - 3 = 0$. Clearly, the root of this polynomial is $\lambda = 3$. Thus, the closed form for the homogeneous equation is of the form $T_n = A * 3^n$.

Since $g(x)$ is a constant, we look for a solution of the form $T_n = A * 3^n + B$. Note that $T_0 = 1$ and $T_1 = 5$. Letting $n = 0, 1$ in the above equation yields

$$n = 0 \Rightarrow 1 = A + B;$$

$$n = 1 \Rightarrow 5 = 3A + B;$$

$$\Rightarrow 4 = 2A \Rightarrow A = 2;$$

$$\Rightarrow 1 = 2 + B \Rightarrow B = -1.$$

Thus $T_n = 2 * 3^n - 1$ is the required closed form. □

In Theorem 6.1.7, we defined $c_m(n)$ to be the number of ways to make a fixed necklace with n beads where each bead may be any of m different colors and no two adjacent beads are the same color. In that theorem, we found that $c_m(3) = m(m-1)(m-2)$ and $c_m(n) = m(m-1)^{n-1} - c_m(n-1)$ for $n \geq 4$. We now determine a closed form for this sequence.

Theorem 6.5.6 *For $n \geq 3$, let $c_m(n)$ be the number of ways to make a fixed necklace with n beads where each bead may be any of m different colors and no two adjacent beads are the same color. This sequence satisfies the closed form,*

$$c_m(n) = (m-1)^n + (-1)^n(m-1).$$

Proof Note that for $n \geq 3$, we have that $c_m(n+1) + c_m(n) = m(m-1)^{n-1}$. First, we solve the homogeneous recurrence $c_m(n+1) + c_m(n) = 0$. The corresponding characteristic polynomial is $x + 1 = 0$. The characteristic root of this is $\lambda = -1$. Thus, the closed form for the homogeneous equation is $h(n) = A(-1)^n$. Since $g(n) = m(m-1)^{n-1}$, we take $p(n) = B(m-1)^n$. So, we look for a solution of the form

$$c_m(n) = A(-1)^n + B(m-1)^n. \tag{6.25}$$

Since Eq. (6.25) has two unknowns, we anticipate needing an additional initial value. In this case,

$$\begin{aligned} c_m(4) = m(m-1)^3 - c_m(3) &= m(m-1)^3 - m(m-1)(m-2) \\ &= m(m-1)\left[(m-1)^2 - (m-2)\right] \\ &= m(m-1)(m^2 - 3m + 3). \end{aligned}$$

Inputting the values for $n = 3$ and $n = 4$ into Eq. (6.25) yields the system:

$$\begin{aligned} c_m(3) = m(m-1)(m-2) &= A(-1)^3 + B(m-1)^3 = B(m-1)^3 - A \\ c_m(4) = m(m-1)(m^2 - 3m + 3) &= A(-1)^4 + B(m-1)^4 = B(m-1)^4 + A. \end{aligned}$$

Adding these two equations together yields:

$$\begin{aligned} m(m-1)(m-2) + m(m-1)(m^2 - 3m + 3) &= B(m-1)^3 + B(m-1)^4 \\ \Rightarrow m(m-1)[(m-2) + (m^2 - 3m + 3)] &= B(m-1)^3[1 + (m-1)] \\ \Rightarrow m(m-1)(m^2 - 2m + 1)] &= Bm(m-1)^3] \\ \Rightarrow B &= 1. \end{aligned}$$

Thus,

$$m(m-1)(m-2) = (m-1)^3 - A$$

$$\begin{aligned}
\Rightarrow A &= (m-1)^3 - m(m-1)(m-2) \\
&= (m-1)\left[(m-1)^2 - m(m-2)\right] \\
&= (m-1)(m^2 - 2m + 1 - (m^2 - 2m)) = m - 1.
\end{aligned}$$

Thus, the closed form of the recurrence is $c_m(n) = (m-1)(-1)^n + (m-1)^n$.

■

Example 6.5.7 Find a closed form for $S_n = \sum_{i=0}^{n} i^2$.

Solution In Exercise 6.1.8, we found the recurrence $S_{n+1} - S_n = (n+1)^2$. The homogenous part of this recurrence is $S_{n+1} - S_n = 0$, which has characteristic polynomial $x - 1 = 0$. The characteristic root is $\lambda = 1$. Thus, every solution to the homogeneous is of the form $h(n) = A$. Since $g(n) = (n+1)^2$ and 1 is a characteristic root, we take $p(n) = Bn + Cn^2 + Dn^3$. So the closed form of the recurrence is

$$S_n = A + Bn + Cn^2 + Dn^3. \tag{6.26}$$

We need four initial values to solve for the coefficients. By direct computation, we have that $S_0 = 1$, $S_1 = 1$, $S_2 = 5$, and $S_3 = 14$. This yields the following system:

$$\begin{aligned}
S_0 &= A = 0, \\
S_1 &= A + B + C + D = 1, \\
S_2 &= A + 2B + 4C + 8D = 5, \\
S_3 &= A + 3B + 9C + 27D = 14.
\end{aligned}$$

Solving this system using a computer algebra system yields $A = 0$, $B = 1/6$, $C = 1/2$, and $D = 1/3$. Thus,

$$S_n = \frac{n}{6} + \frac{n^2}{2} + \frac{n^3}{3}.$$

□

Example 6.5.8 Find a closed form for the recurrence $R_n = 9R_{n-1} - 20R_{n-2} + 4n^2 - 2n + 3$, where $R_0 = R_1 = 1$.

Solution We first find a solution to the homogeneous equation, $R_n = 9R_{n-1} - 20R_{n-2}$. This can be rewritten as $R_{n+2} - 9R_{n+1} + 20R_n = 0$. The characteristic polynomial for this recurrence is $x^2 - 9x + 20 = (x-4)(x-5)$. The characteristic roots for this polynomial are $\lambda_1 = 4$ and $\lambda_2 = 5$. Thus all solutions to the homogeneous equation are of the form $h(n) = A4^n + B5^n$. Because $g(n) = 4n^2 - 2n + 3$, we guess that $p(n) = Cn^2 + Dn + E$. Hence by Theorem 6.5, all possible solutions to the recurrence are of the form

$$R_n = h(n) + p(n) = A4^n + B5^n + Cn^2 + Dn + E. \tag{6.27}$$

To find the coefficients, we must have at least five values of the sequence $\{R_n\}$. Using the two values already known to us, we find that:

$$R_2 = 9R_1 - 20R_0 + 4(2^2) - 2(2) + 3 = 9 - 20 + 16 - 4 + 3 = 4;$$

$$R_3 = 9R_2 - 20R_1 + 4(3^2) - 2(3) + 3 = 9(4) - 20 + 36 - 6 + 3 = 49;$$

$$R_4 = 9R_3 - 20R_2 + 4(4^2) - 2(4) + 3 = 9(49) - 20(4) + 64 - 8 + 3 = 420.$$

Substituting $n = 0, 1, 2, 3, 4$ into Eq. (6.27), yields the following system:

$$n = 0 \Rightarrow A + B + E = 1;$$

$$n = 1 \Rightarrow 4A + 5B + C + D + E = 1;$$

$$n = 2 \Rightarrow 16A + 25B + 4C + 2D + E = 4;$$

$$n = 3 \Rightarrow 64A + 125B + 9C + 3D + E = 49;$$

$$n = 4 \Rightarrow 256A + 625B + 16C + 4D + E = 420.$$

Using a computer algebra system, we find the solution to the above system is

$$A = \frac{-89}{27}, B = 2, C = \frac{1}{3}, D = \frac{14}{9}, \quad \text{and} \quad E = \frac{62}{27}.$$

Thus the solution of the recurrence is

$$R_n = \frac{-89}{27} * 4^n + 2 * 5^n + \frac{n^2}{3} + \frac{14n}{9} + \frac{62}{27}.$$

□

Example 6.5.9 Find a closed form for the recurrence $R_n = 6R_{n-1} - 9R_{n-2} + 3^n$, where $R_0 = R_1 = 1$.

Solution The homogeneous recurrence that corresponds to this sequence is $R_n = 6R_{n-1} - 9R_{n-2}$. This can be rewritten as $R_{n+2} - 6R_{n+1} + 9R_n = 0$. The associated characteristic polynomial is $x^2 - 6x + 9$. This has characteristic roots $\lambda_1 = \lambda_2 = 3$. Thus the solutions to the homogeneous recurrence are of the form $h(n) = A3^n + Bn3^n$. Because $g(n) = 3^n$, an appropriate guess for $p(n)$ would be $p(n) = 3^n$. However, since $h(n)$ contains both 3^n and $n3^n$, we must take $p(n) = Cn^2 3^n$. Thus all possible solutions of the recurrence are of the form

$$R_n = A3^n + Bn3^n + Cn^2 3^n. \tag{6.28}$$

Note that $R_2 = 6(1) - 9(1) + 3^2 = 6$. We now substitute $n = 0, 1, 2$ into Eq. (6.28).

$$n = 0 \Rightarrow 1 = A;$$

$$n = 1 \Rightarrow 1 = 3A + 3B + 3C;$$

$$n = 2 \Rightarrow 6 = 9A + 18B + 36C.$$

This yields the smaller system:

$$3B + 3C = -2$$

$$18B + 36C = -3.$$

$$\Rightarrow 18C = 9 \Rightarrow C = \frac{1}{2} \Rightarrow B = \frac{-7}{6}.$$

Thus the solution of the above recurrence is

$$R_n = 3^n\left(1 - \frac{7n}{6} + \frac{n^2}{2}\right).$$

□

Before proceeding with additional examples, we present an additional theorem to aid in finding the solution of non-homogenous recurrences.

Theorem 6.5.10 *Suppose that $\{R_n\}$ is a sequence satisfying the nonhomogeneous recurrence*

$$a_k R_{n+k} + \cdots + a_1 R_{n+1} + a_0 R_n = g_1(n) + g_2(n).$$

If $p_1(n)$ and $p_2(n)$ are "good guesses" for $g_1(n)$ and $g_2(n)$, respectively, then $p(n) = p_1(n) + p_2(n)$ is a "good guess" for the above recurrence.

Example 6.5.11 Find a closed form for the recurrence $R_n = 2R_{n-1} - R_{n-2} + 2R_{n-3} + 3^n + 2$, where $R_0 = R_1 = R_2 = 1$.

Solution The associated homogeneous recurrence is $R_n = 2R_{n-1} - R_{n-2} + 2R_{n-3}$. This can be rewritten as $R_{n+3} - 2R_{n+2} + R_{n+1} - 2R_n = 0$. The characteristic polynomial is $x^3 - 2x^2 + x - 2 = (x-2)(x-i)(x+i)$. Hence the characteristic roots are $\lambda_1 = 2$, $\lambda_2 = i$, and $\lambda_3 = -i$. From this it follows that the solution of the homogeneous equation is of the form $h(n) = A2^n + Bi^n + C(-i)^n$. Since $g(n) = 3^n + 2$, a valid guess for $p(n)$ is $p(n) = D3^n + E$. Hence, all possible solutions to the recurrence are of the form

$$R_n = A2^n + Bi^n + C(-i)^n + D3^n + E. \tag{6.29}$$

We must use the recurrence to generate one additional value of the sequence. In particular,

$$R_3 = 2R_2 - R_1 + 2R_0 + 3^3 + 2 = 2(1) - 1 + 2(1) + 27 + 2 = 32;$$

Substituting $n = 0, 1, 2, 3$ into Eq. (6.29) yields the system

$$n = 0 \Rightarrow A + B + C + D + E = 1;$$

$$n = 1 \Rightarrow 2A + iB - iC + 3D + E = 1;$$

$$n = 2 \Rightarrow 4A - B + C + 9D + E = 1;$$

$$n = 3 \Rightarrow 8A - iB + iC + 27D + E = 32.$$

In solving this system it may be useful to compare the imaginary parts on each side of the system. This yields $B = C$. Since the terms involving B and C cancel out in the second, third, and fourth equations, this yields a smaller system

$$2A + 3D + E = 1;$$

$$4A + 9D + E = 1;$$
$$8A + 27D + E = 32.$$

Solving this reduced system gives us

$$A = \frac{-31}{2}, D = \frac{31}{6}, \quad \text{and} \quad E = \frac{33}{6}.$$

Inputting these values into the first equation yields

$$B = C = \frac{-31}{12}.$$

Thus the closed form of the recurrence is

$$R_n = \left(\frac{-31}{2}\right) 2^n + \left(\frac{-31}{12}\right) i^n + \left(\frac{-31}{12}\right) (-i)^n + \left(\frac{31}{6}\right) 3^n + \frac{33}{6}.$$

□

We should note that in the previous equation, we only required four linear equations to solve for five unknowns. This is because our solution involved complex numbers. Anytime a system involves complex numbers, you actually have two systems in place. The first system is the real part of the system and the second part is the imaginary part of the system.

Example 6.5.12 Find a closed form for the recurrence $R_n = 11R_{n-1} - 39R_{n-2} + 45R_{n-3} + 2 * 5^n - 4n + \cos\left(\frac{n\pi}{2}\right)$, where $R_0 = R_1 = R_2 = 1$.

Solution The associated homogeneous recurrence is $R_n = 11R_{n-1} - 39R_{n-2} + 45R_{n-3}$. This can be rewritten as $R_{n+3} - 11R_{n+2} + 39R_{n+1} - 45R_n = 0$. The characteristic polynomial is $x^3 - 11x^2 + 39x - 45 = (x-3)^2(x-5)$. Hence the characteristic roots are $\lambda_1 = \lambda_2 = 3$ and $\lambda_3 = 5$. From this it follows that the solution of the homogeneous equation is of the form $h(n) = A3^n + Bn3^n + C5^n$. Since $g(n) = 2 * 5^n - 4n + \cos\left(\frac{n\pi}{2}\right)$, a valid guess for $p(n)$ is

$$p(n) = Dn5^n + En + F + G\cos\left(\frac{n\pi}{2}\right) + H\sin\left(\frac{n\pi}{2}\right).$$

Hence, all possible solutions to the recurrence are of the form

$$R_n = A3^n + Bn3^n + C5^n + Dn5^n + En + F + G\cos\left(\frac{n\pi}{2}\right) + H\sin\left(\frac{n\pi}{2}\right). \tag{6.30}$$

Since we have eight unknowns, we anticipate requiring an additional five values of the sequence. These can be generated using the recurrence:

$$R_3 = 11R_2 - 39R_1 + 45R_0 + 2(125) - 4(3) + 0 = 255;$$
$$R_4 = 11(255) - 39 + 45 + 2(625) - 16 + 1 = 4046;$$

$$R_5 = 11(4046) - 39(255) + 45 + 2(3125) - 20 = 40836;$$
$$R_6 = 11(40836) - 39(4046) + 45(255) + 2(15625) - 24 - 1 = 271602;$$
$$R_7 = 11(271602) - 39(40836) + 45(4046) + 2(78125) - 28 + 0 = 1733310;$$

Substituting $n = 0, \ldots, 7$ into Eq. (6.30) yields the following system:

$$n = 0 \Rightarrow A + C + F + G = 1;$$
$$n = 1 \Rightarrow 3A + 3B + 5C + 5D + E + F + H = 1;$$
$$n = 2 \Rightarrow 9A + 18B + 25C + 50D + 2E + F - G = 1;$$
$$n = 3 \Rightarrow 27A + 81B + 125C + 375D + 3E + F - H = 255;$$
$$n = 4 \Rightarrow 81A + 324B + 625C + 2500D + 4E + F + G = 4046;$$
$$n = 5 \Rightarrow 243A + 1215B + 3125C + 15625D + 5E + F + H = 40836;$$
$$n = 6 \Rightarrow 726A + 4374B + 15625C + 93750D + 6E + F - G = 271602;$$
$$n = 7 \Rightarrow 2187A + 15309B + 78125C + 546875D + 7E + F - H = 1733310.$$

Using a computer algebra system, we find the solution of this system to be

$$A = \frac{-506348}{325}, B = \frac{5362853}{7800}, C = \frac{-84652011}{270400}, D = \frac{1811011}{52000},$$

$$E = \frac{36927287}{10400}, F = \frac{5206203}{4160}, G = \frac{10487547}{16900}, H = \frac{-13497537}{16900}.$$

Hence the closed form of the recurrence is

$$R_n = \left(\frac{-506348}{325}\right)3^n + \left(\frac{5362853}{7800}\right)n3^n + \left(\frac{-84652011}{270400}\right)5^n$$
$$+ \left(\frac{1811011}{52000}\right)n5^n + \left(\frac{36927287}{10400}\right)n + \frac{5206203}{4160}$$
$$+ \left(\frac{10487547}{16900}\right)\cos\left(\frac{n\pi}{2}\right) + \left(\frac{-13497537}{16900}\right)\sin\left(\frac{n\pi}{2}\right).$$

□

Exercise 6.5.13 Find a closed form for the recurrence $R_n = 4R_{n-1} + 6$, where $R_0 = 1$.

Exercise 6.5.14 Find a closed form for $S_n = \sum_{i=0}^{n} i^3$.

Exercise 6.5.15 Find a closed form for $S_n = \sum_{i=0}^{n} i^4$.

Exercise 6.5.16 Find a closed form for $S_n = \sum_{i=0}^{n} i(i-1)$.

Exercise 6.5.17 Find a closed form for the recurrence $R_n = 3R_{n-1} + n^2 + n + 1$, where $R_0 = 1$.

Exercise 6.5.18 Find a closed form for the recurrence $R_n = 6R_{n-1} + 7R_{n-2} + n^2 + 3n + 5$, where $R_0 = R_1 = 1$.

Exercise 6.5.19 Find a closed form for the recurrence $R_n = 5R_{n-1} + 4^n$, where $R_0 = 1$.

Exercise 6.5.20 Find a closed form for the recurrence $R_n = 8R_{n-1} - 15R_{n-2} + 3^n$, where $R_0 = R_1 = 1$.

Exercise 6.5.21 Find a closed form for the recurrence $R_n = 7R_{n-1} + 5^n - n^2$, where $R_0 = 1$.

Exercise 6.5.22 Find a closed form for the recurrence $R_n = 2R_{n-1} + 2^n$, where $R_0 = 1$.

Exercise 6.5.23 Find a closed form for the recurrence $R_n = 8R_{n-1} + \sin(n\pi)$, where $R_0 = 1$.

Exercise 6.5.24 Find a closed form for the recurrence $R_n = 7R_{n-1} - 12R_{n-2} + 5n^2 - 3n + 7$, where $R_0 = R_1 = 1$.

Exercise 6.5.25 Find a closed form for the recurrence $R_n = 10R_{n-1} - 33R_{n-2} + 36R_{n-3} + 4^n - n^2 + 2n + 1$, where $R_0 = R_1 = R_2 = 1$.

Exercise 6.5.26 Find a closed form for the recurrence $R_n = R_{n-1} + R_{n-2} + \cos(n\pi)$, where $R_0 = R_1 = 1$.

Exercise 6.5.27 Find a closed form for the recurrence $R_n = 8R_{n-1} - 16R_{n-2} + 4^n$, where $R_0 = R_1 = 1$.

Exercise 6.5.28 Find a closed form for the recurrence $R_n = 8R_{n-1} - 20R_{n-2} + 32R_{n-3} - 64R_{n-4} + (-1)^n + 4^n + n^2 - \sin(n\pi)$, where $R_0 = R_1 = R_2 = R_3 = 1$.

Chapter 7
Advanced Counting—Inclusion and Exclusion

7.1 The Principle of Inclusion and Exclusion

There are many counting problems that can not be solved easily using the techniques that we have previously developed. For example, consider the number of non-negative integer solutions to

$$x_1 + x_2 + x_3 + x_4 = 25$$

that satisfy *at least one* of the constraints

$$0 \leq x_1 \leq 8;$$
$$0 \leq x_2 \leq 6;$$
$$0 \leq x_3 \leq 5;$$
$$0 \leq x_4 \leq 3.$$

Using our previous techniques, we could possibly use generating functions to find the number of solutions that satisfy at least one of the of the constraints. However, this would require four generating functions. Moreover, some of the solutions obtained may satisfy a second (or third or fourth) constraint. This would also not help us in determining the number of solutions that satisfy *exactly* two of the constraints.

As another example, consider the problem of determining the number of positive integers less than or equal to 50000 that are divisible by at least one of 2, 3, 7, 15, and 19. The set of divisible numbers cannot be partitioned easily into disjoint sets as many numbers (such as 6, 14, and 42) are divisible by two or more of the numbers. While this problem can be solved using brute force analysis (such as writing a computer program to look at each of the possibilities), we would prefer a method that is easier to generalize.

Note that if we were dealing with only two numbers, say 2 and 3, then the above problem can be solved using the techniques of Sect. 2.4 (in particular, Example 2.4.3). Let A be the set of all positive integers less than 50001 that are divisible by 2. Let

R. A. Beeler, *How to Count*, DOI 10.1007/978-3-319-13844-2_7

B be the set of all positive integers less than 50001 divisible by 3. By the Principle of Inclusion and Exclusion, the number of positive integers less than 50001 that are divisible by 2 or 3 is given by:

$$|A \cup B| = |A| + |B| - |A \cap B|.$$

We note that in this case the set $A \cap B$ represents the set of all positive integers less than 50001 that are divisible by *both* 2 and 3, namely the integers divisible by 6. Hence we need only determine how many integers are divisible by 2, 3, and 6. By Corollary 2.4.2, the number of positive integers less than x that are divisible by n is given by $\lfloor \frac{x}{n} \rfloor$. From this it follows that the number of positive integers less than 50000 is given by:

$$\begin{aligned}|A \cup B| &= |A| + |B| - |A \cap B| \\ &= \left\lfloor \frac{50000}{2} \right\rfloor + \left\lfloor \frac{50000}{3} \right\rfloor - \left\lfloor \frac{50000}{6} \right\rfloor \\ &= 25000 + 16666 - 8333 = 33333.\end{aligned}$$

Using an improved version of Theorem 1.3.8, we can solve the more generalized problem. In the more generalized version, we consider the problem of determining the cardinality of the union of n sets, $A_1,...,A_n$. To aid in this, we define the quantity c_m for $m = 1, ..., n$ as

$$c_m = \sum_{i_1,...,i_m} |A_{i_1} \cap \cdots \cap A_{i_m}|.$$

In the above definition, the sum is taken over all distinct $i_1,...,i_m$. This means that:

$$\begin{aligned}c_1 &= \sum_{i=1}^{n} |A_i|, \\ c_2 &= \sum_{i \neq j} |A_i \cap A_j|, \quad \text{and} \\ c_n &= |A_1 \cap \cdots \cap A_n|.\end{aligned}$$

Essentially, c_m can be thought of as the cardinality of a multiset that contains all elements that are a member of *at least* m of the sets, $A_1,...,A_m$. Since we are considering this as a multiset, the element x is counted once for each time it is a member of the intersection of m-sets.

Theorem 7.1.1 (The Generalized Principle of Inclusion and Exclusion) *Let $A_1,...,A_n$ be subsets of some universal (or relevant) set A. Then the following hold:*

(i)

$$\begin{aligned}|A_1 \cup \cdots \cup A_n| \\ = c_1 - c_2 + \cdots + (-1)^{\ell+1} c_\ell + \cdots + (-1)^{n+1} c_n \\ = \sum_{\ell=1}^{n} (-1)^{\ell+1} c_\ell;\end{aligned}$$

(ii)

$$\begin{aligned}|A_1^c \cap \cdots \cap A_n^c| = |(A_1 \cup \cdots \cup A_n)^c| \\ = |A| - c_1 + c_2 + \cdots + (-1)^{\ell} c_\ell + \cdots + (-1)^n c_n \\ = |A| + \sum_{\ell=1}^{n} (-1)^{\ell} c_\ell.\end{aligned}$$

Proof

(i) Let $x \in A_1 \cup \cdots \cup A_n$. It suffices to show that x is counted the same number of times by both sides of the equation. Clearly, x is counted once in $|A_1 \cup \cdots \cup A_n|$. Suppose that x is an element of *exactly* m of the subsets. It follows that:
 (a) x is counted m times by c_1;
 (b) x is counted $\binom{m}{2}$ times by c_2;
 (c) In general, x is counted $\binom{m}{\ell}$ times by c_ℓ.
 From this it follows that the right side of the equation counts x

$$\binom{m}{1} - \binom{m}{2} + \cdots + (-1)^{\ell+1}\binom{m}{\ell} + \cdots + (-1)^{m+1}\binom{m}{m}$$
$$= \sum_{i=1}^{m} (-1)^{i+1}\binom{m}{i} \quad \text{times.}$$

Recall that Corollary 3.3.4 implies that

$$0 = \sum_{i=0}^{m} (-1)^i \binom{m}{i} = 1 + \sum_{i=1}^{m} (-1)^i \binom{m}{i}$$
$$\Rightarrow 1 = -\sum_{i=1}^{m} (-1)^i \binom{m}{i} = \sum_{i=1}^{m} (-1)^{i+1} \binom{m}{i}.$$

Since x is counted the same number of times in each side of the equation, the theorem holds.

(ii) Note that DeMorgan's Law implies that:

$$A_1^c \cap \cdots \cap A_n^c = (A_1 \cup \cdots \cup A_n)^c.$$

Since the complements are assumed to be taken within the set A, we have:

$$\begin{aligned}|A_1^c \cap \cdots \cap A_n^c| &= |(A_1 \cup \cdots \cup A_n)^c| \\ &= |A| - |A_1 \cup \cdots \cup A_n|.\end{aligned}$$

Thus by (i), we have

$$\begin{aligned}|A_1^c \cap \cdots \cap A_n^c| &= |(A_1 \cup \cdots \cup A_n)^c| \\ = |A| - c_1 + c_2 + &\cdots + (-1)^\ell c_\ell + \cdots + (-1)^n c_n.\end{aligned}$$

■

Using the Principle of Inclusion and Exclusion, we can solve the above problems. However, we first solve a simpler example.

Example 7.1.2 Find the number of lattice paths from $(0,0)$ to $(10,10)$ that pass through $(2,4)$ or $(4,7)$ or $(6,8)$.

Solution Let A_1 be the set of all lattice paths from $(0,0)$ to $(10,10)$ that pass through the point $(2,4)$. Let A_2 be the set of all lattice paths from $(0,0)$ to $(10,10)$ that pass through the point $(4,7)$. Let A_3 be the set of all lattice paths from $(0,0)$ to $(10,10)$ that pass through the point $(6,8)$. Stars and Bars implies that

$$|A_1| = \binom{2+4}{4}\binom{10-2+10-4}{10-4} = \binom{6}{4}\binom{14}{6} = 45045;$$

$$|A_2| = \binom{4+7}{7}\binom{10-4+10-7}{10-7} = \binom{11}{7}\binom{9}{3} = 27720;$$

$$|A_3| = \binom{6+8}{8}\binom{10-6+10-8}{10-8} = \binom{14}{8}\binom{6}{2} = 45045;$$

$$|A_1 \cap A_2| = \binom{2+4}{4}\binom{4-2+7-4}{7-4}\binom{10-4+10-7}{10-7} = 12600;$$

$$|A_1 \cap A_3| = \binom{2+4}{4}\binom{6-2+8-4}{8-4}\binom{10-6+10-8}{10-8} = 15750;$$

$$|A_2 \cap A_3| = \binom{4+7}{7}\binom{6-4+8-7}{8-7}\binom{10-6+10-8}{10-8} = 14850;$$

$$|A_1 \cap A_2 \cap A_3| = \binom{6}{4}\binom{5}{3}\binom{3}{1}\binom{6}{2} = 6750.$$

Thus,

$$\begin{aligned}c_1 &= |A_1| + |A_2| + |A_3| = 117810, \\ c_2 &= |A_1 \cap A_2| + |A_1 \cap A_3| + |A_2 \cap A_3| = 43200, \\ \text{and} \quad c_3 &= |A_1 \cap A_2 \cap A_3| = 6750.\end{aligned}$$

By the Principle of Inclusion and Exclusion, we have that:

$$\begin{aligned}|A_1 \cup A_2 \cup A_3| &= c_1 - c_2 + c_3 \\ &= 117810 - 43200 + 6750 = 81360.\end{aligned}$$ □

For most problems, we do not compute the c_m for their own sake. We simply compute the larger sum directly. This is the case with the next problem.

Example 7.1.3 Find the number of positive integers less than or equal to 50000 that are divisible by at least one of 2, 3, 7, 15, and 19.

Solution Let A_i denote the set of positive integers less than or equal to 50000. Note that $A_{lcm(i,j)}$ will be the set of all positive integers divisible by both i and j, where $lcm(i, j)$ is the *least common multiple* of i, j. We note that $A_{lcm(i,j)} = A_i \cap A_j$. Further note that $A_{15} \subset A_3$. Hence, $A_3 \cap A_{15} = A_{15}$. Thus we must determine $|A_i|$ for $i = 2, 3, 6, 7, 14, 15, 19, 21, 30, 38, 42, 57, 105, 114, 133, 210, 266, 285, 399, 570, 798, 1995$, and 3990. Note that

$$|A_2| = \left\lfloor \frac{50000}{2} \right\rfloor = 25000; \quad |A_3| = \left\lfloor \frac{50000}{3} \right\rfloor = 16666;$$

$$|A_2 \cap A_3| = |A_6| = \left\lfloor \frac{50000}{6} \right\rfloor = 8333; \quad |A_7| = \left\lfloor \frac{50000}{7} \right\rfloor = 7142;$$

$$|A_2 \cap A_7| = |A_{14}| = \left\lfloor \frac{50000}{14} \right\rfloor = 3571; \quad |A_{15}| = \left\lfloor \frac{50000}{15} \right\rfloor = 3333;$$

$$|A_{19}| = \left\lfloor \frac{50000}{19} \right\rfloor = 2631; \quad |A_3 \cap A_7| = |A_{21}| = \left\lfloor \frac{50000}{21} \right\rfloor = 2380;$$

$$|A_2 \cap A_{15}| = |A_{30}| = \left\lfloor \frac{50000}{30} \right\rfloor = 1666;$$

$$|A_2 \cap A_{19}| = |A_{38}| = \left\lfloor \frac{50000}{38} \right\rfloor = 1315;$$

$$|A_2 \cap A_3 \cap A_7| = |A_{42}| = \left\lfloor \frac{50000}{42} \right\rfloor = 1190;$$

$$|A_3 \cap A_{19}| = |A_{57}| = \left\lfloor \frac{50000}{3} \right\rfloor = 877;$$

$$|A_7 \cap A_{15}| = |A_{105}| = \left\lfloor \frac{50000}{105} \right\rfloor = 476;$$

$$|A_2 \cap A_3 \cap A_{19}| = |A_{114}| = \left\lfloor \frac{50000}{114} \right\rfloor = 438;$$

$$|A_7 \cap A_{19}| = |A_{133}| = \left\lfloor \frac{50000}{133} \right\rfloor = 375;$$

$$|A_2 \cap A_7 \cap A_{15}| = |A_{210}| = \left\lfloor \frac{50000}{210} \right\rfloor = 238$$

$$|A_2 \cap A_7 \cap A_{19}| = |A_{266}| = \left\lfloor \frac{50000}{266} \right\rfloor = 187;$$

$$|A_{15} \cap A_{19}| = |A_{285}| = \left\lfloor \frac{50000}{285} \right\rfloor = 175;$$

$$|A_3 \cap A_7 \cap A_{19}| = |A_{399}| = \left\lfloor \frac{50000}{399} \right\rfloor = 125;$$

$$|A_2 \cap A_{15} \cap A_{19}| = |A_{570}| = \left\lfloor \frac{50000}{570} \right\rfloor = 87;$$

$$|A_2 \cap A_3 \cap A_7 \cap A_{19}| = |A_{798}| = \left\lfloor \frac{50000}{798} \right\rfloor = 62;$$

$$|A_7 \cap A_{15} \cap A_{19}| = |A_{1995}| = \left\lfloor \frac{50000}{1995} \right\rfloor = 25;$$

$$|A_2 \cap A_7 \cap A_{15} \cap A_{19}| = |A_{3990}| = \left\lfloor \frac{50000}{3990} \right\rfloor = 12.$$

By the Principle of Inclusion and Exclusion, we have:

$$\begin{aligned}
&|A_2 \cup A_3 \cup A_7 \cup A_{15} \cup A_{19}| = \\
&|A_2| + |A_3| + |A_7| + |A_{15}| + |A_{19}| \\
&-|A_6| - |A_{14}| - |A_{30}| - |A_{38}| - |A_{21}| - |A_{15}| - |A_{57}| - |A_{95}| - |A_{133}| - |A_{285}| \\
&+|A_{42}| + |A_{30}| + |A_{114}| + |A_{210}| + |A_{266}| + |A_{570}| + |A_{105}| + |A_{399}| + |A_{285}| \\
&+|A_{1995}| - |A_{210}| - |A_{798}| - |A_{570}| - |A_{3990}| - |A_{1995}| + |A_{3990}| \\
&= 25000 + 16666 + 7142 + 3333 + 2631 - 8333 - 3571 - 1666 - 1315 - 2380 \\
&-3333 - 877 - 526 - 375 - 175 + 1190 + 1666 + 438 + 238 + 187 + 87 + 476 \\
&+125 + 175 + 25 - 238 - 62 - 87 - 12 - 25 + 12 \\
&= 36416.
\end{aligned}$$

□

Note that the previous example can be done in a more compact manner if we realize that $A_{15} \subset A_3$ implies that $A_3 \cup A_{15} = A_3$.

Now, we return to the problem that motivated this section.

Example 7.1.4 Find the number of non-negative integer solutions to

$$x_1 + x_2 + x_3 + x_4 = 25$$

that satisfy *at least one* of the constraints

$$0 \leq x_1 \leq 8;$$
$$0 \leq x_2 \leq 6;$$
$$0 \leq x_3 \leq 5;$$
$$0 \leq x_4 \leq 3.$$

Solution Let A_i denote the set of all non-negative integer solutions to the ith constraint. By the techniques in Chap. 5, we can find the cardinality of each of these sets.

In particular, $|A_1|$ is the coefficient of x^{25} in

$$\left(\frac{1-x^9}{1-x}\right)\left(\frac{1}{1-x}\right)^3.$$

Using a computer algebra, system, we find $|A_1| = 2307$. Similarly, $|A_2|$ is the coefficient of x^{25} in

$$\left(\frac{1-x^7}{1-x}\right)\left(\frac{1}{1-x}\right)^3,$$

$|A_3|$ is the coefficient of x^{25} in

$$\left(\frac{1-x^6}{1-x}\right)\left(\frac{1}{1-x}\right)^3,$$

and $|A_4|$ is the coefficient of x^{25} in

$$\left(\frac{1-x^4}{1-x}\right)\left(\frac{1}{1-x}\right)^3.$$

Using a computer algebra system, we find these coefficients to be $|A_2| = 1946$, $|A_3| = 1736$, and $|A_4| = 1252$.

Similarly, $|A_1 \cap A_2|$ is the coefficient of x^{25} in

$$\left(\frac{1-x^9}{1-x}\right)\left(\frac{1-x^7}{1-x}\right)\left(\frac{1}{1-x}\right)^2,$$

$|A_1 \cap A_3|$ is the coefficient of x^{25} in

$$\left(\frac{1-x^9}{1-x}\right)\left(\frac{1-x^6}{1-x}\right)\left(\frac{1}{1-x}\right)^2,$$

$|A_1 \cap A_4|$ is the coefficient of x^{25} in

$$\left(\frac{1-x^9}{1-x}\right)\left(\frac{1-x^4}{1-x}\right)\left(\frac{1}{1-x}\right)^2,$$

$|A_2 \cap A_3|$ is the coefficient of x^{25} in

$$\left(\frac{1-x^7}{1-x}\right)\left(\frac{1-x^6}{1-x}\right)\left(\frac{1}{1-x}\right)^2,$$

$|A_2 \cap A_4|$ is the coefficient of x^{25} in

$$\left(\frac{1-x^7}{1-x}\right)\left(\frac{1-x^4}{1-x}\right)\left(\frac{1}{1-x}\right)^2,$$

and $|A_3 \cap A_4|$ is the coefficient of x^{25} in

$$\left(\frac{1-x^6}{1-x}\right)\left(\frac{1-x^4}{1-x}\right)\left(\frac{1}{1-x}\right)^2.$$

Using a computer algebra system, we have that $|A_1 \cap A_2| = 1197$, $|A_1 \cap A_3| = 1053$, $|A_1 \cap A_4| = 738$, $|A_2 \cap A_3| = 861$, $|A_2 \cap A_4| = 602$, and $|A_3 \cap A_4| = 528$.

Now for the sets taken three at a time. The cardinality of $A_1 \cap A_2 \cap A_3$ is the coefficient of x^{25} in

$$\left(\frac{1-x^9}{1-x}\right)\left(\frac{1-x^7}{1-x}\right)\left(\frac{1-x^6}{1-x}\right)\left(\frac{1}{1-x}\right),$$

$|A_1 \cap A_2 \cap A_4|$ is the coefficient of x^{25} in

$$\left(\frac{1-x^9}{1-x}\right)\left(\frac{1-x^7}{1-x}\right)\left(\frac{1-x^4}{1-x}\right)\left(\frac{1}{1-x}\right),$$

$|A_1 \cap A_3 \cap A_4|$ is the coefficient of x^{25} in

$$\left(\frac{1-x^9}{1-x}\right)\left(\frac{1-x^6}{1-x}\right)\left(\frac{1-x^4}{1-x}\right)\left(\frac{1}{1-x}\right),$$

and $|A_2 \cap A_3 \cap A_4|$ is the coefficient of x^{25} in

$$\left(\frac{1-x^7}{1-x}\right)\left(\frac{1-x^6}{1-x}\right)\left(\frac{1-x^4}{1-x}\right)\left(\frac{1}{1-x}\right).$$

Using a computer algebra system we find that $|A_1 \cap A_2 \cap A_3| = 378$, $|A_1 \cap A_2 \cap A_4| = 252$, $|A_1 \cap A_3 \cap A_4| = 216$, and $|A_2 \cap A_3 \cap A_4| = 168$.

Finally, $|A_1 \cap A_2 \cap A_3 \cap A_4|$ is the coefficient of x^{25} in

$$\left(\frac{1-x^9}{1-x}\right)\left(\frac{1-x^7}{1-x}\right)\left(\frac{1-x^6}{1-x}\right)\left(\frac{1-x^4}{1-x}\right).$$

Using a computer algebra system (or common sense), we find that $|A_1 \cap A_2 \cap A_3 \cap A_4| = 0$. Using the Principle of Inclusion and Exclusion, we find that:

$$\begin{aligned}|A_1 \cup A_2 \cup A_3 \cup A_4| &= |A_1| + |A_2| + |A_3| + |A_4| \\ &-|A_1 \cap A_2| - |A_1 \cap A_3| - |A_1 \cap A_4| - |A_2 \cap A_3| - |A_2 \cap A_4| - |A_3 \cap A_4| \\ &+|A_1 \cap A_2 \cap A_3| + |A_1 \cap A_2 \cap A_4| + |A_1 \cap A_3 \cap A_4| + |A_2 \cap A_3 \cap A_4| \\ &-|A_1 \cap A_2 \cap A_3 \cap A_4| \\ &= 2307 + 1946 + 1736 + 1252 - 1197 - 1053 \\ &-738 - 861 - 602 - 528 + 378 + 252 + 216 + 168 - 0 \\ &= 3276.\end{aligned}$$

□

At this point, you may realize that is tedious to compute the cardinalities of these sets. So far, we have restricted our attention to cases where we have only considered the union of four or five sets. However, it would be much more tedious to compute this if we were considering hundreds (or even dozens) of sets. In such cases, it may be convenient to write a simple computer program to assist in the computations. In other cases, we may be able to use certain symmetries to assist in finding a solution. This idea will be especially useful in later sections where we are considering the union of an arbitrary number of sets.

Example 7.1.5 Find the number of non-negative integer solutions to

$$x_1 + x_2 + x_3 + x_4 + x_5 = 25$$

such that at least one of the x_i satisfy $0 \leq x_i \leq 5$.

Solution Note that the x_i all have identical constraints and contribute the same amount to the equation $x_1 + x_2 + x_3 + x_4 + x_5 = 25$. Hence, we define the set A_i ($i = 1, ..., 5$) to be the set of all non-negative integer solutions to $x_1+x_2+x_3+x_4+x_5 = 25$ that satisfy i of the constraints. To aid in this, we define a_i to be the number of solutions that satisfy *some fixed set* of i constraints. Hence, a_i is the coefficient of x^{25} in

$$\left(\frac{1-x^6}{1-x}\right)^i \left(\frac{1}{1-x}\right)^{5-i}.$$

Using a computer algebra system, we find that $a_1 = 14896$, $a_2 = 8421$, $a_3 = 3996$, $a_4 = 1296$, and $a_5 = 1$. To compute $|A_i|$ we

(i) Select i of the constraints to satisfy. There are $\binom{5}{i}$ ways to do this.
(ii) Determine the number of ways to satisfy these constraints. By definition, this is given by a_i.

Hence, by the Multiplication Principle we have that $|A_i| = \binom{5}{i}a_i$ for all i. Thus, $|A_1| = 74480$, $|A_2| = 84210$, $|A_3| = 39960$, $|A_4| = 6480$, and $|A_5| = 1$. Thus by the Principle of Inclusion and Exclusion, we have that the number of non-negative integer solutions that satisfy at least one of the constraints is given by

$$|A_1| - |A_2| + |A_3| - |A_4| + |A_5|$$
$$= 74480 - 84210 + 39960 - 6480 + 1 = 23751. \quad \square$$

You may be curious as to why we have not considered any table setting examples using the Principle of Inclusion and Exclusion. We will consider an important example of such a problem in a few sections.

Exercise 7.1.6 Find the number of lattice paths from $(0,0)$ to $(7,9)$ that pass through $(2,4)$ or $(4,5)$ or $(6,6)$.

Exercise 7.1.7 Find the number of lattice paths from $(0,0)$ to $(12,12)$ that pass through neither $(3,5)$ nor $(4,7)$ nor $(8,8)$ nor $(9,11)$.

Exercise 7.1.8 Find the number of lattice paths from $(0,0)$ to $(7,9)$ that pass through $(2,4)$ or $(4,5)$ or $(6,6)$.

Exercise 7.1.9 Find the number of lattice paths from $(0,0)$ to $(12,12)$ that pass through neither $(3,5)$ nor $(4,7)$ nor $(8,8)$ nor $(9,11)$.

Exercise 7.1.10 Find the number of positive integers less than or equal to 25000 that are divisible by at least one of 2, 5, 15, and 23.

Exercise 7.1.11 Find the number of positive integers less than or equal to 30000 that are divisible by none of 3, 7, 15, 17, and 29.

Exercise 7.1.12 Find the number of positive integers less than or equal to 25000 that are divisible by at least one of 2, 5, 15, and 23.

Exercise 7.1.13 Find the number of positive integers less than or equal to 30000 that are divisible by none of 3, 7, 15, 17, and 29.

Exercise 7.1.14 Find the number of positive integers less than or equal to 50000 that are divisible by at least one of 6, 10 or 15.

Exercise 7.1.15 Find the number of positive integers less than or equal to 40000 that are divisible by none of 15, 17, 21, and 35.

Exercise 7.1.16 Find the number of non-negative integer solutions to:

$$x_1 + x_2 + x_3 + x_4 = 35$$

that satisfy *at least one* of the constraints:

$$0 \le x_1 \le 12;$$
$$0 \le x_2 \le 10;$$
$$0 \le x_3 \le 7;$$
$$0 \le x_4 \le 4.$$

Exercise 7.1.17 Find the number of non-negative integer solutions to:

$$x_1 + x_2 + x_3 + x_4 + x_5 + x_6 = 40$$

such that at least one of the x_i satisfy $0 \leq x_i \leq 8$.

Exercise 7.1.18 Find the number of non-negative integer solutions to:

$$x_1 + x_2 + x_3 + x_4 + x_5 + x_6 + x_7 = 50$$

such that at least one of the x_i satisfy $0 \leq x_i \leq 7$.

Exercise 7.1.19 Suppose that we have 25 gumdrops to be distributed to four children (Alice, Bob, Chad, and Diane). Each child has their own set of requirements:

(i) Alice must have either one or five gumdrops.
(ii) Bob must have an even number of gumdrops.
(iii) Chad must have at least four gumdrops.
(iv) Diane must have no more than six gumdrops.

Find the number of ways to distribute the gumdrops such that none of the children's restrictions are satisfied.

Exercise 7.1.20 Determine how many ordered words of length 35 from the alphabet $\{a, b, c, d, e\}$ satisfy at least one of the following:

(i) 'a' is used no more than seven times;
(ii) 'b' is used a multiple of five times;
(iii) 'c' is used an odd number of times;
(iv) 'd' is used at least six times;
(v) 'e' is used zero or a prime number of times.

7.2 Items That Satisfy a Prescribed Number of Conditions

We now return to a problem alluded to in the last section. Namely, the problem of finding the number of integer solutions to

$$x_1 + x_2 + x_3 + x_4 = 25$$

that satisfy *exactly two* of the constraints

$$0 \leq x_1 \leq 8;$$
$$0 \leq x_2 \leq 6;$$
$$0 \leq x_3 \leq 5;$$
$$0 \leq x_4 \leq 3.$$

This problem is more subtle, but is also tractable using a combination of generating functions and the Principle of Inclusion and Exclusion. However, we first need to prove an additional theorem. Essentially, we want to find the number of elements that are in exactly k of n possible sets. In this case, the elements may be members of any k sets. However, we do not care which of the k sets the elements belong. Recall that

$$c_m = \sum_{i_1,\ldots,i_m} |A_{i_1} \cap \cdots \cap A_{i_m}|.$$

Theorem 7.2.1 *Let $A_1,\ldots,A_n \subset A$ be sets. Let B_k be the set of all elements that belong to* exactly *k of these sets. The number of elements that belong to exactly k sets is given by:*

$$|B_k| = \sum_{i=0}^{n-k} (-1)^i \binom{k+i}{i} c_{k+i}.$$

Proof Let $x \in A$. It suffices to show that x is counted the same number of times by each side of the equation. There are three possibilities:

(i) Suppose that x is a member of less than k of the sets $A_1, \ldots, A_n$. By definition, $|B_k|$ counts the number of elements that are in exactly k sets. Hence, x is not counted by the left side. Similarly, c_{k+i} only counts those elements that are in at least $k+i$ sets. Thus, x is not counted by the right side either.
(ii) Suppose that x is in exactly k of the sets $A_1,\ldots,A_n$. Hence it is counted once by $|B_k|$ and once by c_k. It is not counted by any of the other c_{k+i}, as these only count those elements that are in at least $k+i$ sets for $i \geq 1$.
(iii) Suppose that x is in more than k of the sets $A_1,\ldots,A_n$. Without loss of generality, suppose that x is in exactly $k+j$ sets, where $j \geq 1$. By definition, x is not counted by $|B_k|$. On the left side, x is counted by each of the terms up to c_{k+j}. In computing c_{k+i}, x is counted once for each of the ways we can choose $k+i$ sets from the $k+j$ sets that contain x. Hence, x is counted $\binom{k+j}{k+i}$ times by c_{k+i} for $i = 0, \ldots, j$. Thus, the number of times that x is counted in the right side is given by

$$\sum_{i=0}^{j} (-1)^i \binom{k+i}{i}\binom{k+j}{k+i}$$

$$= \binom{k+j}{k} + \sum_{i=1}^{j} (-1)^i \binom{k+j}{k+i}\binom{k+i}{i}$$

$$= \binom{k+j}{k} + \sum_{i=1}^{j} (-1)^i \binom{k+j}{k}\binom{j}{i} \quad \text{(by Theorem 3.4.5)}$$

$$= \binom{k+j}{k}\left(1 + \sum_{i=1}^{j} (-1)^i \binom{j}{i}\right)$$

$$= \binom{k+j}{k}\left(\sum_{i=0}^{j} (-1)^i \binom{j}{i}\right)$$

$$= 0 \text{ (by Corollary 3.3.4.)}$$

In any case, x is counted the same number of times in each side of the equation. Hence the theorem holds. ■

Example 7.2.2 Find the number of non-negative integers solutions to:

$$x_1 + x_2 + x_3 + x_4 = 25$$

that satisfy *exactly two* of the constraints:

$$\begin{aligned} 0 &\le x_1 \le 8; \\ 0 &\le x_2 \le 6; \\ 0 &\le x_3 \le 5; \\ 0 &\le x_4 \le 3. \end{aligned}$$

Solution Let A_i denote the set of all non-negative integer solutions to the ith constraint. In Example 7.1.4, we found that:

$$|A_1 \cap A_2| = 1197, |A_1 \cap A_3| = 1053, |A_1 \cap A_4| = 738, |A_2 \cap A_3| = 861,$$
$$|A_2 \cap A_4| = 602, |A_3 \cap A_4| = 528, |A_1 \cap A_2 \cap A_3| = 378,$$
$$|A_1 \cap A_2 \cap A_4| = 252, |A_1 \cap A_3 \cap A_4| = 216, |A_2 \cap A_3 \cap A_4| = 168,$$
$$\text{and } |A_1 \cap A_2 \cap A_3 \cap A_4| = 0.$$

From this it follows that:

$$c_2 = \sum_{i \neq j} |A_i \cap A_j| = 1197 + 1053 + 738 + 861 + 602 + 528 = 4979,$$

$$c_3 = \sum_{i \neq j \neq \ell} |A_i \cap A_j \cap A_\ell| = 378 + 252 + 216 + 168 = 1014,$$

$$\text{and } c_4 = |A_1 \cap A_2 \cap A_3 \cap A_4| = 0.$$

Thus by Theorem 7.2.1, the number of solutions that satisfy *exactly* two of the constraints is given by:

$$\binom{2}{0} c_2 - \binom{3}{1} c_3 + \binom{4}{2} c_4$$
$$= 1(4979) - 3(1014) + (6)0 = 1937.$$

□

This can also be used to compute the number of positive integers that are divisible by a specific number of integers. This is illustrated in the next example.

Example 7.2.3 Find the number of positive integers less than or equal to 50000 that are divisible by *exactly* three of 2, 3, 7, 15, and 19.

Solution Much of the relevant work is contained in Example 7.1.3. Based on this work, we see that:

$$c_3 = \sum_{i \neq j \neq \ell} |A_i \cap A_j \cap A_\ell|$$

$$= |A_2 \cap A_3 \cap A_7| + |A_2 \cap A_3 \cap A_{15}| + |A_2 \cap A_3 \cap A_{19}| + |A_2 \cap A_7 \cap A_{15}|$$
$$+|A_2 \cap A_7 \cap A_{19}| + |A_2 \cap A_{15} \cap A_{19}| + |A_3 \cap A_7 \cap A_{15}| + |A_3 \cap A_7 \cap A_{19}|$$
$$+|A_3 \cap A_{15} \cap A_{19}| + |A_7 \cap A_{15} \cap A_{19}|$$
$$= 1190 + 1666 + 438 + 238 + 187 + 87 + 476 + 125 + 175 + 25 = 4607,$$

$$c_4 = |A_2 \cap A_3 \cap A_7 \cap A_{15}| + |A_2 \cap A_3 \cap A_7 \cap A_{19}| + |A_2 \cap A_3 \cap A_{15} \cap A_{19}|$$
$$+|A_2 \cap A_7 \cap A_{15} \cap A_{19}| + |A_3 \cap A_7 \cap A_{15} \cap A_{19}|$$
$$= 238 + 62 + 87 + 12 + 25 = 424,$$

$$\text{and} \quad c_5 = |A_2 \cap A_3 \cap A_7 \cap A_{15} \cap A_{19}| = 12.$$

Hence the number of positive integers less than or equal to 50000 that are divisible by exactly three of the numbers is given by:

$$|B_3| = \binom{3}{0}c_3 - \binom{4}{1}c_4 + \binom{5}{2}c_5$$
$$= 1(4607) - 4(424) + 10(12) = 3031. \qquad \square$$

Exercise 7.2.4 Find the number of non-negative integers solutions to:

$$x_1 + x_2 + x_3 + x_4 = 35$$

that satisfy *exactly two* of the constraints:

$$0 \le x_1 \le 12;$$
$$0 \le x_2 \le 10;$$
$$0 \le x_3 \le 7;$$
$$0 \le x_4 \le 4.$$

Exercise 7.2.5 Find the number of positive integers less than or equal to 25000 that are divisible by *exactly* two of 2, 5, 15, and 23.

Exercise 7.2.6 Suppose that we have 25 gumdrops to be distributed to four children (Alice, Bob, Chad, and Diane). Each child has their own set of requirements:

(i) Alice must have either one or five gumdrops.
(ii) Bob must have an even number of gumdrops.
(iii) Chad must have at least four gumdrops.
(iv) Diane must have no more than six gumdrops.

Find the number of ways to distribute the gumdrops such that exactly two of the children's restrictions are satisfied.

Exercise 7.2.7 Determine how many ordered words of length 35 from the alphabet $\{a, b, c, d, e\}$ satisfy exactly three of the following:

(i) 'a' is used no more than seven times;
(ii) 'b' is used a multiple of five times;
(iii) 'c' is used an odd number of times;
(iv) 'd' is used at least six times;
(v) 'e' is used zero or a prime number of times.

Exercise 7.2.8 Find the number of positive integers less than or equal to 30000 that are divisible by *exactly* two of 3, 7, 15, 17, and 29.

Exercise 7.2.9 Let $A_1, \ldots, A_n$ be subsets of a universal set A. Find and prove a formula for the number of elements that are in *at least* k of the sets $A_1, \ldots, A_n$.

7.3 Stirling Numbers of the Second Kind and Derangements Revisited

In this section, we revisit two of our previous problems. Namely, we revisit

(i) The number of ways to distribute n labeled balls into k unlabeled urns such that no urn is empty;
(ii) The number of permutations of $[n]$ that contain no fixed points.

The number of distributions in the first problem is counted by $S(n, k)$, a Stirling number of the second kind. However, we were unable to derive an explicit formula for this quantity. In this section, we will rectify this.

The second problem deals with the number of derangements on $[n]$. While we were able to obtain a formula for this quantity, the solution required generating functions to solve the recurrence. In this section, we hope to achieve an alternate solution.

To achieve an explicit formula for $S(n, k)$, we instead look at the number of distributions of n labeled balls into k labeled urns in which no urn is empty. By Theorem 4.3.10, the number of distributions of n labeled balls into k labeled urns such that no urn is empty is given by $k!S(n, k)$.

Theorem 7.3.1 *For all $n, k \in \mathbb{N}$, the number of distributions of n labeled balls into k labeled urns such that no urn is empty is given:*

$$\sum_{i=0}^{k} (-1)^i \binom{k}{i} (k-i)^n.$$

Proof Let A denote the set of all distributions of n labeled balls into k labeled urns. Note that $|A| = k^n$ by Proposition 4.2.3. Let A_i denote the set of all distributions in which the urn labeled i is empty. Thus by the Principle of Inclusion and Exclusion, the number of distributions is given by

$$|(A_1 \cup \cdots \cup A_k)^c| =$$

$$|A| - \sum_{i=1}^{k} |A_i| + \sum_{i \neq j} |A_i \cap A_j| - \cdots +$$

$$(-1)^\ell \sum_{i_1, \ldots, i_\ell} |A_{i_1} \cap \cdots \cap A_{i_\ell}| + \cdots + (-1)^k |A_1 \cap \cdots \cap A_k|.$$

Note that if at least ℓ urns are empty, then we must distribute n labeled balls into $k - \ell$ labeled urns. There are $(k - \ell)^n$ ways to do this by Proposition 4.2.3. Thus $|A_{i_1} \cap \cdots \cap A_{i_\ell}| = (k - \ell)^n$ for distinct i_j. To compute the sum over all distinct $i_1, \ldots, i_\ell$, note that distributions into specific urns is irrelevant. It is only relevant that we are distributing to ℓ urns. There are $\binom{k}{\ell}$ ways to choose ℓ labeled urns from k labeled urns. Thus,

$$\sum_{i_1, \ldots, i_\ell} |A_{i_1} \cap \cdots \cap A_{i_\ell}| = \binom{k}{\ell} (k - \ell)^n.$$

Note that

$$|A| = k^n = \binom{k}{0} (k-0)^n.$$

Hence the number of distributions is given by

$$\binom{k}{0}(k-0)^n - \binom{k}{1}(k-1)^n + \binom{k}{2}(k-2)^n + \cdots$$

$$+(-1)^i \binom{k}{i}(k-i)^n + \cdots + (-1)^k \binom{k}{k}(k-k)^n$$

$$= \sum_{i=0}^{k} (-1)^i \binom{k}{i} (k-i)^n. \qquad \blacksquare$$

From this it follows that

$$k!S(n,k) = \sum_{i=0}^{k} (-1)^i \binom{k}{i} (k-i)^n.$$

Thus,

$$S(n,k) = \frac{1}{k!}\sum_{i=0}^{k}(-1)^i\binom{k}{i}(k-i)^n.$$

We now proceed with finding a simpler solution to the problem of the derangement.

Theorem 7.3.2 *For $n \in \mathbb{N}$, the number of derangements on $[n]$ is given by:*

$$D_n = n!\sum_{i=0}^{n}\frac{(-1)^i}{i!}.$$

Proof Let A denote the set of all permutations on $[n]$. Let A_i denote the set of permutations in which i is a fixed point. By the Principle of Inclusion and Exclusion, the number of derangements is given by

$$D_n = |(A_1 \cup \cdots \cup A_n)^c| =$$

$$|A| - \sum_{i=1}^{n}|A_i| + \sum_{i \neq j}|A_i \cap A_j| - \cdots$$

$$+(-1)^\ell \sum_{i_1,\ldots,i_\ell}|A_{i_1} \cap \cdots \cap A_{i_\ell}| + \cdots + (-1)^n|A_1 \cap \cdots \cap A_n|.$$

Note that $A_{i_1} \cap \cdots \cap A_{i_\ell}$ is the set of all distributions in which ℓ elements are fixed. Thus $|A_{i_1} \cap \cdots \cap A_\ell|$ counts the number of permutations on $[n-\ell]$. Thus there are $(n-\ell)!$ permutations in this set. Since this is constant for all distinct choices of $i_1, \ldots, i_\ell$, we have that:

$$\sum_{i_1,\ldots,i_\ell}|A_{i_1} \cap \cdots \cap A_{i_\ell}| = \binom{n}{\ell}(n-\ell)!$$

$$= \frac{n!}{\ell!(n-\ell)!}(n-\ell)! = \frac{n!}{\ell!}.$$

Note that

$$|A| = n! = \frac{n!}{0!}.$$

From this it follows that

$$\begin{aligned} D_n &= \frac{n!}{0!} - \frac{n!}{1!} + \frac{n!}{2!} \pm \cdots + \frac{(-1)^i n!}{i!} + \cdots + \frac{(-1)^n n!}{n!} \\ &= n! \left(\frac{1}{0!} - \frac{1}{1!} + \frac{1}{2!} \pm \cdots + \frac{(-1)^i}{i!} + \cdots + \frac{(-1)^n}{n!} \right) \\ &= n! \sum_{i=0}^{n} \frac{(-1)^i}{i!}. \end{aligned}$$

■

Exercise 7.3.3 Confirm the entries in Table 4.5.

Exercise 7.3.4 Find the following limit:

$$\lim_{n \to \infty} \frac{D_n}{n!}.$$

Exercise 7.3.5 Find the number of distributions of n labeled balls into k labeled urns in such a way that *exactly* m urns are empty (which urns are empty is irrelevant).

Exercise 7.3.6 Find the number of distributions of n labeled balls into k unlabeled urns in such a way that *exactly* m urns are empty (which urns are empty is irrelevant).

Exercise 7.3.7 *(The problème des rencontres)* Find $D_{n,k}$, the number of permutations on $[n]$ which have *exactly* k fixed points.

7.4 Problème des Ménage

The ménage problem (problème des ménage) is one of the classic problems in combinatorics. The problem is a table setting problem. Suppose that n married couples are to be seated around a circular dinner table with $2n$ seats numbered 1, ..., $2n$. Sexes must alternate and no one can sit next to their own spouse. In how many visually distinct ways can this be done?

Note that by numbering the seats, we have eliminated the "rotations" and "flips" of the table. Thus we have insured that all table settings are visually distinct. If we consider "rotations" to be identical, then we would divide our final solution by $2n$. If we consider "flips" to be identical, then we divide our final solution by 2.

As a possible solution, we might try a chivalrous solution. Namely, we seat the women first. While a solution along these lines *is* possible, it is beyond the scope of this text. The solution presented here involves elementary methods and is based on the solution by Bogart and Doyle [10]. This method uses the Principle of Inclusion and Exclusion conditioning on how many couples are seated together.

If a couple is seated together, then they occupy two adjacent seats at the table. So, if we have k couples seated together, then we must "reserve" k pairs of adjacent seats for those couples. Hence, we first need to know the number of ways to select these k pairs of seats. This is equivalent to the number of ways to place k non-overlapping dominoes on a circular number line (Fig. 7.1).

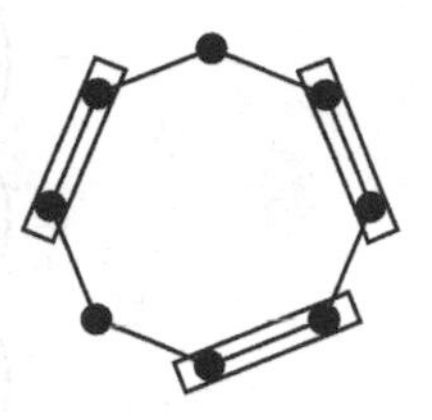

Fig. 7.1 Non overlapping dominoes

Lemma 7.4.1 *The number of ways of placing k non-overlapping indistinguishable dominoes on a circle with m positions numbered* 1, ..., *m is given by*

$$\frac{m}{m-k}\binom{m-k}{k}.$$

Proof This can be counted in three disjoint, exhaustive sets:

(i) The set of placements in which a domino covers the numbers 1 and m. The circle is now broken. Therefore, this problem reduces to finding the number of ways of placing $k-1$ non-overlapping dominoes on $m-2$ spaces. By Exercise 3.5.6, the number of ways of doing this is given by

$$\binom{m-2-(k-1)}{k-1} = \binom{m-k-1}{k-1}.$$

(ii) The set of placements in which a domino covers 1 and 2. By the same argument as above, there are $\binom{m-k-1}{k-1}$ ways of doing this.

(iii) The set of placements in which 1 is not covered by a domino. This again breaks the cycle. Hence, this problem reduces to finding the number of ways of placing k non-overlapping dominoes on $m-1$ spaces. By Exercise 3.5.6, the number of ways of doing this is given by

$$\binom{m-1-k}{k} = \binom{m-k-1}{k}.$$

Thus the number of ways of placing k non-overlapping dominoes on a circle with m numbered positions is given by

$$\begin{aligned}
&2\binom{m-k-1}{k-1} + \binom{m-k-1}{k} \\
=&\frac{2(m-k-1)!}{(k-1)!(m-2k)!} + \frac{(m-k-1)!}{k!(m-2k-1)!} \\
=&\left(\frac{(m-k-1)!}{(k-1)!(m-2k-1)!}\right)\left(\frac{2}{m-2k} + \frac{1}{k}\right) \\
=&\left(\frac{(m-k-1)!}{(k-1)!(m-2k-1)!}\right)\left(\frac{2k+m-2k}{k(m-2k)}\right)
\end{aligned}$$

$$= \left(\frac{(m-k-1)!}{(k-1)!(m-2k-1)!}\right)\left(\frac{m}{k(m-2k)}\right)$$

$$= \left(\frac{(m-k-1)!}{(k-1)!(m-2k-1)!}\right)\left(\frac{m}{k(m-2k)}\frac{m-k}{m-k}\right)$$

$$= \left(\frac{m}{m-k}\right)\left(\frac{(m-k)!}{k!(m-2k)!}\right) = \frac{m}{m-k}\binom{m-k}{k}. \qquad \blacksquare$$

Theorem 7.4.2 (The solution of the ménage problem) *The number of ways of sitting n married couples around a circular dinner table (with 2n numbered seats) in such a way that sexes must alternate and no one sits next to their own spouse is given by*

$$M_n = 2n! \sum_{k=0}^{n} (-1)^k \frac{2n}{2n-k}\binom{2n-k}{k}(n-k)!.$$

Proof Label the couples 1, ..., n. Let A be set of all seatings of n couples in which sexes alternate. Note that $|A| = 2(n!)^2$ by Example 2.3.2. Let A_i be the set of seatings in A that have couple i together. Thus by the Principle of Inclusion and Exclusion, we have

$$M_n = |(A_1 \cup \cdots \cup A_n)^c| =$$

$$|A| - \sum_{i=1}^{n} |A_i| + \sum_{i \neq j} |A_i \cap A_j| \pm \cdots + (-1)^n |A_1 \cap \cdots \cap A_n|.$$

Thus we must count

$$\sum_{i_1,\ldots,i_\ell} |A_{i_1} \cap \cdots \cap A_{i_\ell}|$$

for some arbitrary ℓ. This can be done as follows:

(i) Choose ℓ of the n couples to be seated together. There are $\binom{n}{\ell}$ ways to do this.
(ii) Seat these ℓ couples in such a way that sexes alternate. We define W_ℓ to be the number of settings in which sexes alternate and some fixed set of ℓ couples are adjacent.

Thus by the Multiplication Principle,

$$\sum_{i_1,\ldots,i_\ell} |A_{i_1} \cap \cdots \cap A_{i_\ell}| = \binom{n}{\ell} W_\ell.$$

From this it follows that:

$$M_n = |A| - \binom{n}{1}W_1 + \binom{n}{2}W_2 \pm \cdots (-1)^k\binom{n}{k}W_k + \cdots + (-1)^n\binom{n}{n}W_n.$$

Table 7.1 M_n for small n

n	1	2	3	4	5	6	7	8
M_n	0	0	12	96	3120	115200	5836320	382072320

To complete this problem, we need only compute W_k for any value of k. To compute W_k:

(i) Decide whether to seat the women in the even or the odd numbered seats. There are two ways of doing this.
(ii) Choose k pairs of adjacent seats in which to seat the k couples. This is equivalent to placing k non-overlapping indistinguishable dominoes on a circle with $2n$ positions. The number of way of doing this is $\frac{2n}{2n-k}\binom{2n-k}{k}$ by Lemma 7.4.1.
(iii) Assign the k couples to those k pairs of seats. There are $k!$ ways to do this.
(iv) Seat the remaining $n-k$ women. There are $(n-k)!$ to do this.
(v) Seat the remaining $n-k$ men. There are $(n-k)!$ ways of doing this.

Thus by the Multiplication Principle,

$$W_k = 2\frac{2n}{2n-k}\binom{2n-k}{k}k!(n-k)!(n-k)!$$

$$= \frac{4n}{2n-k}\binom{2n-k}{k}k![(n-k)!]^2.$$

Note that $W_0 = 2(n!)^2 = |A|$. Hence,

$$M_n = \sum_{k=0}^{n}(-1)^k\binom{n}{k}W_k$$

$$= \sum_{k=0}^{n}(-1)^k\binom{n}{k}\frac{4n}{2n-k}\binom{2n-k}{k}k![(n-k)!]^2$$

$$= \sum_{k=0}^{n}(-1)^k\frac{n!}{k!(n-k)!}\frac{4n}{2n-k}\binom{2n-k}{k}k![(n-k)!]^2$$

$$= 2n!\sum_{k=0}^{n}(-1)^k\frac{2n}{2n-k}\binom{2n-k}{k}(n-k)!. \qquad \blacksquare$$

The value of M_n for small values of n are given in Table 7.1.

A related problem is that of the *relaxed ménage*. In the relaxed ménage, there are n married couples to be seated around a circular dinner table with seats numbered $1, \ldots, 2n$. No one can sit next to their own spouse. However, sexes need not alternate. The solution of the relaxed ménage follows in a similar manner to the above. We present the solution to the relaxed ménage in the following theorem.

Theorem 7.4.3 (The solution of the relaxed ménage) *The number of ways of sitting n married couples around a circular dinner table (with $2n$ numbered seats) in such a way that no one sits next to their own spouse is given by*

$$m_n = n\sum_{k=0}^{n}(-1)^k\binom{n}{k}2^{k+1}(2n-k-1)!.$$

Proof Label the couples $1, ..., n$. Let A be set of all settings of $2n$ individuals. Note that $|A| = (2n)!$. Let A_i be the set of settings in A that have couple i together. Thus by the Principle of Inclusion and Exclusion, we have

$$m_n = |\,(A_1 \cup \cdots \cup A_n)^c\,| =$$

$$|A| - \sum_{i=1}^{n}|A_i| + \sum_{i\neq j}|A_i \cap A_j| \pm \cdots + (-1)^n|A_1 \cap \cdots \cap A_n|.$$

Define w_k to be the number of settings in which some fixed set of k couples are adjacent. Note that

$$\sum_{i_1,\ldots,i_\ell}|A_{i_1} \cap \cdots \cap A_{i_\ell}| = \binom{n}{\ell}w_\ell$$

as we choose the ℓ couples (there are $\binom{n}{\ell}$ ways to do this) and we seat the couples in w_ℓ ways. From this it follows that

$$m_n = |A| - \binom{n}{1}w_1 + \binom{n}{2}w_2 \pm \cdots (-1)^k\binom{n}{k}w_k + \cdots + (-1)^n\binom{n}{n}w_n.$$

Therefore, we need only compute w_k for any k. To compute w_k:

(i) Choose k pairs of adjacent seats in which to seat the k couples. This is equivalent to placing k non-overlapping indistinguishable dominoes on a circle with $2n$ positions. The number of way of doing this is $\frac{2n}{2n-k}\binom{2n-k}{k}$ by Lemma 7.4.1.
(ii) Assign the k couples to those k pairs of seats. There are $k!$ ways to do this.
(iii) For each of the k couples, decide if the woman will be seated in the even numbered or odd numbered seat. By the Multiplication Principle, there are 2^k ways of doing this.
(iv) Seat the remaining $2n - 2k$ individuals. There are $(2n - 2k)!$ ways of doing this.

Thus by the Multiplication Principle,

$$\begin{aligned} w_k &= \frac{2n}{2n-k}\binom{2n-k}{k}k!2^k(2n-2k)! \\ &= \frac{2n}{2n-k}\frac{(2n-k)!}{k!(2n-2k)!}k!2^k(2n-2k)! \\ &= n2^{k+1}(2n-k-1)! \end{aligned}$$

Table 7.2 m_n for small n

n	1	2	3	4	5	6	7	8
m_n	0	8	192	11904	1125120	153262080	28507207680	6951513784320

Note that $W_0 = (2n)! = |A|$. Hence,

$$m_n = \sum_{k=0}^{n} (-1)^k \binom{n}{k} w_k$$

$$= \sum_{k=0}^{n} (-1)^k \binom{n}{k} n2^{k+1}(2n-k-1)!$$

$$= n \sum_{k=0}^{n} (-1)^k \binom{n}{k} 2^{k+1}(2n-k-1)! \quad \blacksquare$$

The values of m_n for small n are given in Table 7.2.

Exercise 7.4.4 Confirm the entries in Table 7.1.

Exercise 7.4.5 Confirm the entries in Table 7.2.

Exercise 7.4.6 The solution of the relaxed ménage problem is often given as:

$$m_n = 2^n n! \sum_{k=0}^{n} (-1)^k \binom{n}{k} \frac{2n}{2n-k} \binom{2n-k}{k} (1 * 3 * 5 * \cdots * (2n-2k-1)).$$

Confirm that this answer is equivalent to the one given in Theorem 7.4.3.

Exercise 7.4.7 Suppose that n married couples are to placed in a line in such a way that sexes alternate and no one stands next to their own spouse. How many ways are there to do this?

Exercise 7.4.8 Suppose that n married couples are to placed in a line in such a way that no one stands next to their own spouse (sexes need not alternate). How many ways are there to do this?

Exercise 7.4.9 Find $M_{n,k}$, the number of ways of seating n married couples around a circular dinner table with $2n$ numbered seats in such a way that sexes must alternate and *exactly* k couples are seated together.

Exercise 7.4.10 Find $m_{n,k}$, the number of ways of seating n married couples around a circular dinner table with $2n$ numbered seats in such a way that *exactly* k couples are seated together (sexes need not alternate).

Exercise 7.4.11 Suppose that n married couples are to placed in a line in such a way that sexes alternate. How many ways are there to this such that exactly k men are standing next to their own spouse?

Exercise 7.4.12 Suppose that n married couples are to placed in a line. How many ways are there to this such that exactly k men are standing next to their own spouse (sexes need not alternate)?

Chapter 8
Advanced Counting—Pólya Theory

8.1 Equivalence Relations

In this chapter, we will study Pólya theory. Essentially, Pólya theory allows for "common sense" counting. As a motivating example, suppose that we want to make a necklace with six beads. Each bead may either be white or black. As usual, our goal will be to determine the number of different necklaces that can be made. By Corollary 2.1.4, there are $2^6 = 64$ possibilities. However, we would likely consider two necklaces to be the same if one can be obtained from the other by rotating or flipping the strand. Thus, we would consider many of these possibilities to be equivalent. Figure 8.1 gives three arrangements of beads that can be obtained by rotating or flipping the strand.

We now recall Example 2.1.9. In this example, we determined that the number of distinct ways to label a cube with the elements of [6] was 30. Again, because many of the labelings are identical under rotations of the die, there are fewer than the $6! = 720$ possible labelings that we might expect.

In both cases, the notion of *equivalence* appears. Equivalence is one of the most important notions in mathematics (or, life in general) as it conveys a sense of "sameness." Various examples of equivalence in real life include:

(i) Two restaurants or stores may be considered "equivalent" by consumers if they offer essentially the same goods at essentially the same price.
(ii) Two workers may be considered "equivalent" by management if either can do the job of the other.
(iii) Two vacation spots may be considered "equivalent" by a tourist if they are the same distance from the home of the tourist.

Now, we will formalize the notion of equivalence. To do this, we define a *relation*.

Definition 8.1.1 Let S be a set. A *relation* R on S is a subset of $S \times S$. If $(a, b) \in R$, then we say that *a relates* to b. This situation is denoted $a \sim b$.

As an example of a relation, consider a relation R on the set of real numbers. Suppose that

$$R = \{(x, y) \in \mathbb{R} \times \mathbb{R} : y = x^2\}.$$

R. A. Beeler, *How to Count*, DOI 10.1007/978-3-319-13844-2_8

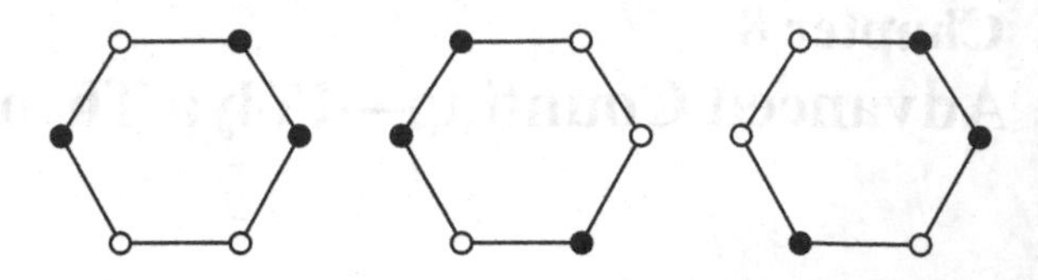

Fig. 8.1 Three equivalent arrangements of the same necklace

The set of all ordered pairs that satisfy this relation gives the graph of the parabola.

With this in mind, we proceed to formally define an *equivalence relation*.

Definition 8.1.2 Let S be a set. The relation R on S is an *equivalence relation* if it satisfies all of the following:

(i) It is *reflexive*. That is, for all $x \in S$, $x \sim x$.
(ii) It is *symmetric*. That is, for all $x, y \in S$, if $x \sim y$, then $y \sim x$.
(iii) It is *transitive*. That is, for all $x, y, z \in S$, if $x \sim y$ and $y \sim z$, then $x \sim z$.

Before discussing equivalence relations further, we give examples that illustrate the reflexive, symmetric, and transitive properties that define an equivalence relation.

Example 8.1.3 Let $S = [3]$. Consider the following relations on S,

(i) $R_1 = \{(2,1),(3,2)\}$;
(ii) $R_2 = \{(1,1),(2,1),(2,2),(3,2),(3,3)\}$;
(iii) $R_3 = \{(1,2),(2,1)\}$;
(iv) $R_4 = \{(1,1),(1,2),(2,1),(2,2)\}$.

For each of the relations R_1, R_2, R_3, and R_4, which of the properties (reflexive, symmetric, and transitive) does R_i satisfy?

Solution *(i)* Since $(1,1) \notin R_1$, R_1 is not reflexive. Because $(2,1) \in R_1$ but $(1,2) \notin R_1$, it follows that R_1 is not symmetric. Similarly, $(3,2),(2,1) \in R_1$, but $(3,1) \notin R_1$. Therefore, the relation is not transitive.

(ii) Since $(1,1),(2,2),(3,3) \in R_2$, the relation is reflexive. Because $(2,1) \in R_2$ but $(1,2) \notin R_2$, it follows that R_2 is not symmetric. Similarly, $(3,2),(2,1) \in R_2$, but $(3,1) \notin R_2$. Therefore, the relation is not transitive.

(iii) The relation R_3 is symmetric as $(1,2),(2,1) \in R_3$. If the relation were transitive, then this would imply that $(1,1) \in R_3$. As this is not the case, the relation is not transitive. As mentioned above, $(1,1) \notin R_3$. Thus, the relation is not reflexive.

(iv) The relation R_4 is transitive for the same reasons as R_3. Since $(1,2),(2,1) \in R_4$ and $(1,1),(2,2) \in R_4$, the relation is transitive. However, $(3,3) \notin R_4$. Therefore, the relation is not reflexive. □

We now present several examples of equivalence relations on various sets. While these proofs are omitted, they follow directly from the above definition. Additional examples may be found in other texts.

(i) By far, the most famous example of an equivalence relation is "equals." This is an equivalence relation on any subset of the complex numbers.
(ii) Let S be the set of all $n \times n$ matrices with real coefficients. Two matrices are *row equivalent* if there is a sequence of elementary row operations (transposing two rows, multiplying a row by a non-zero scalar, or adding a multiple of one row to another), that transform the first matrix into the second.
(iii) Two triangles are *similar* if their interior angles have the same measure.
(iv) Two complex numbers, $a+bi$ and $c+di$, have the same *norm* if $a^2+b^2 = c^2+d^2$.

Example 8.1.4 Let R be an equivalence relation on $\mathbb{Z}$ defined by $n \sim k$ if and only if $13n - 7k$ is even. Show that R is an equivalence relation on $\mathbb{Z}$.

Solution We must show that R is reflexive, symmetric, and transitive.

(i) **(Reflexive)**—Note that $13n - 7n = 6n = 2(3n)$. Thus $13n - 7n$ is even for all $n \in \mathbb{Z}$. It follows that $n \sim n$ for all $n \in \mathbb{Z}$. Thus, R is reflexive.
(ii) **(Symmetric)**—Suppose that $n \sim k$. Then, there exists $\ell \in \mathbb{Z}$ such that $13n - 7k = 2\ell$. From this it follows that

$$13n - 7k - 20n + 20k = 2\ell - 20n + 20k$$
$$\Rightarrow 13k - 7n = 2(\ell - 10n + 10k).$$

Hence, $13k - 7n$ is even. Ergo, $k \sim n$. Therefore, R is symmetric by definition.
(iii) **(Transitive)**—Suppose that $n \sim k$ and $k \sim m$. Thus, there exists $\ell, j \in \mathbb{Z}$ such that $13n - 7k = 2\ell$ and $13k - 7m = 2j$. Adding these two expressions together yields

$$(13n - 7k) + (13k - 7m) = 2\ell + 2j$$
$$\Rightarrow 13n + 6k - 7m = 2\ell + 2j$$
$$13n - 7m = 2\ell + 2j - 6k = 2(\ell + j - 3k).$$

Thus, $13n - 7m$ is an even number. This implies that $n \sim m$. Thus, R is transitive by definition. □

We now present one of the most famous examples of an equivalence relation.

Proposition 8.1.5 *Let $n \in \mathbb{Z}$ such that $n \geq 2$. Let R be a relation on $\mathbb{Z}$ defined by $x \sim y$ if and only if $x - y$ is divisible by n. For all $n \geq 2$, R is an equivalence relation on n.*

Proof It suffices to show that R is reflexive, symmetric, and transitive. Recall that if an integer k is divisible by n, then there exist $m \in \mathbb{Z}$ such that $k = mn$.

(i) **(Reflexive)**—Note that $x - x = 0 = 0n$. Thus, $x - x$ is divisible by n. Hence, the relation is reflexive.
(ii) **(Symmetric)**—Suppose that $x \sim y$. By definition, $x - y$ is divisible by n. Thus, there exists $m \in \mathbb{Z}$ such that $x - y = mn$. It follows that $y - x = (-m)n$. Hence, $y - x$ is likewise divisible by n. Ergo, $y \sim x$. So, the relation is symmetric by definition.
(iii) **(Transitive)**—Suppose that $x \sim y$ and $y \sim z$. By definition, there exists integers $m, \ell \in \mathbb{Z}$ such that $x - y = mn$ and $y - z = \ell n$. Thus,

$$x - z = (x - y) + (y - z) = mn + \ell n = (m + \ell)n.$$

Hence, $x \sim z$. It follows that the relation is transitive. ■

The above equivalence relation is called *congruence modulo n*. In our next definition, we consider the set of all elements that are related to a given element.

Definition 8.1.6 Let R be an equivalence relation on a set S. For $x \in S$, the *equivalence class* of x induced by R is the set

$$cl(x) = \{y \in S : x \sim y\}.$$

For example, consider congruence modulo 4. The equivalence classes are:

$$cl(0) = \{..., -8, -4, 0, 4, 8, ...\},$$
$$cl(1) = \{..., -7, -3, 1, 5, 9, ...\},$$
$$cl(2) = \{..., -6, -2, 2, 6, 10, ...\}, \quad \text{and}$$
$$cl(3) = \{..., -5, -1, 3, 7, 11, ...\}.$$

Notice that each integer is in exactly one of the above equivalence classes. This is generalized in the following theorem.

Theorem 8.1.7 *Let R be an equivalence relation on a set S. Every element of S is in exactly one equivalence class.*

Proof Since R is reflexive, it follows that $x \sim x$ for all $x \in S$. Thus, $x \in cl(x)$. Suppose that there exists $x \in S$ such that $x \in cl(y)$ and $x \in cl(z)$. Thus $y \sim x$ and $z \sim x$. Since R is symmetric, it follows that $x \sim z$. We now have that $y \sim x$ and $x \sim z$. Since R is also transitive, it follows that $y \sim z$. Therefore $cl(y) = cl(z)$. ■

Theorem 8.1.7 defines a key property of equivalence relation. Namely, an equivalence relation on a set S partitions S into equivalence classes. This allows us to give an alternate proof of the fact that every permutation on $[n]$ can be written as a product of disjoint cycles. In this case, the disjoint cycles are the orbits of the permutation. Recall that y is in the orbit of x under σ if there exists $k \in \mathbb{Z}$ such that $\sigma^k(x) = y$.

Theorem 8.1.8 *Let σ be a permutation on $[n]$. Suppose that $x, y \in [n]$. Define a R on $[n]$ by $x \sim y$ if and only if there exists $k \in \mathbb{Z}$ such that $y = \sigma^k(x)$. The relationship R is an equivalence relation. Its equivalence classes are precisely the orbits.*

Proof It suffices to show that R is reflexive, symmetric, and transitive.

(i) **(Reflexive)**—The relationship is reflexive as $\sigma^0(x) = x$. Thus $x \sim x$.
(ii) **(Symmetric)**—Suppose that $x \sim y$. This means that there exists $k \in \mathbb{Z}$ such that $y = \sigma^k(x)$. Since σ is a bijection, it has an inverse. Likewise, σ^k has an inverse, σ^{-k}. Taking the inverse on both sides yields $x = \sigma^{-k}(y)$. Hence $y \sim x$.
(iii) **(Transitive)**—Suppose that $x \sim y$ and $y \sim z$. Thus, there exists $k, m \in \mathbb{Z}$ such that $y = \sigma^k(x)$ and $z = \sigma^m(y)$. Hence, $z = \sigma^m(\sigma^k(x))$. So, $z = \sigma^{k+m}(x)$. Therefore $x \sim z$.

By definition, the equivalence classes of R are the orbits of the elements of $[n]$ under σ. Since equivalence classes partition the set $[n]$, we have that every permutation on $[n]$ can written as a product of disjoint cycles. ■

As an example of equivalence classes, consider our earlier example of a necklace. Our set S will be the set of all 2^6 arrangements of six beads where each bead is either white or black. Two arrangements are related if there is a rotation or flip that transforms the first arrangement to the second. We will show in the next section that this is an equivalence relation. The classes induced by this relation are precisely those arrangements that are visually distinct. In other words, those that cannot be obtained from another by a rotation or a flip. Recall that our goal is to count the number of essentially distinct necklaces. To put this another way, we will count the number of equivalence classes under this relation.

Exercise 8.1.9 Let $S = \mathbb{R}$. Define R on S by $x \sim y$ if and only if $3x + 5y = 7$. Describe the set of points $(x, y) \in S \times S$ such that $x \sim y$.

Exercise 8.1.10 Let $S = \mathbb{R}$. Define R on S by $x \sim y$ if and only if $x^3 = y$. Describe the set of points $(x, y) \in S \times S$ such that $x \sim y$.

Exercise 8.1.11 Let $S = [3]$. Consider the following relations on S,

(i) $R_1 = \{(1, 2), (2, 1), (3, 3)\}$;
(ii) $R_2 = \{(1, 1), (1, 2), (2, 1), (2, 2), (3, 3)\}$;
(iii) $R_3 = \{(1, 2), (2, 3), (1, 3)\}$;
(iv) $R_4 = \{(1, 1), (1, 2), (1, 3), (2, 1), (2, 2), (3, 1), (3, 3)\}$.

For each of the relations R_1, R_2, R_3, and R_4, which of the properties (reflexive, symmetric, and transitive) does R_i satisfy?

Exercise 8.1.12 Show that any relation on the empty set is an an equivalence relation.

Exercise 8.1.13 Show that for any set S, the relation $R = S \times S$ is an equivalence relation.

Exercise 8.1.14 Let S be a non-empty set. Suppose that $R = \emptyset$ (in other words, no elements in S are related). Prove that R is symmetric and transitive, but not reflexive.

Exercise 8.1.15 Let $S = \mathbb{R}$. Define R on S by $x \sim y$ if and only if $x \leq y$. Is this an equivalence relation? If not, what properties does it fail to satisfy?

Exercise 8.1.16 Let S be the set of $n \times m$ matrices with coefficients from $\mathbb{C}$. Prove that row equivalence is an equivalence relation on S. What are the equivalence classes induced by this relation?

Exercise 8.1.17 Let $S = \mathbb{Z}$. Define R on S by $x \sim y$ if and only if $17x - 5y$ is even. Prove that R is an equivalence relation on S. What are the equivalence classes induced by R?

Exercise 8.1.18 Let $S = \mathbb{N}$. Define R on S by $x \sim y$ if and only if $2x + 2y$ is divisible by 4. Prove that R is an equivalence relation. What are the equivalence classes induced by R?

Exercise 8.1.19 Let $S = \mathbb{N}$. Define R on S by $x \sim y$ if and only if $2x + 3y$ is divisible by 5. Is R an equivalence relation? If so, what are the equivalence classes induced by R. If not, what properties does it fail to satisfy?

8.2 Group Actions

One of the most important ideas in abstract algebra is that of a *group*. We will give only a minimal treatment of groups. The interested reader is referred to any of the excellent texts on abstract algebra. In particular, *A First Course in Abstract Algebra* by Fraleigh [22] or *Topics in Algebra* by Herstein [27].

Definition 8.2.1 A *group* is a set G together with a binary operation $*$ such that:

(i) G is *closed* under $*$. That is, for all $x, y \in G$, $x * y \in G$.
(ii) $*$ is *associative* on G. That is, for all $x, y, z \in G$, we have that

$$(x * y) * z = x * (y * z).$$

(iii) There exists an *identity* element in G. That is, there exists $e \in G$ such that for all $x \in G$, $x * e = e * x = x$.
(iv) Every element in G has an *inverse*. That is, for all $x \in G$, there exists $x^{-1} \in G$ such that $x * x^{-1} = x^{-1} * x = e$.

Recall that S_n denotes the set of permutations, or bijective functions, on $[n]$ (see Sect. 2.7). In this case, our binary operation is function composition. As noted in Exercise 1.4.8, the composition of two bijections is likewise a bijection. The identity element is the permutation that maps every element to itself. Similarly, if $\sigma \in S_n$, then σ is a bijection. Thus it has an inverse which is also a bijection. As for associativity, this proof can be found in any text in abstract algebra. Thus, S_n is a group under function composition.

Definition 8.2.2 The nth *symmetric group* is the group of permutations on $[n]$ under function composition. This is denoted S_n.

We will be concerned with *subgroups* of S_n. Let G be a group and H be a subset of the elements of G. If the elements of H form a group under the binary operation of

G, then H is a *subgroup* of G. For instance, $\{e, (1,2)(3)(4), (1)(2)(3,4), (1,2)(3,4)\}$ is a subgroup of S_4. Lagrange's Theorem gives a useful relationship between the orders of groups and their subgroups.

Theorem 8.2.3 *(Lagrange's Theorem) Let G be a group and let H be a subgroup of G. The cardinality of H divides the cardinality of G.*

Proof Let H be a subgroup of G. For all $g \in G$, define $gH = \{gh : h \in H\}$. Define R to be a relation on G, where $g_1 \sim g_2$ if and only if $g_1 \in g_2H$. We claim that R is an equivalence relation.

(i) **(Reflexive)**—Since H is a group, it follow that $e \in H$. Thus, $g_1 = g_1 * e$, where $e \in H$. Thus, $g_1 \sim g_1$.
(ii) **(Symmetric)**—Suppose $g_1 \sim g_2$. Thus, $g_1 \in g_2H$. It follows that $g_1 = g_2 * h$ for some $h \in H$. Since H is a group, h has an inverse, h^{-1} in H. Thus, $g_1 = g_2 * h$ implies that $g_1 * h^{-1} = g_2$. Therefore, $g_2 \sim g_1$.
(iii) **(Transitive)**—Suppose that $g_1 \sim g_2$ and $g_2 \sim g_3$. This means that $g_1 = g_2 * h_1$ and $g_2 = g_3 * h_2$ for some $h_1, h_2 \in H$. It follows that

$$g_1 = g_2 * h_1 = (g_3 * h_2) * h_1 = g_3 * (h_2 * h_1).$$

Since H is closed under *, it follows that $h_2 * h_1 \in H$. Thus, $g_1 \in g_3H$. So, $g_1 \sim g_3$.

Since R is an equivalence relation, it partitions G into equivalence classes. Suppose that $cl(g_1)$,...,$cl(g_k)$ are the distinct equivalence classes induced by this relation. Further note that $|g_iH| = |g_jH| = |H|$ for all $i, j \in [k]$. Since every element of G is represented in exactly one of these classes, we have that $|G| = k|H|$. Thus, the cardinality of H divides the cardinality of G. ■

While there are several problem specific groups that appear, there are two families of groups that appear repeatedly in many combinatorial problems. The first is the *cyclic group* of order n. This is the set of all rotations of the regular n-gon. This group is denoted C_n, where $C_n = \{e, \rho, ..., \rho^{n-1}\}$.

The second is the nth *dihedral group* is the set of all symmetries on the regular n-gon. The nth dihedral group is denoted D_n. An example of D_4 is given in Fig. 8.2. The dihedral group can be expressed as the product of two elements, ρ and τ. These elements satisfy the relationships $\rho^n = \tau^2 = e$ and $\tau\rho^k = \rho^{n-k}\tau$ for $k \in \mathbb{Z}$. This group is particularly useful for studying the number of distinct necklaces with n beads.

The example of the cyclic and the dihedral group have something in common. Namely, we defined the elements of the group by writing them as a product of a subset of the elements of the group. This notion is generalized in the following definition.

Definition 8.2.4 A group G is *generated* by a set $S = \{g_1, ..., g_n\}$ if every element in G can be expressed as a product of the elements of S. If this is the case, then we say that g_1,...,g_n are the *generators* of G and write $G = < g_1, ..., g_n >$.

Based on this definition, we can write $C_n = < \rho >$ and $D_n = < \rho, \tau >$. As another example, consider the group of permutations on the Rubik's Cube. This

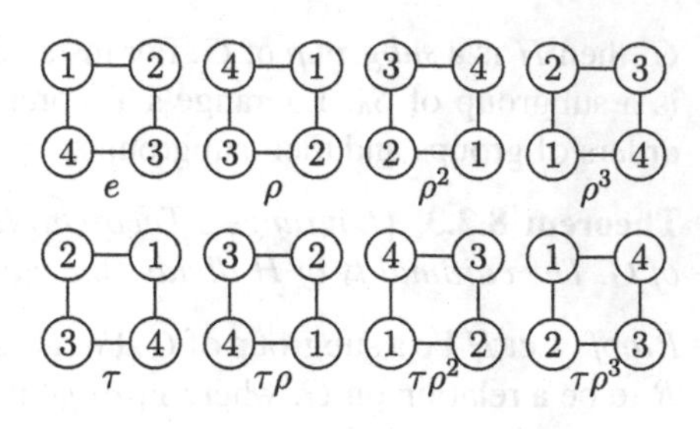

Fig. 8.2 The fourth dihedral group

group is generated by six permutations. Each of these generators correspond to the rotation of one of the six faces of the Rubik's Cube. Often, it is useful to write a group in terms of its generators. Many computer algebra programs (such as Sage) can list all the elements of the group, provided that the generators are known. As usual, we will express our permutations as products of disjoint cycles, as was described in Sect. 2.7. For instance, the permutations in D_4 can be expressed as:

$$e = (1)(2)(3)(4), \quad \rho = (1,4,3,2), \quad \rho^2 = (1,3)(2,4), \quad \rho^3 = (1,2,3,4),$$
$$\tau = (1,2)(3,4), \quad \tau\rho = (1,3)(2)(4), \quad \tau\rho^2 = (1,4)(2,3), \quad \tau\rho^3 = (1)(2,4)(3).$$

Example 8.2.5 Suppose that G is the group of all permutations on the faces of a cube. Find a generating set for this group. What are the elements of the group?

Solution Without loss of generality, we may assume that the faces of the cube are labeled like a standard six-sided die. In other words, the left face is labeled 1, the back face is labeled 2, the bottom face is labeled 3, the top face is labeled 4, the front face is labeled 5, and the right face is labeled 6.

The permutations of the cube can be described in terms of two generators. The first generator is the horizontal rotation of the faces labeled 1, 2, 5, and 6. This corresponds to the cycle $\tau = (1,5,6,2)(3)(4)$. The second generator is the vertical rotation of the faces labeled with 2, 3, 4, and 5. This corresponds to the cycle $\rho = (1)(2,3,5,4)(6)$.

To find the elements of this group, we employ a computer algebra system. This gives us the 24 elements of the group as:

$$e = (1)(2)(3)(4)(5)(6), \quad \tau_1 = (1)(2,3,5,4)(6), \quad \tau_2 = (1)(2,4,5,3)(6),$$
$$\tau_3 = (1)(2,5)(3,4)(6), \quad \tau_4 = (1,2)(3,4)(5,6), \quad \tau_5 = (1,2,3)(4,6,5),$$
$$\tau_6 = (1,2,4)(3,6,5), \quad \tau_7 = (1,2,6,5)(3)(4), \quad \tau_8 = (1,3,2)(4,5,6),$$
$$\tau_9 = (1,3,6,4)(2)(5) \quad \tau_{10} = (1,3)(2,5)(4,6), \quad \tau_{11} = (1,3,5)(2,6,4),$$
$$\tau_{12} = (1,4,2)(3,5,6), \quad \tau_{13} = (1,4,6,3)(2)(5), \quad \tau_{14} = (1,4)(2,5)(3,6),$$
$$\tau_{15} = (1,4,5)(2,6,3), \quad \tau_{16} = (1,5,6,2)(3)(4), \quad \tau_{17} = (1,5,4)(2,3,6),$$
$$\tau_{18} = (1,5,3)(2,4,6), \quad \tau_{19} = (1,5)(2,6)(3,4), \quad \tau_{20} = (1,6)(3,4)(2)(5),$$
$$\tau_{21} = (1,6)(2,3)(4,5), \quad \tau_{22} = (1,6)(2,4)(3,5), \quad \tau_{23} = (1,6)(2,5)(3)(4). \quad \square$$

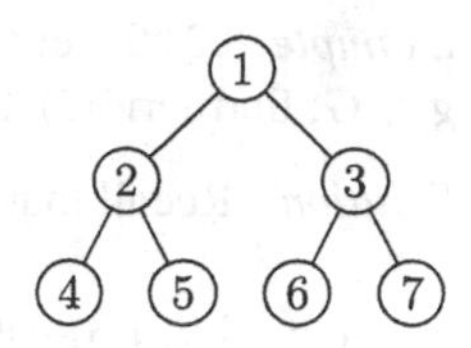

Fig. 8.3 A simple mobile

As a second example, we consider the group of symmetries on the graph given in Fig. 8.3.

Example 8.2.6 Find the group of symmetries on the graph in Fig. 8.3.

Solution To find the group of symmetries, think of this figure as a mobile. The point labeled '1' is where the mobile hangs from the ceiling. Hence, the mobile rotates around this point. As the mobile rotates, the points '2' and '3' may switch positions. However, if they switch positions then the points anchored to them must also switch positions. This corresponds to the generator $a = (1)(2,3)(4,6)(5,7)$. Similarly, the points beneath '2' may rotate around that pivot point and switch positions. This corresponds to the generator $b = (1)(2)(3)(4,5)(6)(7)$. We claim that the group of symmetries is generated by $< a, b >$. There are eight elements generated by $< a, b >$, namely:

$$(1)(2)(3)(4)(5)(6)(7),\quad (1)(2)(3)(4)(5)(6,7),\quad (1)(2)(3)(4,5)(6)(7),$$
$$(1)(2)(3)(4,5)(6,7),\quad (1)(2,3)(4,6)(5,7),\quad (1)(2,3)(4,6,5,7),$$
$$(1)(2,3)(4,7,5,6),\quad (1)(2,3)(4,7)(5,6).$$

□

You may wonder why $c = (1)(2)(3)(4)(5)(6,7)$ was not included in our list of generators. In this case, c was generated by our original elements, as was revealed by our computer algebra system. In fact, $c = aba$. However, if we failed to realize that, then there would be no harm in including superfluous generators into the computer algebra system. Thus, when in doubt, err on the side of having too many generators. Of course, the groups in both of the above examples are small enough that their elements can be listed by brute force.

We now proceed to the notion of a group acting on a set. Let X be a set and let G be a group of permutations mapping X to X. This means that for each $x \in X$ and $g \in G$, we have that $g(x) \in X$. This is a *group action* on the set X. When studying how a group G acts on a set X, one of the most important things to consider is the set of elements in X that $g \in G$ leaves unchanged. This is called the set of *invariants* of g.

Definition 8.2.7 Let G be a group acting on a set X. If $g \in G$, then $Inv(g) = \{x \in X : g(x) = x\}$ is the set of *invariants* of g.

Example 8.2.8 Let G be the group of symmetries on the faces of the cube. For each $g \in G$, find $Inv(G)$ and $cyc(g)$.

Solution Recall that in Example 8.2.5, we found the elements of G to be:

$$\begin{array}{lll}
e = (1)(2)(3)(4)(5)(6), & \tau_1 = (1)(2,3,5,4)(6), & \tau_2 = (1)(2,4,5,3)(6), \\
\tau_3 = (1)(2,5)(3,4)(6), & \tau_4 = (1,2)(3,4)(5,6), & \tau_5 = (1,2,3)(4,6,5), \\
\tau_6 = (1,2,4)(3,6,5), & \tau_7 = (1,2,6,5)(3)(4), & \tau_8 = (1,3,2)(4,5,6), \\
\tau_9 = (1,3,6,4)(2)(5) & \tau_{10} = (1,3)(2,5)(4,6), & \tau_{11} = (1,3,5)(2,6,4), \\
\tau_{12} = (1,4,2)(3,5,6), & \tau_{13} = (1,4,6,3)(2)(5), & \tau_{14} = (1,4)(2,5)(3,6), \\
\tau_{15} = (1,4,5)(2,6,3), & \tau_{16} = (1,5,6,2)(3)(4), & \tau_{17} = (1,5,4)(2,3,6), \\
\tau_{18} = (1,5,3)(2,4,6), & \tau_{19} = (1,5)(2,6)(3,4), & \tau_{20} = (1,6)(3,4)(2)(5), \\
\tau_{21} = (1,6)(2,3)(4,5), & \tau_{22} = (1,6)(2,4)(3,5), & \tau_{23} = (1,6)(2,5)(3)(4).
\end{array}$$

The set of invariants for each element of G is precisely the set of fixed points in its cycle decomposition. Therefore,

$$\begin{array}{lll}
Inv(e) = \{1,2,3,4,5,6\}, & Inv(\tau_1) = \{1,6\}, & Inv(\tau_2) = \{1,6\}, \\
Inv(\tau_3) = \{1,6\}, & Inv(\tau_4) = \emptyset, & Inv(\tau_5) = \emptyset, \\
Inv(\tau_6) = \emptyset, & Inv(\tau_7) = \{3,4\}, & Inv(\tau_8) = \emptyset, \\
Inv(\tau_9) = \{2,5\} & Inv(\tau_{10}) = \emptyset, & Inv(\tau_{11}) = \emptyset, \\
Inv(\tau_{12}) = \emptyset, & Inv(\tau_{13}) = \{2,5\}, & Inv(\tau_{14}) = \emptyset, \\
Inv(\tau_{15}) = \emptyset, & Inv(\tau_{16}) = \{3,4\}, & Inv(\tau_{17}) = \emptyset, \\
Inv(\tau_{18}) = \emptyset, & Inv(\tau_{19}) = \emptyset, & Inv(\tau_{20}) = \{2,5\}, \\
Inv(\tau_{21}) = \emptyset, & Inv(\tau_{22}) = \emptyset, & Inv(\tau_{23}) = \{3,4\}.
\end{array}$$

Similarly, the cycle index for each permutation is the number of cycles in the permutation. Ergo,

$$\begin{array}{lll}
cyc(e) = 6, & cyc(\tau_1) = 3, & cyc(\tau_2) = 3, \\
cyc(\tau_3) = 4, & cyc(\tau_4) = 3, & cyc(\tau_5) = 2, \\
cyc(\tau_6) = 2, & cyc(\tau_7) = 3, & cyc(\tau_8) = 2, \\
cyc(\tau_9) = 3 & cyc(\tau_{10}) = 3, & cyc(\tau_{11}) = 2, \\
cyc(\tau_{12}) = 2, & cyc(\tau_{13}) = 3, & cyc(\tau_{14}) = 3, \\
cyc(\tau_{15}) = 2, & cyc(\tau_{16}) = 3, & cyc(\tau_{17}) = 2, \\
cyc(\tau_{18}) = 2, & cyc(\tau_{19}) = 3, & cyc(\tau_{20}) = 4, \\
cyc(\tau_{21}) = 3, & cyc(\tau_{22}) = 3, & cyc(\tau_{23}) = 4.
\end{array}$$

□

Example 8.2.9 Suppose that G is the group of all symmetries on the graph in Fig. 8.3. For each $g \in G$, find $Inv(g)$ and $cyc(g)$.

Solution In Example 8.2.6, we determined that the elements of G were:

$$\begin{array}{ll} e = (1)(2)(3)(4)(5)(6)(7), & \tau_1 = (1)(2)(3)(4)(5)(6,7), \\ \tau_2 = (1)(2)(3)(4,5)(6)(7), & \tau_3 = (1)(2)(3)(4,5)(6,7), \\ \tau_4 = (1)(2,3)(4,6)(5,7), & \tau_5 = (1)(2,3)(4,6,5,7), \\ \tau_6 = (1)(2,3)(4,7,5,6), & \tau_7 = (1)(2,3)(4,7)(5,6). \end{array}$$

Again, the set of invariants for each permutation is precisely the set of fixed points. Hence,

$$\begin{array}{ll} Inv(e) = \{1,2,3,4,5,6,7\}, & Inv(\tau_1) = \{1,2,3,4,5\}, \\ Inv(\tau_2) = \{1,2,3,6,7\}, & Inv(\tau_3) = \{1,2,3\}, \\ Inv(\tau_4) = \{1\}, & Inv(\tau_5) = \{1\}, \\ Inv(\tau_6) = \{1\}, & Inv(\tau_7) = \{1\}. \end{array}$$

Similarly, the cycle index for each permutation is the number of disjoint cycles in each permutation. Therefore,

$$\begin{array}{llll} cyc(e) = 7, & cyc(\tau_1) = 6, & cyc(\tau_2) = 6, & cyc(\tau_3) = 5, \\ cyc(\tau_4) = 4, & cyc(\tau_5) = 3, & cyc(\tau_6) = 3, & cyc(\tau_7) = 4. \end{array}$$

□

The invariants of a permutation $g \in G$ is the set of elements of X that are fixed under g. A related idea is the *stabilizer* of an element $x \in X$.

Definition 8.2.10 Let X be a set and let G be a group acting on X. The *stabilizer* of $x \in X$, denoted $st(x)$, is the set of all $g \in G$ such that $g(x) = x$. That is, $st(x) = \{g \in G : g(x) = x\}$.

Example 8.2.11 Let $X = [6]$ and $G = D_6$. For each $x \in X$, find $st(x)$.

Solution The elements of D_6 are generated by the elements $\rho = (1,2,3,4,5,6)$ and $\tau = (1)(2,6)(3,5)(4)$. Thus, the 12 elements of D_6 are:

$$\begin{array}{lll} e = (1)(2)(3)(4)(5)(6), & \rho = (1,2,3,4,5,6), & \rho^2 = (1,3,5)(2,4,6), \\ \rho^3 = (1,4)(2,5)(3,6), & \rho^4 = (1,5,3)(2,6,4), & \rho^5 = (1,6,5,4,3,2), \\ \tau = (1)(2,6)(3,5)(4), & \tau\rho = (1,2)(3,6)(4,5), & \tau\rho^2 = (1,3)(2)(4,6)(5), \\ \tau\rho^3 = (1,4)(2,3)(5,6), & \tau\rho^4 = (1,5)(2,4)(3)(6), & \tau\rho^5 = (1,6)(2,5)(3,4). \end{array}$$

To determine the stabilizer for $x \in [6]$, simply find the permutations which have x as a fixed point. Therefore,

$$\begin{array}{lll} st(1) = \{e,\tau\}, & st(2) = \{e,\tau\rho^2\}, & st(3) = \{e,\tau\rho^4\}, \\ st(4) = \{e,\tau\}, & st(5) = \{e,\tau\rho^2\}, & st(6) = \{e,\tau\rho^4\}. \end{array}$$

□

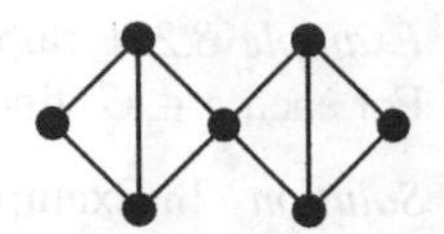

Fig. 8.4 The graph for Exercise 8.2.18

Exercise 8.2.12 What are the elements of D_5? For each $g \in D_5$, find $Inv(g)$ and $cyc(g)$. For each $x \in [5]$, find $st(x)$.

Exercise 8.2.13 What are the elements of S_4? For each $g \in S_4$, find $Inv(g)$ and $cyc(g)$. For each $x \in [4]$, find $st(x)$.

Exercise 8.2.14 Suppose that G is the group of all permutations on the vertices of a cube. Find a generating set for this group. What are the elements of the group? For each $g \in G$, find $Inv(g)$ and $cyc(g)$. If X is the set of vertices of the cube, then find $st(x)$ for each $x \in X$.

Exercise 8.2.15 Suppose that G is the group of all permutations on the edges of a cube. Find a generating set for this group. What are the elements of the group? For each $g \in G$, find $Inv(g)$ and $cyc(g)$. If X is the set of edges of the cube, then find $st(x)$ for each $x \in X$.

Exercise 8.2.16 Consider the 4×4 grid with squares labeled 1 through 16. Let G be the group of permutations obtained by rotating the entire grid. What are the elements of the group? For each $g \in G$, find $Inv(g)$ and $cyc(g)$. For each $x \in [16]$, find $st(x)$.

Exercise 8.2.17 Consider the 4×4 grid with squares labeled 1 through 16. Suppose that when the inner 2×2 grid is rotated clockwise, the outer ring is rotated counterclockwise (and vice versa). What are the elements of the group? For each $g \in G$, find $Inv(g)$ and $cyc(g)$. For each $x \in [16]$, find $st(x)$.

Exercise 8.2.18 Consider the graph given in Fig. 8.4. Let G be the group of all symmetries on the graph. What are the elements of the group? For each $g \in G$, find $Inv(g)$ and $cyc(g)$. If X is the set of vertices of this graph, then find $st(x)$ for all $x \in X$.

Exercise 8.2.19 Consider the graph given in Fig. 8.5. Let G be the group of all symmetries on the graph. What are the elements of the group? For each $g \in G$, find $Inv(g)$ and $cyc(g)$. If X is the set of vertices of this graph, then find $st(x)$ for all $x \in X$.

Exercise 8.2.20 Consider the graph given in Fig. 8.6. Let G be the group of all symmetries on the graph. What are the elements of the group? For each $g \in G$, find $Inv(g)$ and $cyc(g)$. If X is the set of vertices of this graph, then find $st(x)$ for all $x \in X$.

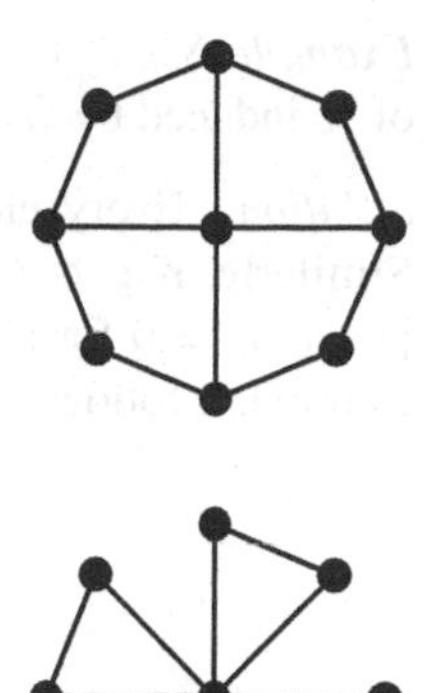

Fig. 8.5 The graph for Exercise 8.2.19

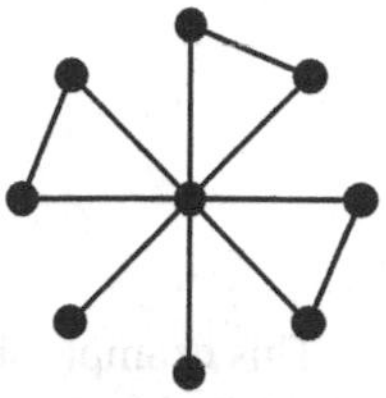

Fig. 8.6 The graph for Exercise 8.2.20

Exercise 8.2.21 Let G be a group. Two elements, $x, y \in G$ are said to be *conjugate* if there exists $g \in G$ such that $g^{-1}xg = y$. Let R be the set of all conjugate pairs (x, y), where $x, y \in G$.

(i) Prove that R is an equivalence relation on G.
(ii) If G is *abelian* (in other words, $x * y = y * x$ for all $x, y \in G$), then what are the equivalence classes induced by R?
(iii) If $G = S_n$, then show that the equivalence class of $\pi \in G$ is precisely the set of permutations that have the same cycle type as π. Hint: Show that conjugacy preserves cycle length.

8.3 Burnside's Lemma

Burnside's Lemma counts the number of distinct equivalent classes induced by a group G acting on a set X. These equivalence classes are the orbits of the permutations of G. In other words, two elements $x, y \in X$ are in the same equivalence class induced by G if there exists $\pi \in G$ such that $\pi(x) = y$. We begin by stating Burnside's Lemma. However, we will not prove this until later in this section. Instead, we will present several examples of how Burnside's Lemma is used.

Theorem 8.3.1 *(Burnside's Lemma) Let G be a group acting on a set X. If R is the equivalence relation on X induced by G, then the number of equivalence classes induced by R is given by*

$$\frac{1}{|G|}\sum_{\pi \in G} |Inv(\pi)|.$$

Example 8.3.2 Let $X = [5]$ and $G = C_5$. Find the number of equivalence classes of X induced by G.

Solution Every element of X is invariant under the identity, so $|Inv(e)| = 5$. Similarly, if $g \in C_5$ and $g \neq e$, then every element of X is moved by g. Thus, $|Inv(g)| = 0$ for all $g \in G$ such that $g \neq e$. Finally, $|C_5| = 5$. Using Burnside's Lemma, it follows that the number of equivalence classes in X induced by G is

$$\frac{1}{|G|}\sum_{\pi \in G} |Inv(\pi)|$$
$$= \frac{1}{5}(5+0+0+0+0) = 1. \qquad \square$$

This example shows that that under C_5, there is exactly one equivalence class on [5]. This class must contain all of the elements of [5]. This makes sense because any element of X can be rotated to any other position. In fact, using a slightly generalized version of the argument above, we can show that there is one equivalence class under C_n acting on the set $[n]$.

Example 8.3.3 Suppose that $X = \{1, 2, 3, 4\}$ and $G = \{e, \pi_1, \pi_2, \pi_3\}$ where

$$e = (1)(2)(3)(4), \quad \pi_1 = (1,2)(3)(4),$$
$$\pi_2 = (1)(2)(3,4), \quad \pi_3 = (1,2)(3,4).$$

Find the number of orbits of G.

Solution The set of invariants for each permutation is precisely the set of fixed points in each permutation. Thus,

$$|Inv(e)| = |\{1,2,3,4\}| = 4, \quad |Inv(\pi_1)| = |\{1,2\}| = 2,$$
$$|Inv(\pi_2)| = |\{3,4\}| = 2, \quad |Inv(\pi_3)| = |\emptyset| = 0.$$

Burnside's Lemma yields the number of orbits or equivalence classes under G. Namely,

$$\frac{1}{|G|}\sum_{\pi \in G} |Inv(\pi)|$$
$$= \frac{1}{4}(4+2+2+0) = 2. \qquad \square$$

There are two orbits in the above example. The first orbit is the orbit of 1, namely $\{1, 2\}$. The second orbit is the orbit of 3, namely $\{3, 4\}$.

Example 8.3.4 Let $X = [6]$ and $G = D_6$. Find the number of orbits in D_6.

Solution The elements of D_6 were given Example 8.2.11. From this, it is easy to find the set of invariants for each element of D_6. This is given below:

$$|Inv(e)| = |\{1,2,3,4,5,6\}| = 6, \quad |Inv(\rho)| = |\emptyset| = 0,$$
$$|Inv(\rho^2)| = |\emptyset| = 0, \quad |Inv(\rho^3)| = |\emptyset| = 0,$$
$$|Inv(\rho^4)| = |\emptyset| = 0, \quad |Inv(\rho^5)| = |\emptyset| = 0,$$
$$Inv(\tau) = |\{1,4\}| = 2, \quad |Inv(\tau\rho)| = |\emptyset| = 0,$$
$$|Inv(\tau\rho^2)| = |\{2,5\}| = 2, \quad |Inv(\tau\rho^3)| = |\emptyset| = 0,$$
$$|Inv(\tau\rho^4)| = |\{3,6\}| = 2, \quad |Inv(\tau\rho^5)| = |\emptyset| = 0.$$

The number of orbits is then given by Burnside's Lemma:

$$\frac{1}{|G|}\sum_{\pi \in G} |Inv(\pi)|$$
$$= \frac{1}{12}(6+2+2+2) = 1. \qquad \square$$

Example 8.3.5 Suppose that G is the group of rotations on the faces of the cube (see Example 8.2.5). How many orbits are in G?

Solution Previously, we found the set of permutations in G as well as the set of invariants for each permutation. This is given below:

$$|Inv(e)| = |\{1,2,3,4,5,6\}| = 6, \quad |Inv(\tau_1)| = |\{1,6\}| = 2,$$
$$|Inv(\tau_2)| = |\{1,6\}| = 2, \quad |Inv(\tau_3)| = |\{1,6\}| = 2,$$
$$|Inv(\tau_4)| = |\emptyset| = 0, \quad |Inv(\tau_5)| = |\emptyset| = 0,$$
$$|Inv(\tau_6)| = |\emptyset| = 0, \quad |Inv(\tau_7)| = |\{3,4\}| = 2,$$
$$|Inv(\tau_8)| = |\emptyset| = 0, \quad |Inv(\tau_9)| = |\{2,5\}| = 2$$
$$|Inv(\tau_{10})| = |\emptyset| = 0, \quad |Inv(\tau_{11})| = |\emptyset| = 0,$$
$$|Inv(\tau_{12})| = |\emptyset| = 0, \quad |Inv(\tau_{13})| = |\{2,5\}| = 2,$$
$$|Inv(\tau_{14})| = |\emptyset| = 0, \quad |Inv(\tau_{15})| = |\emptyset| = 0,$$
$$|Inv(\tau_{16})| = |\{3,4\}| = 2, \quad |Inv(\tau_{17})| = |\emptyset| = 0,$$
$$|Inv(\tau_{18})| = |\emptyset| = 0, \quad |Inv(\tau_{19})| = |\emptyset| = 0,$$
$$|Inv(\tau_{20})| = |\{2,5\}| = 2, \quad |Inv(\tau_{21})| = |\emptyset| = 0,$$
$$|Inv(\tau_{22})| = |\emptyset| = 0, \quad |Inv(\tau_{23})| = |\{3,4\}| = 2.$$

Burnside's Lemma yields the number of orbits in this group, namely

$$\frac{1}{|G|}\sum_{\pi \in G} |Inv(\pi)|$$
$$= \frac{1}{24}(6+2+2+2+2+2+2+2+2+2) = 1. \qquad \square$$

Example 8.3.6 Let G be the group of symmetries on the graph in Fig. 8.3. Find the number of orbits in G.

Solution Previously, we found the elements in this group in Example 8.2.6 and the invariants for each elements. Thus,

$$|Inv(e) = \{1,2,3,4,5,6,7\}| = 7, \quad |Inv(\tau_1)| = |\{1,2,3,4,5\}| = 5,$$
$$|Inv(\tau_2)| = |\{1,2,3,6,7\}| = 5, \quad |Inv(\tau_3)| = |\{1,2,3\}| = 3,$$
$$|Inv(\tau_4)| = |\{1\}| = 1, \quad |Inv(\tau_5)| = |\{1\}| = 1,$$
$$|Inv(\tau_6)| = |\{1\}| = 1, \quad |Inv(\tau_7)| = |\{1\}| = 1.$$

Hence, the number of orbits is given by Burnside's Lemma:

$$\frac{1}{|G|}\sum_{\pi \in G} |Inv(\pi)|$$
$$= \frac{1}{8}(7+5+5+3+1+1+1+1) = 3. \qquad \square$$

Here, the three equivalence classes are $\{1\}$, $\{2,3\}$, and $\{4,5,6,7\}$. Again, this makes sense as 1 must remain fixed under any valid permutation, 2 can only swap with 3 (and vice versa), while any of the other elements can occupy the position of 4 (or 5, 6, or 7).

We present two final examples before proceeding with the proof of Burnside's Lemma. Additional examples of Burnside's Lemma will be given in the next section. Recall, that in Example 2.1.9 we determined that there are 30 ways to label the faces of a cube with distinct elements of [6]. In the following example, we give an alternative derivation using Burnside's Lemma.

Example 8.3.7 Find the number of distinguishable ways to label the faces of a cube with distinct elements of [6].

Solution In this case, the set X is the set of all permutations on the set [6]. Thus, $|Inv(\pi)|$ is no longer the number of fixed points in the permutation π. Instead, it is the number of permutations of [6] which remain unchanged under π. The identity element of G will keep every permutation fixed. Thus, $|Inv(e)| = 6! = 720$. Every other element of G contains a cycle of length at least two. Hence, if $x \in X$ and $g \in G$ such that $g \neq e$, $g(x) \neq x$. For this reason, $|Inv(g)| = 0$ for all $g \in G$ such that $g \neq e$. The number of ways to label the cube is then given by Burnside's Lemma:

$$\frac{1}{|G|}\sum_{\pi \in G} |Inv(\pi)|$$
$$= \frac{1}{24}(720) = 30. \qquad \square$$

Example 8.3.8 Find the number of distinguishable ways to label the vertices of a pentagon with the elements of [5] if

(i) The group acting on the vertices of the pentagon is C_5;
(ii) The group acting on the vertices of the pentagon is D_5.

Solution In either case, the set X is the set of all $5! = 120$ permutations on the set $[5]$. As in the previous example, $|Inv(\pi)|$ is the number of permutations on $[5]$ that are unchanged under π.

(i) Suppose that the group acting on the set of permutations is C_5. Note that the identity element will keep every permutation fixed. This means that $|Inv(e)| = 5! = 120$. Any other element of C_5 is a cycle of length 5. Thus, if $x \in X$ and $g \in C_5$ such that $g \neq e$, then $g(x) \neq x$. It follows that $|Inv(g)| = 0$ for all $g \in C_5$ such that $g \neq e$. Burnside's Lemma yields the number of distinct ways to label the pentagon:

$$\frac{1}{|C_5|}\sum_{\pi \in C_5} |Inv(\pi)|$$
$$= \frac{1}{5}(120) = 24.$$

(ii) Suppose that the group acting on the set of permutations is D_5. If $g \in D_5$ is a rotation or the identity, then that case is covered above. Suppose that g is a "flip." Any such flip, is a line of symmetry passing through one of the vertices of the pentagon. For each of these five symmetries, the cycle type is $[2, 2, 1]$. Any labeling of the two cycles in this permutation will not be invariant. Thus, Burnside's Lemma gives the number of distinct ways to label the pentagon:

$$\frac{1}{|D_5|}\sum_{\pi \in D_5} |Inv(\pi)|$$
$$= \frac{1}{10}(120) = 12. \qquad \square$$

We will now prove Burnside's Lemma. To accomplish this, we will need to establish a relationship between the stabilizing set of $x \in X$ and the equivalence class of x induced by a group G acting on X.

Theorem 8.3.9 *Let G be a group acting on a set X. For each $x \in X$, $|G| = |st(x)||cl(x)|$.*

Proof Suppose that $cl(x) = \{y_1, ..., y_n\}$. For each $i \in [n]$, there exists $\pi_i \in G$ such that $\pi_i(x) = y_i$. Since $y_i \neq y_j$ for $i \neq j$, it follows that $\pi_i \neq \pi_j$ for all $i \neq j$. This establishes a bijection between the set $P = \{\pi_1, ..., \pi_n\}$ and the set $cl(x)$. Thus, $|P| = |cl(x)|$.

We want to show that for every $\sigma \in G$, σ can be written uniquely as a product $\sigma = p * s$, where $p \in P$ and $s \in st(x)$. Let $\sigma \in G$ and note that if $\sigma \in st(x)$, then $\sigma = \sigma * e$, where $e \in P$. By definition, if $\sigma \notin st(x)$, then $\sigma(x) \in cl(x)$. So for some $y_i \in cl(x)$, we have that $\sigma(x) = y_i = \pi_i(x)$. Thus, $\pi_i^{-1} * \sigma = e \in st(x)$. From this it follows that,

$$\sigma = (\pi_i * \pi_i^{-1}) * \sigma = \pi_i * (\pi_i^{-1} * \sigma).$$

Hence, $\sigma = p * s$, where $\pi_i \in P$ and $s \in st(x)$.

We now show that this representation is unique. Assume that $\sigma = p_1 * s_1 = p_2 * s_2$, where $p_1, p_2 \in P$ and $s_1, s_2 \in st(x)$. Our goal is to show that $p_1 = p_2$ and $s_1 = s_2$. Since $p_1 * s_1 = p_2 * s_2$, it follow that $p_1(s_1(x)) = p_2(s_2(x))$. Since $s_1, s_2 \in st(x)$ we have that $p_1(s_1(x)) = p_1(x)$ and $p_2(s_2(x)) = p_2(x)$. Thus, p_1 and p_2 map x to the same element of $cl(x)$. This implies that $p_1 = p_2$. To show that $s_1 = s_2$, note that $p_1 * s_1 = p_2 * s_2 = p_1 * s_2$. Taking the inverse of p_1 on both sides yields $s_1 = s_2$. ■

With this in mind, we are ready to prove Burnside's Lemma.

Proof Let G be a group acting on a set X. Consider the set of ordered pairs $S = (x, \pi)$, where $\pi(x) = x$. We will count this set in two different ways.

The cardinality of S can be counted in $|G|$ disjoint, exhaustive sets. Here, the ith set is the set of all ordered pairs (x, π_i) such that $\pi_i(x) = x$. Thus, the cardinality of the ith set is $|Inv(\pi_i)|$ by definition. Hence, $|S| = \sum_{\pi \in G} |Inv(\pi)|$ by the Addition Principle.

The cardinality of S can also be counted in $|X|$ disjoint, exhaustive sets. Here, the jth set is the set of all ordered pairs (x_j, π) such that $\pi(x_j) = x_j$. From this it follows that the cardinality of the jth set is $|st(x_j)|$ by definition. The Addition Principle yields $|S| = \sum_{x \in X} |st(x)|$. From this it follows that:

$$\sum_{\pi \in G} |Inv(\pi)| = \sum_{x \in X} |st(x)|. \tag{8.1}$$

From Theorem 8.3.9, we have that $|st(x)| = \frac{|G|}{|cl(x)|}$. Applying this to Eq. (8.1) gives us

$$\sum_{\pi \in G} |Inv(\pi)| = |G| \sum_{x \in X} \frac{1}{|cl(x)|}.$$

Equivalently,

$$\frac{1}{|G|} \sum_{\pi \in G} |Inv(\pi)| = \sum_{x \in X} \frac{1}{|cl(x)|}. \tag{8.2}$$

The left side of Eq. (8.2) is the formula given in the statement of Burnside's Lemma. To complete the proof, it suffices to show that the right side counts the number of equivalences classes in G.

Suppose that $cl(x) = \{x_1, ..., x_t\}$. Thus,

$$\begin{aligned} &\frac{1}{|cl(x_1)|} + \cdots + \frac{1}{|cl(x_t)|} \\ &= \underbrace{\frac{1}{t} + \cdots + \frac{1}{t}}_{t \text{ times}} = 1. \end{aligned}$$

This implies that each equivalence class induced by G contributes 1 to the sum $\sum_{x \in X} \frac{1}{|cl(x)|}$. Hence both sides of Eq. (8.2) count the number of equivalence classes induced by G. ■

Exercise 8.3.10 Find the number of orbits in D_8.

Exercise 8.3.11 Find the number of orbits in D_9.

Exercise 8.3.12 Find the number of orbits in S_5.

Exercise 8.3.13 Suppose that G is the group acting on [6] with elements given below:

$$e = (1)(2)(3)(4)(5)(6), \quad \pi_1 = (1,2,3)(4)(5)(6), \quad \pi_2 = (1,3,2)(4)(5)(6),$$
$$\pi_3 = (1)(2)(3)(4,5,6), \quad \pi_4 = (1,2,3)(4,5,6), \quad \pi_5 = (1,3,2)(4,5,6),$$
$$\pi_6 = (1)(2)(3)(4,6,5), \quad \pi_7 = (1,2,3)(4,6,5), \quad \pi_8 = (1,3,2)(4,6,5).$$

Find the number of orbits in G.

Exercise 8.3.14 Suppose that G is the group of all permutations on the vertices of a cube. Find the number of orbits in G. Hint: See Exercise 8.2.14.

Exercise 8.3.15 How many distinguishable ways are there to label the vertices of the cube with distinct elements of [8]?

Exercise 8.3.16 How many distinguishable ways are there to label the vertices of the octagon with distinct elements of [8] if

(i) The group acting on the vertices of the octagon is C_8;
(ii) The group acting on the vertices of the octagon is D_8.

Exercise 8.3.17 How many distinguishable ways are there to label the vertices of the 11-gon with distinct elements of [11] if

(i) The group acting on the vertices of the 11-gon is C_{11};
(ii) The group acting on the vertices of the 11-gon is D_{11}.

Exercise 8.3.18 Suppose that G is the group of all permutations on the edges of a cube. Find the number of orbits in G. Hint: See Exercise 8.2.15.

Exercise 8.3.19 How many distinguishable ways are there to label the edges of the cube with distinct elements of [12]?

Exercise 8.3.20 Consider the 4×4 grid with squares labeled 1 through 16. Let G be the group of permutations obtained by rotating the entire grid. How many orbits are in G? Hint: See Exercise 8.2.16.

Exercise 8.3.21 Consider the 4×4 grid with squares labeled 1 through 16. Suppose that when the inner 2×2 grid is rotated clockwise, the outer ring is rotated counterclockwise (and vice versa). Find the number of orbits in G. Hint: See Exercise 8.2.17.

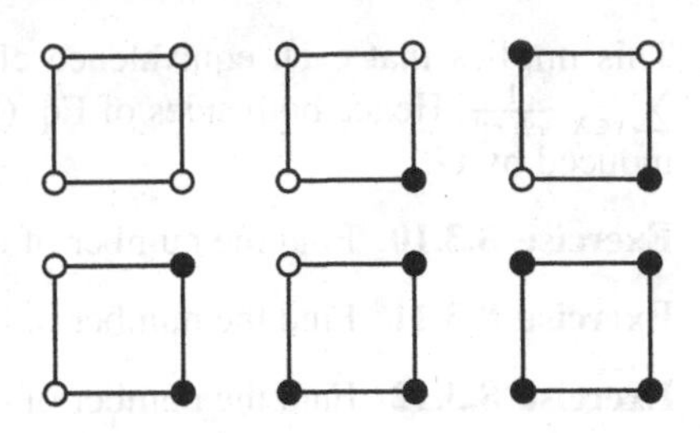

Fig. 8.7 Non-equivalent arrangements of four beads of two colors

Exercise 8.3.22 Consider the graph given in Fig. 8.4. Let G be the group of all symmetries on the graph. How many orbits are in G? Hint: See Exercise 8.2.18.

Exercise 8.3.23 Consider the graph given in Fig. 8.4. How many distinguishable ways are there to label the vertices of this graph with distinct elements of [7]?

Exercise 8.3.24 Consider the graph given in Fig. 8.5. Let G be the group of all symmetries on the graph. How many orbits are in G? Hint: See Exercise 8.2.19.

Exercise 8.3.25 Consider the graph given in Fig. 8.5. How many distinguishable ways are there to label the vertices of this graph with distinct elements of [9]?

Exercise 8.3.26 Consider the graph given in Fig. 8.6. Let G be the group of all symmetries on the graph. How many orbits are in G? Hint: See Exercise 8.2.20.

Exercise 8.3.27 Consider the graph given in Fig. 8.6. How many distinguishable ways are there to label the vertices of this graph with distinct elements of [9]?

8.4 Equivalent Colorings

The most common application of Burnside's Lemma is to equivalent colorings. Recall that we started this chapter with a discussion of the problem of making a necklace with n beads, where each bead can either be white or black. When counting the number of distinguishable necklaces, we count only those necklaces that cannot be obtained by rotating or flipping another. For instance, Fig. 8.7 gives all non-equivalent arrangements of four beads with two different colors.

A *coloring* is a function that maps a set S to a set of colors, $C = \{1, ..., m\}$. For example, suppose that we are making a necklace with six beads, each of which can either be white or black. In this case, the set S is the set of locations on the necklace. Without loss of generality, we assume this is the set [6]. The set of colors is {black, white}. Two colorings, c_1 and c_2, are *equivalent* under G if there is $\pi \in G$ such that $\pi(c_1) = c_2$. When applying Burnside's Lemma to equivalent colorings, X is the set of all $|C|^{|S|}$ possible colorings of S. We now consider a motivating example.

Example 8.4.1 Suppose that we are making a necklace with six beads, where each bead must be either black or white. Determine the number of distinct necklaces.

Solution The beads of the necklace can be flipped or rotated. Thus, the group acting on this set is the sixth dihedral group, D_6. In Example 8.2.11, we found the elements of D_6 to be:

$$\begin{aligned} e &= (1)(2)(3)(4)(5)(6), & \rho &= (1,2,3,4,5,6), & \rho^2 &= (1,3,5)(2,4,6), \\ \rho^3 &= (1,4)(2,5)(3,6), & \rho^4 &= (1,5,3)(2,6,4), & \rho^5 &= (1,6,5,4,3,2), \\ \tau &= (1)(2,6)(3,5)(4), & \tau\rho &= (1,2)(3,6)(4,5), & \tau\rho^2 &= (1,3)(2)(4,6)(5), \\ \tau\rho^3 &= (1,4)(2,3)(5,6), & \tau\rho^4 &= (1,5)(2,4)(3)(6), & \tau\rho^5 &= (1,6)(2,5)(3,4). \end{aligned}$$

The set X is the 2^6 possible ways to color $[6]$ with the colors black and white. We now determine the number of invariants for each element in D_6. Of course, every coloring is invariant under the identity. So, $|Inv(e)| = 2^6 = 64$.

Note that ρ maps 1 to 2. For a coloring to be invariant under ρ, 1 and 2 must be mapped to the same color. By the same logic, 2 and 3 must be mapped to the same color. Continuing in this manner, it follows that every element of $[6]$ must have the same color. Hence, there are precisely two colorings that are invariant under ρ. These invariant colorings are precisely the coloring that colors everything white and the coloring that colors everything black. Similarly, $|Inv(\rho^5)| = 2$.

By a similar argument, ρ^2 maps 1 to 3 and 3 to 5. Thus, 1, 3, and 5 must all receive the same color. Similarly, 2, 4, and 6 must all receive the same color, which may be different than the color received by 1, 3, and 5. Thus, there are four colorings that are invariant under ρ^2. By the same logic, $|Inv(\rho^4)| = 4$.

In ρ^3, 1 and 4 must be mapped to the same color. Similarly, 2 and 5 must receive the same color, which may be different than the color received by 1 and 4. Finally, 3 and 6 must receive the same color, which is not dependent on the colors already assigned. Thus there are $2^3 = 8$ colorings that are invariant under ρ^3. Similarly,

$$|Inv(\tau\rho)| = |Inv(\tau\rho^3)| = |Inv(\tau\rho^5)| = 8.$$

For a coloring to be invariant under τ, 2 and 6 must be mapped to the same color. Further, 3 and 5 must be mapped to the same color. Any color may be applied to 1. Finally, any color may be applied to 4. Thus there $2^4 = 16$ colorings that are invariant under τ. Similarly, $|Inv(\tau\rho^2)| = |Inv(\tau\rho^4)| = 16$.

With the above in mind, the number of acceptable necklaces is given by Burnside's Lemma:

$$\begin{aligned} &\frac{1}{|G|}\sum_{\pi\in G}|Inv(\pi)| \\ &= \frac{1}{12}(64 + 2(2) + 2(4) + 4(8) + 3(16)) = 13. \end{aligned}$$

□

The example above motivates an important observation. Suppose that we want to find a coloring that is invariant under $\pi \in G$. Like all permutations, π can be written as a product of disjoint cycles, in other words, $\pi = c_1 ... c_k$, where $k = cyc(\pi)$ (the cycle index of π). For a coloring to be invariant under π, any two elements of c_i must receive the same color. Thus the number of invariants under π is given by $|C|^k$, where k is the cycle index of π. From this observation, the following proposition is immediate.

Proposition 8.4.2 *Suppose that G is a group acting on a set of colorings, C. If $\pi \in G$, then the number of colorings invariant under π is given by*

$$|Inv(\pi)| = |C|^{cyc(\pi)}.$$

The beauty of this is that we only need to know the cycle index of each permutation in G. Hence, we do not need to have each element of G written out explicitly. Using the above proposition, we give a version of Burnside's Lemma incorporating the cycle index. This version follows immediately from our original version of Burnside's Lemma and the above proposition.

Theorem 8.4.3 *(Burnside's Lemma—Cycle index version) Suppose that S is a set and C is an m-set of colors. Let X be the set of all $m^{|S|}$ colorings of S. If G is a group acting on X, then the number of distinct colorings is given by*

$$\frac{1}{|G|} \sum_{\pi \in G} m^{cyc(\pi)}.$$

We will now give several examples using this version of Burnside's Lemma.

Example 8.4.4 Suppose that we wish to paint the faces of a cube with one of three colors. In how many distinguishable ways can this be done?

Solution In this case, the group acting on the set of colorings is the set of rotations on the faces of the cube. We determined the elements of this group in Example 8.2.5. Later, we found the cycle index for each of these permutations in Example 8.2.8:

$$\begin{array}{lll}
cyc(e) = 6, & cyc(\tau_1) = 3, & cyc(\tau_2) = 3, \\
cyc(\tau_3) = 4, & cyc(\tau_4) = 3, & cyc(\tau_5) = 2, \\
cyc(\tau_6) = 2, & cyc(\tau_7) = 3, & cyc(\tau_8) = 2, \\
cyc(\tau_9) = 3 & cyc(\tau_{10}) = 3, & cyc(\tau_{11}) = 2, \\
cyc(\tau_{12}) = 2, & cyc(\tau_{13}) = 3, & cyc(\tau_{14}) = 3, \\
cyc(\tau_{15}) = 2, & cyc(\tau_{16}) = 3, & cyc(\tau_{17}) = 2, \\
cyc(\tau_{18}) = 2, & cyc(\tau_{19}) = 3, & cyc(\tau_{20}) = 4, \\
cyc(\tau_{21}) = 3, & cyc(\tau_{22}) = 3, & cyc(\tau_{23}) = 4.
\end{array}$$

From this table, we see that there are eight permutations with index 2, 12 permutations with index 3, three permutations with index 4, and one with index 6. Thus,

Burnside's Lemma gives the number of distinguishable m-colorings of the faces of the cube to be:

$$\frac{1}{|G|}\sum_{\pi \in G} m^{cyc(\pi)}$$

$$= \frac{1}{24}\left(m^6 + 3m^4 + 12m^3 + 8m^2\right).$$

For the particular case when $m = 3$, there are 57 distinct colorings of the faces of the cube. □

Example 8.4.5 Suppose that we are building a mobile based on the design in Fig. 8.3. At each of the seven points of this figure, we can place either a star, a moon, a sun, or a planet. How many different mobile designs are possible?

Solution This problem amounts to determining the number of distinguishable 4-colorings of the graph in Fig. 8.3. Fortunately, we have already determined the group of permutations (see Example 8.2.6) and the cycle index for each permutation (see Example 8.2.9). This is given below:

$$cyc(e) = 7, \quad cyc(\tau_1) = 6,$$
$$cyc(\tau_2) = 6, \quad cyc(\tau_3) = 5,$$
$$cyc(\tau_4) = 4, \quad cyc(\tau_5) = 3,$$
$$cyc(\tau_6) = 3, \quad cyc(\tau_7) = 4.$$

From this, we see that there are two permutations with index 3, two permutations with index 4, one permutation with index 5, two permutations with index 6, and one with index 7. Thus, Burnside's Lemma gives the number of distinguishable m-colorings on the graph in Fig. 8.3:

$$\frac{1}{|G|}\sum_{\pi \in G} m^{cyc(\pi)}$$

$$= \frac{1}{8}\left(m^7 + 2m^6 + m^5 + 2m^4 + 2m^3\right).$$

For the particular case when $m = 4$, there are 3280 different mobiles that can be made. □

Example 8.4.6 How many distinguishable ways are there to five-color vertices of the graph in Fig. 8.8?

Solution Any two vertices in this graph can be swapped in this graph. Thus, the group of symmetries is the sixth symmetric group, S_6. However, we do not wish to list all 720 permutations in this group to determine the cycle index. Fortunately, the number of permutations in S_n which have index k is given by the Stirling numbers of the first kind (see Sect. 2.7). Using Table 2.3, we find that $s(6,1) = 120$, $s(6,2) = 274$,

Fig. 8.8 The graph for Example 8.4.6

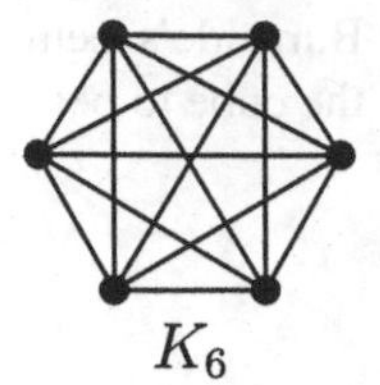

$s(6,3) = 225$, $s(6,4) = 85$, $s(6,5) = 15$, and $s(6,6) = 1$. Burnside's Lemma then gives the number of distinguishable m-colorings on this graph:

$$\begin{aligned}\frac{1}{|G|}\sum_{\pi\in G} m^{cyc(\pi)} &= \frac{1}{720}\sum_{k=1}^{6} s(6,k)m^k \\ &= \frac{1}{720}\left(120m + 274m^2 + 225m^3 + 85m^4 + 15m^5 + m^6\right).\end{aligned}$$

In particular, when $m = 5$, there are 210 distinguishable colorings. □

We can generalize the graph in Fig. 8.8. Suppose that we have n vertices, all of which are mutually adjacent. The resulting graph is the *complete graph on n vertices*, denoted K_n. So, the graph in Fig. 8.8 is K_6. In general, the group of symmetries on K_n is S_n.

Example 8.4.6 also reveals the value of knowing the number of permutation in a group that have a given cycle index. Two of the more common groups that appear in these problems are the cyclic group and the dihedral group. Based on the above comments, we are motivated to determine the number of permutations in C_n and D_n that have index k.

To accomplish this, we need to introduce one of the most important functions in number theory and abstract algebra. The *Euler ϕ-function*, also known as Euler's totient function, counts the number of positive integers less than or equal to n that share no common factor (except 1) with n. If two positive integers only share one as a common factor, then they are *coprime*. The Euler ϕ-function is denoted $\phi(n)$. The value of $\phi(n)$ can be readily computed using the following theorem.

Theorem 8.4.7 *For $n = 1$, $\phi(1) = 1$. If $n \in \mathbb{N}$ and $n \geq 2$, then*

$$\phi(n) = n\prod\left(1 - \frac{1}{p}\right),$$

where the product is taken over all distinct primes dividing n.

Proof Clearly, $\phi(1) = 1$. If n is prime, then every positive integer less than n is coprime with n. Thus $\phi(p) = p - 1$ for all primes.

Suppose that $n = p^m$, where p is prime. Then, $p, 2p, \ldots, (p^{m-1} - 1)p$, and $p^{m-1}p$ are the only positive integers less than or equal to n that share a common factor with

n. There are p^{m-1} such numbers. However, there are p^m positive integers less than or equal to n. By the Subtraction Principle,

$$\phi(p^m) = p^m - p^{m-1} = p^m \left(1 - \frac{1}{p}\right).$$

Suppose that n and m are coprime. We claim that $\phi(nm) = \phi(n)\phi(m)$. To illustrate this, write every positive integer less than or equal to nm in a rectangular array as follows:

$$\begin{array}{lllll}
1 & 2 & \cdots & n-1 & n \\
n+1 & n+2 & \cdots & 2n-1 & 2n \\
\vdots & \vdots & \ddots & \vdots & \vdots \\
(m-2)n+1 & (m-2)n+2 & \cdots & (m-2)n+n-1 & (m-1)n \\
(m-1)n+1 & (m-1)n+2 & \cdots & (m-1)n+n-1 & mn
\end{array}$$

Mark out any column whose heading is not coprime with n. This leaves $\phi(n)$ columns. Similarly, mark out the ith row if i is not coprime with m. This leaves $\phi(m)$ rows. Hence, there are $\phi(n)\phi(m)$ elements left in the array. Ergo, $\phi(nm) = \phi(n)\phi(m)$.

By the Fundamental Theorem of Arithmetic, every positive integer greater than 1 can be written uniquely as a product of prime powers. In other words,

$$n = p_1^{m_1} \ldots p_t^{m_t},$$

where $m_1, \ldots, m_t \geq 1$ and $p_1 \leq \ldots \leq p_t$. From this it follows that

$$\begin{aligned}
\phi(n) &= \phi\left(p_1^{m_1} \ldots p_t^{m_t}\right) \\
&= \phi\left(p_1^{m_1}\right) \ldots \phi\left(p_t^{m_t}\right) \\
&= p_1^{m_1}\left(1 - \frac{1}{p_1}\right) \ldots p_t^{m_t}\left(1 - \frac{1}{p_t}\right) \\
&= p_1^{m_1} \ldots p_t^{m_t}\left(1 - \frac{1}{p_1}\right) \ldots \left(1 - \frac{1}{p_t}\right) \\
&= n \prod \left(1 - \frac{1}{p}\right),
\end{aligned}$$

where the product is taken over all primes dividing n. ■

Example 8.4.8 Determine $\phi(n)$ for:

(i) $n = 67$;
(ii) $n = 81$;
(iii) $n = 47320000$.

Solution

(i) Note that 67 is prime. Therefore there are 66 positive integers less than or equal to 67 that only have 1 as a common factor with 67. This is confirmed by Theorem 8.4.7,

$$\phi(67) = 67\left(1 - \frac{1}{67}\right) = 67 - 1 = 66.$$

(ii) The prime power factorization of 81 is $81 = 3^4$. Using Theorem 8.4.7 we have that

$$\phi(81) = 81\left(1 - \frac{1}{3}\right) = 81 - 27 = 54.$$

(iii) The prime power factorization of 47320000 is $47320000 = 2^5 * 5^3 * 7 * 13^2$. Theorem 8.4.7 yields

$$\phi(47320000) = 2^5 * 5^3 * 7 * 13^2\left(1 - \frac{1}{2}\right)\left(1 - \frac{1}{5}\right)\left(1 - \frac{1}{7}\right)\left(1 - \frac{1}{13}\right)$$

$$= 2^4 * 5^2 * 13(2-1)(5-1)(7-1)(13-1) = 2^4 * 5^2 * 13 * (2^2)(2 * 3) * (2^2 * 3)$$

$$= 2^9 * 3^2 * 5^2 * 13 = 1497600.$$

□

An important fact regarding the Euler ϕ-function is given in the following proposition which we present without proof.

Proposition 8.4.9 *If k is coprime with n, then $C_n =< \rho^k >$. Thus, the number of generators in C_n is given by $\phi(n)$.*

Using the Euler ϕ-function, we can give a formula for the number of permutations in the nth cyclic group which have cycle index k. Define $c(n, k)$ to be the number of permutations in C_n which have cycle index k.

Theorem 8.4.10 *If k does not divide n, then $c(n, k) = 0$. If k divides n, then*

$$c(n, k) = \phi\left(\frac{n}{k}\right).$$

Proof Suppose that C_n is generated by ρ. Let k be a divisor of n. Note that $H =< \rho^k >$ is a subgroup of C_n. Lagrange's Theorem implies that $|H| = n/k$. Further, $\rho^t H$ has the same cardinality as H for $t \in \mathbb{N}$. As in the proof of Lagrange's Theorem, these sets partition C_n. Hence, each cycle in ρ^k has length n/k. Thus, $cyc(\rho^k) = k$. Equivalently, the cycle index must divide n. It follows that $c(n, k) = 0$ when k does not divide n.

Suppose that $\pi \in C_n$ has cycle index k. Let $K = \{e, \rho, ..., \rho^{n/k-1}\}$. Consider the function $f : H \to K$ defined by $f(\rho^{mk}) = \rho^m$. Note that this mapping is a bijection. It follows that K is a cyclic group with n/k elements. Further, any integer coprime with n/k will generate K by Proposition 8.4.9. The number of such generators is given by $\phi(n/k)$ by that same proposition. Thus, $c(n, k) = \phi(n/k)$. ■

The values of $c(n, k)$ for small values of n and k is given in Table 8.1.

Table 8.1 The number of permutations in C_n with cycle index k

$n \setminus k$	1	2	3	4	5	6	7	8	9	10	11	12
1	1											
2	1	1										
3	2	0	1									
4	2	1	0	1								
5	4	0	0	0	1							
6	2	2	1	0	0	1						
7	6	0	0	0	0	0	1					
8	4	2	0	1	0	0	0	1				
9	6	0	2	0	0	0	0	0	1			
10	4	4	0	0	1	0	0	0	0	1		
11	10	0	0	0	0	0	0	0	0	0	1	
12	4	2	2	2	0	1	0	0	0	0	0	1

Example 8.4.11 Suppose that we are painting the front of circular place mats. Each place mat is divided into sixteen indistinguishable wedges. Each wedge must be painted with one of four available colors. How many different place mats can be made?

Solution As only one side of the place mat is painted, the group acting on this set is C_{16}. Since 16 is divisible by 1, 2, 4, 8, and 16, we have that:

$$c(16,1) = \phi\left(\frac{16}{1}\right) = 16\left(1 - \frac{1}{2}\right) = 8,$$

$$c(16,2) = \phi\left(\frac{16}{2}\right) = 8\left(1 - \frac{1}{2}\right) = 4,$$

$$c(16,4) = \phi\left(\frac{16}{4}\right) = 4\left(1 - \frac{1}{2}\right) = 2,$$

$$c(16,8) = \phi\left(\frac{16}{8}\right) = 2\left(1 - \frac{1}{2}\right) = 1,$$

$$\text{and} \quad c(16,16) = \phi\left(\frac{16}{16}\right) = 1.$$

For all other values of k, $c(16,k) = 0$ by Theorem 8.4.10. By Burnside's Lemma, the number of different place mats that can be painted with m colors is

$$\frac{1}{|G|}\sum_{\pi \in G} m^{cyc(\pi)}$$

$$= \frac{1}{16}\sum_{k=1}^{16} c(16,k)m^k$$

$$= \frac{1}{16}\left(8m + 4m^2 + 2m^4 + m^8 + m^{16}\right).$$

In particular, if $m = 4$, then 268439590 different place mats are possible. □

The number of permutations in the dihedral group that have cycle index k is particularly valuable as it allows us to find the number of distinct necklaces. Define $d(n,k)$ to be the number of permutations in D_n that have cycle index k. A method for computing $d(n,k)$ is given in the following theorem.

Theorem 8.4.12 *For $n \in \mathbb{Z}^+$, $n \geq 3$, and $k \leq n$, $d(n,k) = c(n,k) + d'(n,k)$.*

(i) If n is odd, then $d'(n,k) = n$ when $k = (n+1)/2$ and $d'(n,k) = 0$ otherwise.
(ii) If n is even, then $d'(n,k) = n/2$ when $k = n/2$ or $k = (n+2)/2$ and $d'(n,k) = 0$ otherwise.

Proof The set of permutations in D_n can be partitioned into two disjoint, exhaustive sets:

(i) The set of permutations in D_n that only involve rotating the elements of $[n]$. These permutations form a cyclic subgroup with n elements. The number of such permutations that have index k is given by $c(n,k)$ by Theorem 8.4.10.
(ii) The set of permutations in D_n that involve a reflection across a line of symmetry. Suppose that the number of such permutations that have index k is counted by $d'(n,k)$.

By the Addition Principle, $d(n,k) = c(n,k) + d'(n,k)$. We must now determine the values of $d'(n,k)$. This involves two cases:

(i) Suppose that n is odd. This will fix one point and transpose the remaining $n-1$ points in pairs (see the left side of Fig. 8.9). Thus, the cycle index of such a permutation is $1 + (n-1)/2 = (n+1)/2$. Since this reflection passes through one vertex, there are n choices for the line of symmetry. Hence, $d'(n,k) = n$ when $k = (n+1)/2$ and $d'(n,k) = 0$ otherwise.
(ii) Suppose that n is even. If the line of symmetry passes through two edges, then all n points are swapped in pairs (see the middle of Fig. 8.9). Such a reflection will have cycle index $n/2$. There are n such edges. However, when one edge is chosen, the edge on the opposite side is also chosen. Thus, there are $n/2$ such lines of symmetry. Ergo, $d'(n,k) = n/2$ when $k = n/2$.
Suppose instead that the line of symmetry passes through two vertices. Those two vertices will be fixed by the permutation. The remaining vertices will be transposed in pairs (see the right side of Fig. 8.9). It follows that any such line of symmetry will have index $2 + (n-2)/2 = (n+2)/2$. Because the vertices for this line of symmetry are chosen in pairs, there are $n/2$ such reflections. Hence $d'(n,k) = n/2$ when $k = (n+2)/2$.
So, when n is even, $d'(n,k) = n/2$ when $k = n/2, (n+2)/2$ and $d'(n,k) = 0$ otherwise. ■

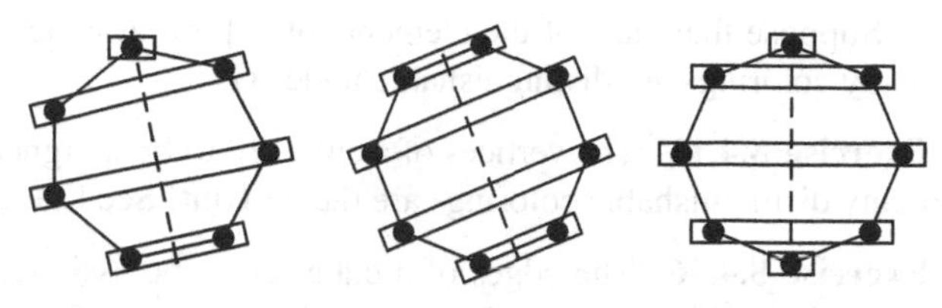

Fig. 8.9 The lines of symmetry on regular polygons

Table 8.2 The number of permutations in D_n with cycle index k

$n \backslash k$	1	2	3	4	5	6	7	8	9	10	11	12
3	2	3	1									
4	2	3	2	1								
5	4	0	5	0	1							
6	2	2	4	3	0	1						
7	6	0	0	7	0	0	1					
8	4	2	0	5	4	0	0	1				
9	6	0	2	0	9	0	0	0	1			
10	4	4	0	0	6	5	0	0	0	1		
11	10	0	0	0	0	11	0	0	0	0	1	
12	4	2	2	2	0	7	6	0	0	0	0	1

The values of $d(n,k)$ for small values of n and k are given in Table 8.2.

Example 8.4.13 How many distinct 12 bead necklaces can be made with beads of five different colors?

Solution The group acting on the set of m-colorings of the necklace is D_{12}. The number of permutations in D_{12} with cycle index k is given by Table 8.2. Namely, $d(12,1) = 4$, $d(12,2) = 2$, $d(12,3) = 2$, $d(12,4) = 2$, $d(12,6) = 7$, $d(12,7) = 6$, and $d(12,12) = 1$. Therefore, Burnside's Lemma gives that the number of distinct m-colorings of this necklace is

$$\frac{1}{24}\sum_{k=1}^{12} d(12,k)m^k$$

$$= \frac{1}{24}\left(4m + 2m^2 + 2m^3 + 2m^4 + 7m^6 + 6m^7 + m^{12}\right).$$

In particular, when $m = 5$, the number of distinct colorings is 10196680. □

Exercise 8.4.14 Suppose that G is the group acting on [6] with elements given below:

$$e = (1)(2)(3)(4)(5)(6), \quad \pi_1 = (1,2,3)(4)(5)(6), \quad \pi_2 = (1,3,2)(4)(5)(6),$$

$$\pi_3 = (1)(2)(3)(4,5,6), \quad \pi_4 = (1,2,3)(4,5,6), \quad \pi_5 = (1,3,2)(4,5,6),$$

$$\pi_6 = (1)(2)(3)(4,6,5), \quad \pi_7 = (1,2,3)(4,6,5), \quad \pi_8 = (1,3,2)(4,6,5).$$

Suppose that each of the elements of [6] are assigned one of three colors. How many colorings are distinguishable under G?

Exercise 8.4.15 The vertices of a cube are to be assigned one of four colors. How many distinguishable colorings are there? Hint: See Exercise 8.2.14.

Exercise 8.4.16 The edges of a cube are to be assigned one of five colors. How many distinguishable colorings are there? Hint: See Exercise 8.2.15.

Exercise 8.4.17 Consider the 4×4 grid with squares labeled 1 through 16. Let G be the group of permutations obtained by rotating the entire grid. How many distinguishable ways are there to two-color the grid? Hint: See Exercise 8.2.16.

Exercise 8.4.18 Consider the 4×4 grid with squares labeled 1 through 16. Suppose that when the inner 2×2 grid is rotated clockwise, the outer ring is rotated counterclockwise (and vice versa). How many distinguishable three-colorings are there? Hint: See Exercise 8.2.17.

Exercise 8.4.19 Consider the graph given in Fig. 8.4. How many distinguishable 3-colorings of the vertices are there? Hint: See Exercise 8.2.18.

Exercise 8.4.20 Consider the graph given in Fig. 8.5. How many distinguishable 4-colorings of the vertices are there? Hint: See Exercise 8.2.19.

Exercise 8.4.21 Consider the graph given in Fig. 8.6. How many distinguishable 7-colorings of the vertices are there? Hint: See Exercise 8.2.20.

Exercise 8.4.22 How many distinguishable ways are there to five-color the vertices of the complete graph on seven vertices?

Exercise 8.4.23 How many distinguishable ways are there to seven-color the vertices of the complete graph on eight vertices?

Exercise 8.4.24 Compute $\phi(n)$ for each of the following values of n:

(i) $n = 73$;
(ii) $n = 78125$;
(iii) $n = 262758417075$.

Exercise 8.4.25 Confirm the entries in Table 8.1.

Exercise 8.4.26 Suppose that we want to make circular place mats. The front of the place mat is divided into 18 indistinguishable wedges. Each wedge is to be painted with one of three colors. How many different place mats can be made?

Exercise 8.4.27 Suppose that we want to make circular place mats. The front of the place mat is divided into 30 indistinguishable wedges. Each wedge is to be painted with one of four colors. How many different place mats can be made?

Exercise 8.4.28 Confirm the entries in Table 8.2.

Exercise 8.4.29 Suppose that we want to make necklaces that will have nine beads on the strand. Each bead is to be either red, white, or blue. How many distinguishable necklaces can we make?

Exercise 8.4.30 Suppose that we want to make necklaces that will have ten beads on the strand. Each bead is to be either purple, orange, blue, or gold. How many distinguishable necklaces can we make?

8.5 Pólya Enumeration

Let us return to our original problem of making a necklace with six beads, each of which is either black or white. In that problem, we assumed that we had unlimited beads of each color to place on the necklace. It is perhaps more realistic to consider the problem where we only have a set number of beads of each color. For this problem, we will assume that we are making distinguishable necklaces that have four black beads and two white beads.

Again, we will be using Burnside's Lemma for the calculation. In this case, X is the set of all colorings with four black beads and two white beads. However, we do not wish to go back to the original formulation of Burnside's Lemma. Instead, we would prefer something closer to what was used in the last section. The solution requires us to keep track of how many of each color we have used. Hence, we will incorporate the method of generating functions (see Chap. 5) into Burnside's Lemma to yield Pólya's Enumeration Theorem.

To motivate this, we will consider a few permutations in D_6 and their invariant colorings. As before, the permutation $\rho = (1,2,3,4,5,6)$ moves 1 to 2 and so on. Therefore, the only colorings that are invariant under ρ are the colorings where everything is either all white or all black. However, neither of these colorings are in our set X. This corresponds to finding the coefficient of b^4w^2 in $b^6 + w^6$. In this case, the variable b keeps track of the number of times the color black will be used in the coloring (similarly for the variable w). The exponent on each term implies that the respective color must be used six times. In this instance, the coefficient of b^4w^2 in $b^6 + w^6$ is 0. Thus there are no invariant colorings under this permutation. A similar argument holds for ρ^5.

Similarly, ρ^2 has two cycles of length three. Each element of a cycle must receive the same color to be invariant under ρ^2. Therefore, the first cycle must receive either three black beads or three white beads. This corresponds to a factor of $b^3 + w^3$. An analogous argument holds for the second cycle. Thus, the corresponding term is $(b^3 + w^3)^2$. The coefficient of b^4w^2 in this term is again 0. A similar argument holds for ρ^4.

As for ρ^3, this has three cycles each of length two. By a similar argument to above, the corresponding term will be $(b^2 + w^2)^3$. The coefficient of b^4w^2 in this term is 3. A similar argument holds for $\tau\rho$, $\tau\rho^3$, and $\tau\rho^5$. Since e has six cycles of length one, the corresponding term will be $(b^1 + w^1)^6$. The coefficient of b^4w^2 is $\binom{6}{2} = 15$. Hence, there are 15 invariant colorings under this permutation.

Finally, τ has two cycles of length one and two cycles of length two. Therefore, the corresponding term is $(b^1 + w^1)^2(b^2 + w^2)^2$. The coefficient of b^4w^2 in this term is 3. This means that there are three colorings invariant under this permutation. A similar argument holds for $\tau\rho^2$ and $\tau\rho^4$.

Using the above information and Burnside's Lemma, we have that the number of distinguishable colorings of the necklace with four black beads and two white beads is given by:

$$\frac{1}{12}(15 + 2(0) + 2(0) + 4(3) + 3(3)) = 3.$$

The above example illustrates the importance of the cycle type when considering this type of coloring. As with generating functions, there is a monomial associated with each permutation. This has a variable for each distinct cycle length. Suppose that G is a subgroup of S_n. If a permutation $\pi \in G$ has $k_i(\pi)$ cycles of length i, then the corresponding term has a factor of $x_i^{k_i(\pi)}$. Thus, the *cycle index monomial* associated with the permutation π is

$$\prod_{i=1}^{n} x_i^{k_i(\pi)}.$$

The cycle index monomial of π will be denoted $cim(\pi)$. Note that this gives us an alternate way to represent the cycle type of π.

Example 8.5.1 For each element of D_6, determine the associated cycle index monomial.

Solution In Example 8.2.11, we found the elements of D_6. Therefore, the cycle index monomial for each of the elements is:

$$\begin{array}{lll} cim(e) = x_1^6 & cim(\rho) = x_6, & cim(\rho^2) = x_3^2, \\ cim(\rho^3) = x_2^3, & cim(\rho^4) = x_3^2, & cim(\rho^5) = x_6, \\ cim(\tau) = x_1^2x_2^2, & cim(\tau\rho) = x_2^3, & cim(\tau\rho^2) = x_1^2x_2^2, \\ cim(\tau\rho^3) = x_2^3, & cim(\tau\rho^4) = x_1^2x_2^2, & cim(\tau\rho^5) = x_2^3. \end{array}$$

□

We now define the *cycle index polynomial* of a group G. Essentially, this gives an "average" of the cycle index monomials over all permutations in the group G. Let Z_G denote the cycle index polynomial of G. Note that

$$Z_G(x_1, ..., x_n) = \frac{1}{|G|}\sum_{\pi \in G} cim(\pi).$$

Equivalently, we can write

$$Z_G(x_1, ..., x_n) = \frac{1}{|G|}\sum_{\pi \in G}\left(\prod_{i=1}^{n} x_i^{k_i(\pi)}\right),$$

where $k_i(\pi)$ is the number of cycles of length i in π.

Example 8.5.2 Find the cycle index polynomial associated with D_6.

Solution In Example 8.5.1, we found the cycle index monomial associated with each permutation in D_6. In order to find the cycle index polynomial, we need only average these monomials. Hence,

$$Z_{D_6}(x_1, ..., x_6) = \frac{1}{12}\left(x_1^6 + 3x_1^2x_2^2 + 4x_2^3 + 2x_3^2 + 2x_6\right). \qquad \square$$

The form of the cycle index polynomial should suggest a few possibilities. First, it allows for us to substitute in expressions for the x_i. For instance, if we let $x_i = b^i + w^i$, then the coefficient on b^4w^2 in the resulting generating function would give the number of distinguishable necklaces with four black beads and two white beads. For this reason, the cycle index polynomial is often referred to as a *pattern inventory*. In this case, the function $x_i = b^i + w^i$ acts as a *weighted coloring*. In general, a *weighting function* is any function that assigns exponents, or weights, to the colors being used. Second, if we let $x_i = m$, then the cycle index polynomial is precisely the number of distinguishable m-colorings induced by the group G as given by Burnside's Lemma. These observations will lead directly to Pólya's Enumeration Theorem.

Theorem 8.5.3 *(Pólya's Enumeration Theorem) Suppose that S is a set and $C = \{c_1, ..., c_m\}$ is a set of colors. Let X be the set of all m-colorings of S. Let $f(c_1, ..., c_m)$ be a weighting function on the set of colors. Let G be a group acting on X, where G is a subgroup of S_n. The number of distinct colorings in which c_i is used ℓ_i times is given by the coefficient of $\prod_{i=1}^m c_i^{\ell_i}$ in the generating function*

$$Z_G\left(f(c_1, ..., c_m), f(c_1^2, ..., c_m^2), ..., f(c_1^n, ..., c_m^n)\right),$$

where $Z_G(x_1, ..., x_n)$ is the cycle index polynomial of G.

The proof of Pólya's Enumeration Theorem follows immediately from the use of generating functions and Burnside's Lemma. Before proceeding with examples, we note that the most common weighting function is the one that assigns the same weight to each color. In other words,

$$f(c_1, ..., c_m) = c_1 + \cdots + c_m$$

is the weighting function that will be used for most problems. In many subgroups of S_n, there are no permutations which have a cycle of length of k. In which case, we do not list x_k in our notation for the cycle index polynomial.

Example 8.5.4 Suppose that we want to paint the faces of a cube such that three faces are painted red, two faces are painted white, and one face is painted blue. In how many distinguishable ways can this be done?

Solution In Example 8.2.5, we found the elements of G, the group of rotations on the cube. The cycle index monomial associated with each of these permutations is given below.

$$\begin{array}{lll} cim(e) = x_1^6, & cim(\tau_1) = x_1^2x_4, & cim(\tau_2) = x_1^2x_4, \\ cim(\tau_3) = x_1^2x_2^2, & cim(\tau_4) = x_2^3, & cim(\tau_5) = x_3^2, \\ cim(\tau_6) = x_3^2, & cim(\tau_7) = x_1^2x_4, & cim(\tau_8) = x_3^2, \\ cim(\tau_9) = x_1^2x_4, & cim(\tau_{10}) = x_2^3, & cim(\tau_{11}) = x_3^2, \\ cim(\tau_{12}) = x_3^2, & cim(\tau_{13}) = x_1^2x_4, & cim(\tau_{14}) = x_2^3, \\ cim(\tau_{15}) = x_3^2, & cim(\tau_{16}) = x_1^2x_4, & cim(\tau_{17}) = x_3^2, \\ cim(\tau_{18}) = x_3^2, & cim(\tau_{19}) = x_2^3, & cim(\tau_{20}) = x_1^2x_2^2, \\ cim(\tau_{21}) = x_2^3, & cim(\tau_{22}) = x_2^3, & cim(\tau_{23}) = x_1^2x_2^2. \end{array}$$

Therefore, the cycle index polynomial associated with G is

$$Z_G(x_1, x_2, x_3, x_4) = \frac{1}{24}\left(x_1^6 + 6x_1^2x_4 + 3x_1^2x_2^2 + 8x_3^2 + 6x_2^3\right).$$

To obtain our pattern inventory, we evaluate $Z_G(x_1, x_2, x_3, x_4)$ at $x_i = r^i + w^i + b^i$. According to Pólya's Enumeration Theorem, the number of acceptable colorings is given by the coefficient of r^3w^2b in this generating function. Using a computer algebra system, we find this coefficient to be 3. Hence, there are three distinguishable colorings that color three faces red, two faces white, and one face blue. □

Example 8.5.5 Suppose that we want to color the vertices of the graph in Fig. 8.3. How many distinguishable ways are there to color the vertices if three vertices are to be colored orange, two vertices are to be colored purple, and two vertices are to be colored green?

Solution In Example 8.2.6, we found the elements of G, the group of symmetries of this graph. Thus, the cycle index monomials associated with each permutation are

$$\begin{array}{ll} cim(e) = x_1^7, & cim(\tau_1) = x_1^5x_2, \\ cim(\tau_2) = x_1^5x_2, & cim(\tau_3) = x_1^3x_2^2, \\ cim(\tau_4) = x_1x_2^3, & cim(\tau_5) = x_1x_2x_4, \\ cim(\tau_6) = x_1x_2x_4, & cim(\tau_7) = x_1x_2^3. \end{array}$$

Thus, the cycle index polynomial associated with G is

$$Z_G(x_1, x_2, x_4) = \frac{1}{8}\left(x_1^7 + 2x_1^5x_2 + x_1^3x_2^2 + 2x_1x_2^3 + 2x_1x_2x_4\right).$$

To obtain the pattern inventory, we evaluate $Z_G(x_1, x_2, x_4)$ at $x_i = o^i + p^i + g^i$. Pólya's Enumeration Theorem yields that the number of acceptable colorings is given by the coefficient of $o^3p^2g^2$ in this generating function. Using a computer algebra system, we find this coefficient to be 42. □

We now turn our attention to some of the more commonly used groups. Namely, the cyclic group C_n and the dihedral group D_n. For these, we will need the cycle index polynomial. Fortunately, the form of the cycle index polynomial for these groups follows immediately from the proofs of Theorems 8.4.10 and 8.4.12. As for the symmetric group, it is typically easier to use a computer algebra system to generate the cycle index polynomial.

Theorem 8.5.6

(i) The cycle index polynomial of the nth cyclic group, C_n, is given by

$$Z_{C_n}(x_1, \ldots, x_n) = \frac{1}{n} \sum_{d|n} \phi(d) x_d^{n/d}.$$

(ii) The cycle index polynomial of the nth dihedral group, D_n, is given by

$$Z_{D_n}(x_1, \ldots, x_n) = \frac{Z_{C_n}(x_1, \ldots, x_n)}{2} + \begin{cases} \frac{1}{2} x_1 x_2^{(n-1)/2} & n \text{ is odd} \\ \frac{1}{4}\left(x_1^2 x_2^{(n-2)/2} + x_2^{n/2}\right) & n \text{ is even.} \end{cases}$$

Example 8.5.7 Suppose that we want to paint circular place mats which are divided into fifteen identical wedges. Suppose that we want to do this in such a way that we use five colors, each of which we use three times. In how many distinguishable ways can this be done?

Solution By Theorem 8.5.6, the cycle index polynomial for C_n is given by

$$Z_{C_{15}}(x_1, \ldots, x_n) = \frac{1}{15}\left(x_1^{15} + 2x_3^5 + 4x_5^3 + 8x_{15}\right).$$

Evaluating this polynomial with $x_i = c_1^i + c_2^i + c_3^i + c_4^i + c_5^i$ yields the pattern inventory. The number of acceptable colorings is given by the coefficient of $c_1^3 c_2^3 c_3^3 c_4^3 c_5^3$ in this generating function by Pólya's Enumeration Theorem. Using a computer algebra system, we find this coefficient to be 11211216. □

Example 8.5.8 Suppose that we want to make a necklace with five blue beads, four red beads, and three purple beads. How many distinguishable necklaces can we make?

Solution By Theorem 8.5.6 (or a computer algebra system), the cycle index polynomial of D_{12} is

$$Z_{D_{12}}(x_1, \ldots, x_{12}) = \frac{1}{24}\left(x_1^{12} + 6x_2^5 x_1^2 + 7x_2^6 + 2x_3^4 + 2x_4^3 + 2x_6^2 + 4x_{12}\right).$$

For the pattern inventory, we substitute $x_i = b^i + r^i + p^i$ into this polynomial. By Pólya's Enumeration Theorem, the number of acceptable colorings is given by the coefficient of $b^5 r^4 p^3$ in this polynomial. Using a computer algebra system, we find this number to be 1170. □

The reader may wonder why we did not do any examples involving coloring the complete graph. In the complete graph, any vertex may be swapped with any other. Thus in a restricted coloring, there is only one distinguishable coloring. So for instance, suppose that we are dealing with the graph in Fig. 8.8. Further, suppose that we want to color four vertices black and two vertices white. Because any two vertices can be swapped, it does not matter which vertices we color black and which we color white. Hence, we can rearrange the vertices so that the first four vertices are black and the last two vertices are white. Thus, there is only one way to do this.

Exercise 8.5.9 Suppose that G is the group acting on [6] with elements given below:

$$e = (1)(2)(3)(4)(5)(6), \quad \pi_1 = (1,2,3)(4)(5)(6), \quad \pi_2 = (1,3,2)(4)(5)(6),$$
$$\pi_3 = (1)(2)(3)(4,5,6), \quad \pi_4 = (1,2,3)(4,5,6), \quad \pi_5 = (1,3,2)(4,5,6),$$
$$\pi_6 = (1)(2)(3)(4,6,5), \quad \pi_7 = (1,2,3)(4,6,5), \quad \pi_8 = (1,3,2)(4,6,5).$$

Suppose that one of the elements of [6] is to be colored white, two elements are to be colored black, and three are to be colored green. How many colorings are distinguishable under G?

Exercise 8.5.10 The vertices of a cube are to be assigned one of four colors. Each of the four colors is to used twice. How many distinguishable colorings are there? Hint: See Exercise 8.2.14.

Exercise 8.5.11 The edges of a cube are to be assigned one of three colors, namely black, white, and red. How many distinguishable colorings are there if black is to be used five times, white is to be used four times, and red is to be used three times? Hint: See Exercise 8.2.15.

Exercise 8.5.12 Consider the 4×4 grid with squares labeled 1 through 16. Let G be the group of permutations obtained by rotating the entire grid. How many distinguishable ways are there to two-color the grid if the first color is used ten times and the second color is used six times? Hint: See Exercise 8.2.16.

Exercise 8.5.13 Consider the 4×4 grid with squares labeled 1 through 16. Suppose that when the inner 2×2 grid is rotated clockwise, the outer ring is rotated counterclockwise (and vice versa). Each square is to be assigned one of three colors, namely orange, purple, and white. How many distinguishable colorings are there if seven squares are to colored orange, six squares are to be colored purple, and three squares are to be colored white? Hint: See Exercise 8.2.17.

Exercise 8.5.14 Consider the graph given in Fig. 8.4. How many distinguishable ways are there to color the vertices if three vertices are to colored blue and four vertices are to be colored gold? Hint: See Exercise 8.2.18.

Exercise 8.5.15 Consider the graph given in Fig. 8.5. How many distinguishable 3-colorings of the vertices are there if every color is to be used three times? Hint: See Exercise 8.2.19.

Exercise 8.5.16 Consider the graph given in Fig. 8.6. The vertices are to be colored blue, red, black, or white. How many distinguishable colorings of the vertices are there if blue is to be used three times and each of the other colors is to be used twice? Hint: See Exercise 8.2.20.

Exercise 8.5.17 Suppose that we want to make circular place mats. The front of the place mat is divided into 18 indistinguishable wedges. Each wedge is to be painted with one of three colors. How many different place mats can be made if each color is to be used six times?

Exercise 8.5.18 Suppose that we want to make circular place mats. The front of the place mat is divided into 30 indistinguishable wedges. Each wedge is to be painted with one of four colors, namely orange, purple, blue, and gold. How many different place mats can be made if orange is to be used ten times, purple is to be used nine times, blue is to be used six times, and gold is to be used five times?

Exercise 8.5.19 Suppose that we want to make necklaces that will have nine beads on the strand. Each bead is to be either red, white, or blue. How many distinguishable necklaces can we make if red is to be used five times, white is to be used twice, and blue is to be used twice?

Exercise 8.5.20 Suppose that we want to make necklaces that will have ten beads on the strand. Each bead is to be either purple, white, green, or orange. How many distinguishable necklaces can we make if purple is to be used three times, white is to be used three times, green is to be used twice, and orange is to be used twice?

Chapter 9
Application: Probability

9.1 Basic Discrete Probability

Consider an experiment in which we know all of the possible outcomes. However, we do not know which of the possible outcomes will occur. Such an experiment is a *random experiment*. Examples of such an experiment include flipping a coin, rolling a die, or counting the number of customers that enter a store during the next hour. *Probability* describes how likely any given outcome in a random experiment will happen. The *sample space* S is the set of all possible outcomes in a random experiment. We are only interested in the case where there is a bijection between S and a subset of $\mathbb{N}$. In this case, S is referred to as a *discrete sample space*. This is in contrast to a *continuous* space where the possibilities can exist for all real numbers. Examples of random experiments with a continuous sample space include the time it takes to accomplish a task, heights of individuals, and when a radioactive particle will discharge.

A *random variable* assigns a real number to each outcome in the sample space. For example, suppose that we flip a coin 50 times and observe the pattern of heads. The sample space is all 2^{50} binary strings of length 50. The random variable X could count the number of heads obtained in the 50 tosses. Thus, the random variable maps S to the set $\{0, 1, ..., 50\}$.

Note that some phenomenon are purely deterministic. For instance, suppose that a bowling ball is dropped off of a fifty foot tall cliff. The time it takes to reach the ground can be determined using elementary physics. While this is a random variable, it is simply one that is always a constant.

Suppose that each outcome in the sample space S is assigned a probability. This induces a probability on the random variable X which maps S to $\mathbb{R}$. For example, consider the above example of flipping a coin fifty times. Assuming a fair coin, each of the 2^{50} possible outcomes is equally likely. Again, we define the random variable X to be the function that counts the number of heads in each string. Suppose that we wanted the probability of 15 heads. This could be computed by determining the number of ways to get 15 heads, divided by the total number of possibilities. There are $\binom{50}{15}$ ways to arrange 15 heads and 35 tails by Stars and Bars. Thus, the probability of 15 heads is given by $\binom{50}{15}/2^{50}$.

R. A. Beeler, *How to Count*, DOI 10.1007/978-3-319-13844-2_9

In general, the probability that a random variable X equals x is the sum of probabilities of values of S that map to x under the random variable X. This probability is denoted $p(x)$. This will be referred to as the *probability distribution function* of x. If S is a discrete sample space, X is a random variable associated with S, and $p(x)$ is a probability function on X, then $p(x)$ satisfies the following:

(i) $p(x) \geq 0$ for all $x \in X$;
(ii) $\sum_{x \in X} p(x) = 1$.

An important notion is that of *independence*. Two events, $x, y \in S$ are said to be *independent* if the outcome of one does not influence the other. As an example, consider flipping a coin twice in succession. Since the coin has no memory, the coin flips can be considered independent. However, not all events are independent. For example, consider the following events:

(i) x: "You get a flat tire on the way to class;"
(ii) y: "You are late to class."

These events are *not independent* because if you get a flat tire, then it will increase your chances of being late. As before, two sets of events are *disjoint* if the occurrence of one precludes the possibility of the second occurring.

An *event* is any subset of the sample space. For instance, consider the problem of drawing a five card poker hand (see Sect. 3.2). The sample space consists of all $\binom{52}{5}$ five card hands. An event A could be drawing three of a kind. If all outcomes in S are equally likely, then the probability of A is $|A|$ divided by $|S|$. In other words, $p(A) = |A|/|S|$. Referring to Example 3.2.1, there are 54912 ways to draw three of a kind. Thus, the probability of drawing this hand is $54912/\binom{52}{5} \approx .021128$.

The assumption that all possibilities are equally likely is a standard assumption that covers many common probability problems. For instance, many "games of chance" are based on this. These games include flipping a fair coin, rolling a fair die, and drawing cards from a deck.

If two events are independent, then the outcome of one event does not influence the other. Hence, we have an analog to the Multiplication Principle. Namely, $p(A \cap B) = p(A)p(B)$. Often, probability texts will use this as the definition of independence. This is discussed more thoroughly in the next few paragraphs.

Probabilitists often use the symbol '$\cap$' a bit differently than how we have used it in previous chapters. To illustrate this, consider the problem of flipping a fair coin twice. In particular, we are concerned with the following event sets:

(i) A: "The first toss lands on heads;"
(ii) B: "The second toss lands on heads."

As we have noted, the coin has no memory. Hence, the two tosses are independent and the associated events are independent. We note that $A \cap B$ is the event that both tosses land on heads. The sample space for this is given by:

$$S = \{(T, T), (T, H), (H, T), (H, H)\}.$$

Hence, the probability of both tosses landing on heads is one fourth.

Another way of thinking of this is to consider the two events separately. In this setting, we have two sample spaces, one for each of the two coin tosses. In this setting, $S_1 = \{T, H\}$ and $S_2 = \{T, H\}$. Clearly, $S = S_1 \times S_2$. Further, the Multiplication Principle applies in this case. Hence:

$$|S| = |S_1 \times S_2| = |S_1||S_2| = 2(2) = 4.$$

Thus, when we look at $A \cap B$ in a probabilistic setting, what we actually consider is closer to:

$$p(A \cap B) = \frac{|A \times B|}{|S_1 \times S_2|}.$$

As another example, consider the problem of drawing a single card from a standard deck of fifty-two cards. Consider the following events:

(i) A: "The card is an Ace;"
(ii) B: "The card is a Heart."

In this case, the intersection truly is the intersection of the two sets because $A \cap B$ is the event that the card is the Ace of Hearts. For this reason, we expect that the probability is one out of 52. Note that:

$$p(A) = \frac{4}{52} = \frac{1}{13} \quad \text{and} \quad p(B) = \frac{13}{52} = \frac{1}{4}.$$

Further,

$$\frac{1}{52} = p(A \cap B) = p(A)p(B) = \frac{1}{13}\left(\frac{1}{4}\right).$$

Hence, the two events are independent, even though we are drawing just one card. In a sense, we could consider this as drawing one card from each of two decks and achieve an equivalent result.

Considering the probability as a set has some advantages. The principal advantage is that it allows us to apply set theoretic rules to a probabilistic settings. In particular, this allows us to give analogs to the Subtraction Principle and the Principle of Inclusion and Exclusion.

Theorem 9.1.1 *Let S be a discrete sample space. For any $A, B \subseteq S$, we have the following:*

(i) $p(A) = 1 - p(A^c)$;
(ii) $p(A \cup B) = p(A) + p(B) - p(A \cap B)$.

Proof

(i) As always, we consider the complement to be taken within the set S. In other words, $A^c = S - A$. Note that:

$$1=\sum_{x\in S}p(x)\Rightarrow 1=\sum_{x\in A}p(x)+\sum_{x\in A^c}p(x)$$
$$\Rightarrow \sum_{x\in A}p(x)=1-\sum_{x\in A^c}p(x)\Rightarrow p(A)=1-p(A^c).$$

(ii) Note that:

$$p(A\cup B)=\frac{|A\cup B|}{|S|}=\frac{|A|+|B|-|A\cap B|}{|S|}$$
$$=\frac{|A|}{|S|}+\frac{|B|}{|S|}-\frac{|A\cap B|}{|S|}=p(A)+p(B)-p(A\cap B).$$

■

We begin with several rudimentary examples. Unless otherwise noted, we will assume that a n-sided die has its sides labeled $1,\ldots,n$.

Example 9.1.2 Find the following probabilities:

(i) A fair coin is tossed five times and lands on heads each time;
(ii) A fair coin lands on heads *and* a fair six-sided die lands on 2 or 5;
(iii) A fair coin lands on heads *or* a fair six-sided die lands on 2 or 5.

Solution Ignoring esoteric and exceedingly improbable possibilities, such as the coin balancing on edge, the sample space of a coin toss is $S_1=\{T,H\}$. Similarly, the sample space for the roll of a six-sided die is $S_2=\{1,2,3,4,5,6\}$. Further, since the coin and die are "fair," we can assume that each outcome in each of the sample spaces is equally likely.

(i) Note that the sample space for five coin tosses is $S_1\times S_1\times S_1\times S_1\times S_1$. Hence the cardinality is $|S_1|^5=32$. Exactly one of these events will result in five heads in a row, namely (H,H,H,H,H). Ergo, the probability is $\frac{1}{32}$.
(ii) Let A be the event where the coin lands on heads. Let B be the event that the die lands on 2 or 5. The coin flip does not influence the outcome of the die roll. Hence, we can assume the events to be independent. Note that $p(A)=\frac{1}{2}$ and $p(B)=\frac{2}{6}=\frac{1}{3}$. Since the events are independent, we have:

$$p(A\cap B)=p(A)p(B)=\frac{1}{2}\left(\frac{1}{3}\right)=\frac{1}{6}.$$

(iii) In this example, we must compute $p(A\cup B)$. This can be done as follows:

$$p(A\cup B)=p(A)+p(B)-p(A\cap B)=\frac{1}{2}+\frac{1}{3}-\frac{1}{6}=\frac{2}{3}.$$

□

Example 9.1.3 Assume that all cards in a standard fifty-two card deck are equally likely. In five-card poker, find the probability of drawing a "junk" hand.

Solution The sample space in this case is the set of all five-card poker hands. The cardinality of this set is given by $\binom{52}{5}=2598960$. By Example 3.2.5, the number of ways of getting a "junk" hand is 1302540. Hence, the probability of this hand is given by:

$$\frac{1302540}{2598960} \approx .501177.$$

□

It is somewhat surprising that over half of all poker hands are "junk." Hence, if you are playing with one other person, then the lowest pair will win over half the time.

Example 9.1.4 Suppose that all permutations on $[n]$ are equally likely. Find the probability that a randomly chosen permutation will have no fixed points.

Solution The sample space in this problem is the set of all permutations on $[n]$. The cardinality of this set is given by $n!$. Our event is selecting a permutation in which there are no fixed points. In other words, we must find the set of all derangements on $[n]$. By Theorem 7.3.2, the cardinality of this set is given by:

$$D_n = n! \sum_{i=0}^{n} \frac{(-1)^i}{i!}.$$

From this it follows that the probability that the chosen permutation has no fixed points is given by:

$$\frac{D_n}{n!} = \sum_{i=0}^{n} \frac{(-1)^i}{i!}.$$

□

Example 9.1.5 (The Birthday Paradox) Find the probability that in a room with k randomly chosen individuals that at least two will have the same birthday.

Solution We begin by making certain assumptions. First we will assume that all birthdays are equally likely. For this reason, we will ignore Leap Day (February 29). Further, we assume that the each person's birthday is independent of everyone else's. In a room full of *randomly* chosen individuals, this is a reasonable enough assumption. Before proceeding, we note that if the second assumption were violated, then the probability that two persons share the same birthday would actually *increase*. If the first assumption were violated, then it would decrease the probability, but only marginally. This being the case, we can be assured that this solution will at least give a lower bound for the probability.

Note that the sample space for this probability is the set of all k-tuples chosen from $[365]$ with replacement allowed. Hence, there are 365^k elements in this sample space. Let A be the event where at least two people in the room share the same birthday. It is actually easier to compute the cardinality of A^c, the event where everyone in the room has a *different* birthday. The cardinality of A^c can be computed by observing that there are 365 choices for the first person's birthday, 364 choices for the second

Table 9.1 Birthday probabilities for certain small k

k	$p(A)$	k	$p(A)$
2	.0027397	24	.5383443
3	.0082042	25	.5686997
4	.0163559	26	.5982408
5	.0271356	27	.6268593
6	.0404625	28	.6544615
7	.0562357	29	.6809685
8	.0743353	30	.7063162
9	.0946238	31	.7304546
10	.1169482	35	.8143832
11	.1411414	40	.8912318
12	.1670248	45	.9409759
13	.1944103	50	.9703736
14	.2231025	55	.9862623
15	.2529013	60	.9941227
16	.2836040	65	.9976831
17	.3150077	70	.9991596
18	.3469114	75	.9997199
19	.3791185	80	.9999143
20	.4114384	85	.9999760
21	.4436883	90	.9999938
22	.4756953	95	.9999986
23	.5072972	100	.9999997

person's birthday,..., and $365 - k + 1$ choices for the last person's birthday. Hence, $|A^c| = P(365, k)$. From this it follows that:

$$p(A^c) = \frac{P(365,k)}{365^k} \Rightarrow$$

$$p(A) = 1 - p(A^c) = 1 - \frac{P(365,k)}{365^k}.$$

□

This problem is referred to as the Birthday Paradox because it is somewhat counterintuitive. For instance, Table 9.1 shows that with a room of only twenty-three people, the odds are better that fifty percent that at least two people will share the same birthday. While the Pigeonhole Principle guarantees that a room with 366 people will have at least two people that share the same birthday, a much smaller room will virtually guarantee the same result. For instance, in a room with sixty individuals the probability that at least two share the same birthday is practically one, as shown in Table 9.1.

As with other combinatorial problems, it may be useful (and possibly necessary) to use generating functions (see Chap. 5) to solve certain problems.

Example 9.1.6 Find the probability that a non-negative integer solution to:

$$x_1 + x_2 + x_3 + x_4 = 22$$

satisfies the constraints:

$$0 \le x_1 \le 7, \quad 0 \le x_2 \le 10,$$

$$0 \le x_3 \le 5, \quad 0 \le x_2 \le 8.$$

Solution We begin by finding the cardinality of the sample space. The sample space is the set of all non-negative integer solutions to $x_1 + x_2 + x_3 + x_4 = 22$. The cardinality of this set is the coefficient of x^{22} in the generating function:

$$\underbrace{(1+x+\cdots)}_{x_1}\underbrace{(1+x+\cdots)}_{x_2}\underbrace{(1+x+\cdots)}_{x_3}\underbrace{(1+x+\cdots)}_{x_4}.$$

Using a computer algebra system, we find that this coefficient is 2300.

The event is selecting an element of the set of solutions that satisfy the above constraints. Hence the cardinality of this set is given by the coefficient of x^{22} in the generating function:

$$\underbrace{(1+x+\cdots+x^7)}_{x_1} * \underbrace{(1+x+\cdots+x^{10})}_{x_2}$$

$$* \underbrace{(1+x+\cdots+x^5)}_{x_3} * \underbrace{(1+x+\cdots+x^8)}_{x_4}.$$

Using a computer algebra system, we find that this coefficient is 154.

Hence the probability that a non-negative integer solution to $x_1+x_2+x_3+x_4 = 22$ satisfies the constraints is given by $\frac{154}{2300} \approx .06696$. □

Example 9.1.7 Five fair, six-sided dice are rolled. Find the probability that their sum is 15, 16, or 17.

Solution It is convenient to think of the five dice as being distinguishable from each other. Since the sample space for the roll of the ith die is $S_i = \{1, 2, 3, 4, 5, 6\}$, it follows that the sample space for rolling five distinguishable die is $S_1 \times S_2 \times S_3 \times S_4 \times S_5$. Thus the cardinality of the sample space is $6^5 = 7776$. The number of ways that the sum of the five dice can be fifteen is given by the coefficient of x^{15} in the generating function:

$$(x + x^2 + x^3 + x^4 + x^5 + x^6)^5.$$

Using a computer algebra system, we find that this coefficient is 651. Similarly, the number of ways to roll a sixteen and a seventeen are given by the coefficients of x^{16} and x^{17}, respectively. The coefficient of x^{16} in this generating function is 735. The coefficient on x^{17} in the generating function is 780. Therefore, the probability of rolling a fifteen, sixteen, or seventeen is:

$$\frac{651 + 735 + 780}{7776} = \frac{2166}{7776} \approx .278549.$$

□

Note that Example 9.1.7 can also be accomplished using a *probability generating function*. In a probability generating function, all of the coefficients are probabilities. For instance, the coefficient on x^i corresponds to the probability that i is rolled. Hence the probability generating function for the sum of five fair six-sided dice is:

$$\left(\frac{1}{6}x + \frac{1}{6}x^2 + \frac{1}{6}x^3 + \frac{1}{6}x^4 + \frac{1}{6}x^5 + \frac{1}{6}x^6\right)^5.$$

One advantage of using probability generating functions is that they allow for the case where every outcome is not equally likely. This possibility is considered in our next example.

Example 9.1.8 Suppose that four unfair, six-sided dice are rolled. The respective probabilities for each die are as follows:

*Die**Roll*	1	2	3	4	5	6
1	.2	.3	.25	.05	.1	.1
2	0	.02	.03	.02	.03	.9
3	.1	.15	.25	.25	.15	.1
4	.4	.05	.05	.05	.05	.4

Find the probability that the sum of the dice is 14.

Solution The probability that the sum of the dice is fourteen is given by the coefficient of x^{14} in the probability generating function:

$$\underbrace{(.2x + .3x^2 + .25x^3 + .05x^4 + .1x^5 + .1x^6)}_{\text{Die 1}}$$

$$* \underbrace{(.02x^2 + .03x^3 + .02x^4 + .03x^5 + .9x^6)}_{\text{Die 2}}$$

$$* \underbrace{(.1x + .15x^2 + .25x^3 + .25x^4 + .15x^5 + .1x^6)}_{\text{Die 3}}$$

$$*\underbrace{(.4x + .05x^2 + .05x^3 + .05x^4 + .05x^5 + .4x^6)}_{\text{Die 4}}.$$

Using a computer algebra system, we find this coefficient to be approximately 0.1014075. Hence, this gives the probability of rolling a 14 as the sum of the dice. □

Exercise 9.1.9 Find the probability of drawing a pair in poker.

Exercise 9.1.10 Find the probability of drawing a full house in poker.

Exercise 9.1.11 Suppose that tickets for the movie theater cost five dollars and the cashier has no change. Of the $2n$ people in line, n people have five dollar bills and n people have ten dollar bills. If all people stay in the random order that they arrived in, then find the probability that the cashier never runs out of change.

Exercise 9.1.12 Suppose that there are n married couples to be seated at a circular dinner table. Find the probability that sexes alternate and no one sits next to their own spouse.

Exercise 9.1.13 Suppose that there are n married couples to be seated at a circular dinner table in such a way that sexes must alternate. Find the probability that no one sits next to their own spouse.

Exercise 9.1.14 Six fair, four-sided dice are rolled. Find the probability that their sum is 12, 13, or 14.

Exercise 9.1.15 A fair four-sided, six-sided, eight-sided, ten-sided, twelve-sided, and twenty-sided die are each rolled once. Find the probability that the sum is 26, 27, or 28.

Exercise 9.1.16 Find the probability that a non-negative integer solution to:

$$x_1 + x_2 + x_3 + x_4 = 27$$

satisfies the constraints:

$$0 \le x_1 \le 8, \quad 0 \le x_2 \le 14,$$
$$0 \le x_3 \le 6, \quad 0 \le x_2 \le 11.$$

Exercise 9.1.17 Given a fair four-sided die and a fair ten-sided die, find the following probabilities:

(i) The four-sided die lands on three on each of six consecutive tosses;
(ii) The four-sided die lands on three *and* the ten-sided die lands on 2, 3, or 7;
(iii) The four-sided die lands on three *or* the ten-sided die lands on 2, 3, or 7.

Exercise 9.1.18 Suppose that five unfair, eight-sided dice are rolled. The respective probabilities for each die are as follows:

*Die**Roll*	1	2	3	4	5	6	7	8
1	.1	.1	.1	.1	.1	.1	.1	.3
2	.2	.2	.2	.2	.05	.05	.05	.05
3	.05	.1	.15	.25	.25	.15	.1	.05
4	.2	.05	.2	.05	.2	.05	.2	.05
5	.03	.03	.03	.03	.03	.03	.03	.79

Find the probability that the sum of the dice is 19.

9.2 The Expected Value and Variance

Suppose that X is a discrete random variable with sample space S, then the *expected value* of X, denoted $E(X)$, is defined as:

$$E(X) = \sum_{x \in X} xp(x).$$

Usually we think of the expected value as the *average* or *mean* of the random variable. The mean of a random variable is often denoted μ.

As an example, consider rolling a fair die a very large number of times. If we roll the die enough times, then we would expect that it would land on each number approximately one-sixth of the time. In the expected value formula, x is the *weight* of each roll. Thus the expected value of the die roll is:

$$E(X) = \frac{1}{6}(1) + \frac{1}{6}(2) + \frac{1}{6}(3) + \frac{1}{6}(4) + \frac{1}{6}(5) + \frac{1}{6}(6) = 3.5.$$

Under no circumstances should the expected value be confused with the *most likely value*. For instance, the expected value for the roll of a fair, six-sided die is 3.5. However, 3.5 is not one of the values in the sample space.

We now give additional properties of the expected value.

Theorem 9.2.1 *(Properties of the Expected Value) Suppose that X_1 and X_2 are random variables.*

(i) Suppose that $a, b \in \mathbb{R}$ and that $Y = aX_1 + b$. It follows that $E(Y) = aE(X_1) + b$.
(ii) The expected value of a sum is the sum of the expected values. In other words, $E(X_1 + X_2) = E(X_1) + E(X_2)$.
(iii) If X_1 and X_2 are independent, then $E(X_1X_2) = E(X_1)E(X_2)$.

Proof

(i) Note that

$$E(Y) = E(aX_1 + b) = \sum_{x \in X_1} (ax + b)p(x)$$
$$= a \sum_{x \in X_1} xp(x) + b \sum_{x \in X_1} p(x)$$
$$= aE(X_1) + b.$$

(ii) We proceed in a similar manner to (i). Similarly, we can assume that they have a joint probability distribution $p(x_1, x_2)$. Here, $p(x_1, x_2)$ denotes the probability that $X_1 = x_1$ and $X_2 = x_2$.

$$E(X_1 + X_2) = \sum_{x_1 \in X_1} \sum_{x_2 \in X_2} (x_1 + x_2)p(x_1, x_2)$$
$$= \sum_{x_1 \in X_1} \sum_{x_2 \in X_2} x_1 p(x_1, x_2) + \sum_{x_1 \in S_1} \sum_{x_2 \in X_2} x_2 p(x_1, x_2).$$

Note that if we sum the joint distribution over one variable, we get the distribution function for the other variable. In other words,

$$\sum_{x_1 \in X_1} p(x_1, x_2) = p(x_2).$$

We apply this to the above expression to complete the proof.

$$E(X_1 + X_2) = \sum_{x_1 \in X_1} \sum_{x_2 \in X_2} x_1 p(x_1, x_2) + \sum_{x_1 \in X_1} \sum_{x_2 \in X_2} x_2 p(x_1, x_2)$$
$$= \sum_{x_1 \in X_1} x_1 \sum_{x_2 \in X_2} p(x_1, x_2) + \sum_{x_2 \in X_2} x_2 \sum_{x_1 \in X_1} p(x_1, x_2)$$
$$= \sum_{x_1 \in X_1} x_1 p(x_1) + \sum_{x_2 \in X_2} x_2 p(x_2) = E(X_1) + E(X_2).$$

(iii) We use the same assumptions as in (ii). However, we make the additional assumption that X_1 and X_2 are independent. Since X_1 and X_2 are independent, we have that $p(x_1, x_2) = p(x_1)p(x_2)$. Therefore,

$$E(X_1 X_2) = \sum_{x_1 \in X_1} \sum_{x_2 \in X_2} x_1 x_2 p(x_1, x_2)$$
$$= \sum_{x_1 \in X_1} \sum_{x_2 \in X_2} x_1 x_2 p(x_1)p(x_2) = \sum_{x_1 \in X_1} x_1 p(x_1) \sum_{x_2 \in X_2} x_2 p(x_2)$$
$$= E(X_1)E(X_2).$$

■

We now proceed with an example of how Theorem 9.2.1 can be applied to find the expected value.

Example 9.2.2 Suppose that we are playing a game involving rolling a fair eight-sided die.

(i) Find the expected value of the die.
(ii) Suppose that we compute our score by multiplying the die roll by two and adding three. What is the expected value of our score?
(iii) To compute the number of spaces that we move, we roll the die twice and sum the two rolls. What is the expected value for the number of spaces that we move?
(iv) The size of the board is a $n \times k$ rectangle. The values of n and k are determined by rolling the die twice at the beginning of the game. Find the expected value for the number of spaces on the board.

Solution

(i) The sample space is $S = \{1, ..., 8\}$. Our random variable X is the value of the roll. For this reason, $X = S$. Since the die is fair, $p(x) = 1/8$ for all $x \in X$. Thus,

$$E(X) = \sum_{x \in X} xp(x) = \frac{1}{8}(1 + \cdots + 8) = 4.5.$$

(ii) The score is a random variable $Y = 2X + 3$. Thus, $E(Y) = 2E(X) + 3$ by Theorem 9.2.1. From (i), $E(X) = 4.5$. It follows that $E(Y) = 2(4.5) + 3 = 12$.
(iii) The sum of two rolls is a random variable $X_1 + X_2$, where X_i is the result of the roll of the ith die. Thus,

$$E(X_1 + X_2) = E(X_1) + E(X_2) = 4.5 + 4.5 = 9.$$

(iv) The number of spaces on the board is nk. The values of n and k are random variables obtained by rolling a fair eight-sided die. So, we replace n by X_1 and we replace k by X_2. It follows that

$$E(X_1 X_2) = E(X_1)E(X_2) = 4.5^2 = 20.25.$$

□

We now define the *variance* of a random variable. The variance of a random variable X, denoted $V(X)$, is defined as

$$V(X) = E((X - \mu)^2),$$

where $\mu = E(X)$. The term $(X - \mu)^2$ is the square distance between X and the mean. Thus, $V(X)$ is a measure of the spread or deviation of the random variable. As an example, consider the variance of a fair six-sided die. From above, we have that $E(X) = 3.5$. Thus,

$$
\begin{aligned}
V(X) &= E((X-\mu)^2) = \sum_{x \in X}(x-\mu)^2 p(x) = \frac{1}{6}\sum_{x \in X}(x-3.5)^2 \\
&= \frac{1}{6}\big((1-3.5)^2 + (2-3.5)^2 + (3-3.5)^2 + (4-3.5)^2 \\
&\quad + (5-3.5)^2 + (6-3.5)^2\big) \\
&= \frac{1}{6}(6.25 + 2.25 + .25 + .25 + 2.25 + 6.25) = \frac{17.5}{6} \approx 2.9167.
\end{aligned}
$$

Again, it is valuable to find properties of the variance.

Theorem 9.2.3 *(Properties of the Variance) Suppose that X and Y are independent random variables.*

(i) $V(X) = E(X^2) - \mu^2$.
(ii) For $a, b \in \mathbb{R}$, $V(aX + b) = a^2 V(X)$.
(iii) $V(X + Y) = V(X) + V(Y)$.

Proof

(i) This involves a simple computation using Theorem 9.2.1.

$$
\begin{aligned}
V(X) &= E((X-\mu)^2) = E(X^2 - 2\mu X + \mu^2) \\
&= E(X^2) - 2\mu E(X) + E(\mu^2) \\
&= E(X^2) - 2\mu^2 + \mu^2 = E(X^2) - \mu^2.
\end{aligned}
$$

(ii) This follows in a similar manner to (i). Note that the expected value of $aX + b$ is $a\mu + b$ by Theorem 9.2.1.

$$
\begin{aligned}
V(aX+b) &= E((aX + b - (a\mu + b))^2) \\
&= E((aX+b)^2 - 2(a\mu + b)(aX + b) + (a\mu + b)^2) \\
&= E((aX+b)^2) - 2(a\mu + b)E(aX + b) + (a\mu + b)^2 \\
&= E(a^2X^2 + 2abX + b^2) - 2(a\mu + b)^2 + (a\mu + b)^2 \\
&= a^2E(X^2) + 2ab\mu + b^2 - a^2\mu^2 - 2ab\mu - b^2 \\
&= a^2E(X)^2 - a^2\mu^2 = a^2(E(X^2) - \mu^2) = a^2V(X).
\end{aligned}
$$

(iii) Suppose that the expected value of X is μ_x and that the expected value of Y is μ_y. By Theorem 9.2.1, $E(X + Y) = \mu_x + \mu_y$. Our computation will be similar to (i) and (ii).

$$
\begin{aligned}
V(X+Y) &= E(((X+Y) - (\mu_x + \mu_y))^2) \\
&= E((X+Y)^2 - 2(\mu_x + \mu_y)(X + Y) + (\mu_x + \mu_y)^2)
\end{aligned}
$$

$$= E((X+Y)^2) - 2(\mu_x + \mu_y)^2 + (\mu_x + \mu_y)^2$$
$$= E(X^2 + 2XY + Y^2) - (\mu_x^2 + 2\mu_x\mu_y + \mu_y^2)$$
$$= E(X^2) + 2E(XY) + E(Y^2) - \mu_x^2 - 2\mu_x\mu_y - \mu_y^2.$$

Since X and Y are independent, we have that $E(XY) = E(X)E(Y) = \mu_x\mu_y$. From this, it follows that

$$V(X+Y) = E(X^2) + 2E(XY) + E(Y^2) - \mu_x^2 - 2\mu_x\mu_y - \mu_y^2$$
$$= E(X^2) + 2\mu_x\mu_y + E(Y^2) - \mu_x^2 - 2\mu_x\mu_y - \mu_y^2$$
$$= E(X^2) - \mu^2 + E(Y^2) - \mu_y^2 = V(X) + V(Y).$$

■

We now proceed with an example of how Theorem 9.2.3 can be used for computation.

Example 9.2.4 Suppose that we are playing a game involving rolling a fair six-sided die and a fair eight-sided die. Let X be the random variable for the value of the six-sided die. Let Y be the random variable for the value of the eight-sided die.

(i) Find $V(X)$ and $V(Y)$.
(ii) Determine $V(5Y+4)$.
(iii) Determine $V(X+Y)$.

Solution

(i) Above, we determined that $V(X) = 17.5/6$. Perhaps the most efficient way of determining the variance is the formula $V(Y) = E(Y^2) - (E(Y))^2$. In Example 9.2.2, we found that $E(Y) = 4.5$. Thus, it suffices to compute $E(Y^2)$. This can be done as follows:

$$E(Y^2) = \sum_{y \in Y} y^2 p(y)$$
$$= \frac{1}{8}(1^2 + \cdots + 8^2) = \frac{204}{8} = 25.5.$$

Applying this to Theorem 9.2.3 yields

$$V(Y) = E(Y^2) - (E(Y))^2 = 25.5 - 4.5^2 = 5.25.$$

(ii) By Theorem 9.2.3, we have

$$V(5Y+4) = 5^2 V(Y) = 25(5.25) = 131.25.$$

(iii) Applying Theorem 9.2.3 gives us

$$V(X+Y) = V(X) + V(Y) = 17.5/6 + 5.25 = 49/6.$$

□

Exercise 9.2.5 Suppose that we are rolling an unfair six-sided die with probabilities listed below.

x	1	2	3	4	5	6
$p(x)$	.2	.1	.25	.05	.3	.1

Find the expected value and the variance for the roll of the die.

Exercise 9.2.6 Suppose that we are playing a game involving rolling a fair ten-sided die.

(i) Find the expected value and variance of the die.
(ii) Suppose that we compute our score by multiplying the die roll by five and subtracting three. What is the expected value of our score? What is the variance?
(iii) To compute the number of spaces that we move, we roll the die twice and sum the two rolls. What is the expected value for the number of spaces that we move? What is the variance?
(iv) The size of the board is a $n \times k$ rectangle. The values of n and k are determined by rolling the die twice at the beginning of the game. Find the expected value for the number of spaces on the board. What is the variance?

Exercise 9.2.7 Suppose that we are playing a game involving rolling a fair twelve-sided die.

(i) Find the expected value and variance of the die.
(ii) Suppose that we compute our score by multiplying the die roll by seven and adding five. What is the expected value of our score? What is the variance?
(iii) To compute the number of spaces that we move, we roll the die twice and sum the two rolls. What is the expected value for the number of spaces that we move? What is the variance?
(iv) The size of the board is a $n \times k$ rectangle. The values of n and k are determined by rolling the die twice at the beginning of the game. Find the expected value for the number of spaces on the board. What is the variance?

Exercise 9.2.8 Suppose that we are rolling two unfair eight-sided dice with probabilities listed below

$Die \backslash Roll$	1	2	3	4	5	6	7	8
1	.1	.1	.1	.1	.1	.1	.2	.2
2	.4	0	.2	.05	.15	.1	0	.1

Let X be the random variable assigned to the value of the first die. Let Y be the random variable assigned to the value of the second die.

(i) Find $E(X)$ and $E(Y)$.
(ii) Find $E(4X + 3)$ and $E(5Y - 8)$.

(iii) Find $V(X)$ and $V(Y)$.
(iv) Find $V(4X + 3)$ and $V(5Y - 8)$.
(v) Find $E(X + Y)$ and $V(X + Y)$.

Exercise 9.2.9 Suppose that we are rolling two unfair seven-sided dice with probabilities listed below

Die\Roll	1	2	3	4	5	6	7
1	.1	.3	.1	.1	.1	.1	.2
2	.3	.15	.2	.1	.15	.1	0

Let X be the random variable assigned to the value of the first die. Let Y be the random variable assigned to the value of the second die.

(i) Find $E(X)$ and $E(Y)$.
(ii) Find $E(5X + 7)$ and $E(3Y - 2)$.
(iii) Find $V(X)$ and $V(Y)$.
(iv) Find $V(5X + 7)$ and $V(3Y - 2)$.
(v) Find $E(X + Y)$ and $V(X + Y)$.

9.3 The Binomial Distribution

In probability, there are several distributions that appear repeatedly. Provided that a random variable satisfies the conditions of the distribution, we can use the distribution to find the probability that a random variable achieves a value or a set of values. While there are many distributions that are discussed in a course on probability, we will focus on a handful that have a combinatorial flavor to them. The interested reader is referred to Morris DeGroot's *Probability and Statistics* [18] for a more comprehensive treatment.

We begin with one of the simplest distributions, namely the *Bernoulli distribution*. The Bernoulli distribution is used in situations or experiments in which the only possible outcomes are "success" and "failure." Such an experiment is often referred to as a *Bernoulli trial*. A canonical example of a Bernoulli trial is a coin toss, as the only possibilities (barring some exceedingly improbable outcomes) are heads and tails. Because there are only two outcomes for a Bernoulli random variable, the sample space for the random variable is $S = \{\text{success, failure}\}$. In this case, the random variable takes on the value '1' if the experiment was successful and '0' if the experiment failed. We assume that the probability that $x = 1$ is p and the probability that $x = 0$ is $1 - p$. In this case, we say that the random variable X has a Bernoulli distribution with parameter p. It is easy to see that the probability distribution function for a Bernoulli random variable is $p(x) = p^x(1 - p)^{1-x}$.

Theorem 9.3.1 *Suppose that X is a Bernoulli random variable with parameter p. The expected value of X is $E(X) = p$. The variance of X is $V(X) = p(1 - p)$.*

Proof We begin by computing the expected value.

$$E(X) = \sum_{x \in \{0,1\}} xp(x)$$
$$= 0(p^0(1-p)^1) + 1(p^1(1-p)^0) = p.$$

Similarly, we can compute the variance. To do this, we first compute $E(X^2)$ and apply Theorem 9.2.3.

$$E(X^2) = \sum_{x \in \{0,1\}} x^2 p(x)$$
$$= 0(p^0(1-p)^1) + 1(p^1(1-p)^0) = p.$$

Since $\mu = p$, it follows that

$$V(X) = E(X^2) - \mu^2 = p - p^2 = p(1-p).$$

■

Example 9.3.2 Suppose that X is a Bernoulli random variable with parameter p. Compute $E(X)$ and $V(X)$ for each of the following values of p: *(i)* $p = .5$; *(ii)* $p = .3$; *(iii)* $p = .6$.

Solution For each of the above values of p, we simply apply Theorem 9.3.1.

(i) For $p = .5$, we have that $E(X) = p = .5$ and $V(X) = p(1-p) = .5^2 = .25$.
(ii) Similarly, if $p = .3$, then $E(X) = .3$ and $V(X) = .3(.7) = .21$.
(iii) Finally, when $p = .6$ we have that $E(X) = .6$ and $V(X) = .6(.4) = .24$. □

Our next distribution follows directly from the Bernoulli distribution. Suppose that we conduct n independent Bernoulli trials and count the number of successes. In this case, the random variable is the number of successes. A random variable that counts the number of successes in n independent Bernoulli trials is said to have a *binomial distribution* with parameters n and p. For a binomial distribution, we must have at least zero successes. Clearly, we can have no more than n successes. For this reason, random variable for a binomial distribution takes on values from the set $\{0, 1, ..., n\}$. Before proceeding with an example, we need to develop the probability distribution function, expected value, and variance for a random variable with a binomial distribution.

Theorem 9.3.3 *Suppose that X is a binomial random variable with parameters n and p. The probability that $X = x$ is given by $p(x) = \binom{n}{x} p^x (1-p)^{n-x}$. The expected value for X is $E(X) = np$. The variance of X is $V(X) = np(1-p)$.*

Proof We begin by developing the probability distribution. There are $\binom{n}{x}$ ways to select x of the n trials to be successful. Each of these x trials has probability of success p. Therefore, the probability that a fixed set of x trials are all successful is p^x. The remaining $n - x$ trials must all be failures, each with probability $1 - p$. It follows that the probability that these $n-x$ trials are all failures is $(1-p)^{n-x}$. Thus, the probability of exactly x successes is given by $p(x)=\binom{n}{x}p^x(1-p)^{n-x}$.

Suppose that X is binomial random variable. By definition, we can write $X = X_1+\cdots+X_n$, where the X_i are independent Bernoulli random variables. By Theorem 9.3.1, $E(X_i)=p$ and $V(X_i)=p(1-p)$ for $i=1,\ldots,n$. Theorem 9.2.1 yields

$$E(X)=E(X_1+\cdots+X_n)=E(X_1)+\cdots+E(X_n)$$
$$=\underbrace{p+\cdots+p}_{n\text{ times}}=np.$$

Since the X_i are independent, we can apply Theorem 9.2.3 to obtain the variance:

$$V(X)=V(X_1+\cdots+X_n)=V(X_1)+\cdots+V(X_n)$$
$$=\underbrace{p(1-p)+\cdots+p(1-p)}_{n\text{ times}}=np(1-p).$$

■

Example 9.3.4 Suppose that X is a binomial random variable with parameters $n = 16$ and $p=.6$.

(i) Determine $E(X)$ and $V(X)$.
(ii) Find the probability of exactly four successes.
(iii) Determine the probability of between seven and ten successes.
(iv) What is the probability of at least one success?

Solution

(i) Using Theorem 9.3.3, we have that

$$E(X)=np=16(.6)=9.6$$
$$\text{and } V(X)=np(1-p)=16(.6)(.4)=3.84.$$

(ii) Again, we apply Theorem 9.3.3 to obtain the probability of exactly four successess. This is given by

$$p(4)=\binom{n}{4}p^4(1-p)^{n-4}=\binom{16}{4}(.6^4)(.4^{12})\approx .003957.$$

(iii) Using the Addition Principle, it suffices to compute

$$p(7)+p(8)+p(9)+p(10)=\sum_{x=7}^{10}\binom{16}{x}(.6^x)(.4^{16-x})\approx .612841.$$

(iv) This can be obtained in a similar manner to (iii). However, the computation is easier if we observe that "at least one success" is the complement of "zero successes." So if $A =$ "at least one success," then

$$p(A) = 1 - p(A^c) = 1 - p(0) = 1 - \binom{16}{0}(.6^0)(.4^{12}) \approx .999983. \qquad \square$$

A distribution that is closely related to the binomial distribution is the *multinomial distribution*. In a multinomial distribution, there are n independent trials. Each trial has one of k possible outcomes. The ith outcome occurs with probability p_i, where $p_1 + \cdots + p_k = 1$. Here, X_i is a random variable that counts the number of times that outcome i has occurred. In this case, we are concerned with a *random vector*, $\overrightarrow{X} = (X_1, ..., X_k)$. Each of the X_i can be thought of as a binomial random variable. Therefore, it immediately follows from Theorem 9.3.3 that $E(X_i) = np_i$ and $V(X_i) = np_i(1 - p_i)$. Using a similar argument to Theorem 9.3.3, we have the probability distribution function for a multinomial random vector is

$$p(x_1, ..., x_k) = \binom{n}{x_1, ..., x_k} p_1^{x_1} ... p_k^{x_k}.$$

The proof of this is left as an exercise for the reader.

Example 9.3.5 Suppose that the probabilities associated with a multinomial distribution has probabilities $p_1 = .1$, $p_2 = .3$, $p_3 = .35$, and $p_4 = .25$. Find the probability that after eight trials, we have that $X_1 = 2$, $X_2 = 2$, $X_3 = 2$, and $X_4 = 2$.

Solution We apply the above formula to find the answer:

$$p(2, 2, 2, 2) = \binom{8}{2, 2, 2, 2}(.1^2)(.3^2)(.35^2)(.25^2) \approx .138915. \qquad \square$$

Exercise 9.3.6 Suppose that X is a Bernoulli random variable with parameter p. Compute $E(X)$ and $V(X)$ for each of the following values of p: *(i)* $p = .75$; *(ii)* $p = .45$; *(iii)* $p = .1$.

Exercise 9.3.7 Prove that $\sum_{x=0}^{n} \binom{n}{x} p^x (1 - p)^{n-x} = 1$.

Exercise 9.3.8 Suppose that X is a binomial random variable with parameters $n = 12$ and $p = .3$.

(i) Determine $E(X)$ and $V(X)$.
(ii) Find the probability of exactly three successes.
(iii) Determine the probability of between five and eight successes.

(iv) What is the probability of at least one success?

Exercise 9.3.9 Suppose that X is a binomial random variable with parameters $n = 18$ and $p = .7$.

(i) Determine $E(X)$ and $V(X)$.
(ii) Find the probability of exactly three successes.
(iii) Determine the probability of between five and eight successes.
(iv) What is the probability of at least one success?

Exercise 9.3.10 Suppose that $\overrightarrow{X}$ is a multinomial random vector. Prove that the probability that $\overrightarrow{X} = (x_1, ..., x_k)$ is given by

$$p(x_1, ..., x_k) = \binom{n}{x_1, ..., x_k} p_1^{x_1} ... p_k^{x_k}.$$

Exercise 9.3.11 Prove that

$$\sum_{x_1,...,x_k \geq 0} \binom{n}{x_1, ..., x_k} p_1^{x_1} ... p_k^{x_k} = 1.$$

Exercise 9.3.12 Suppose that the probabilities associated with a multinomial distribution has probabilities $p_1 = .2$, $p_2 = .4$, $p_3 = .15$, and $p_4 = .25$. Find the probability that after ten trials, we have that $X_1 = 3$, $X_2 = 5$, $X_3 = 0$, and $X_4 = 2$.

Exercise 9.3.13 Suppose that the probabilities associated with a multinomial distribution has probabilities $p_1 = .1$, $p_2 = .25$, $p_3 = .05$, $p_4 = .2$, and $p_5 = .4$. Find the probability that after 15 trials, we have that $X_1 = 2$, $X_2 = 3$, $X_3 = 1$, $X_4 = 5$, and $X_5 = 4$.

9.4 The Geometric Distribution

Suppose that we are allowed to conduct infinitely many independent Bernoulli trials, each with parameter p. We stop our experiment as soon as we have our first success. Let the random variable X denote the number of failures before our first success. In this case, the random variable X is said to have a *geometric distribution* with parameter p. The sample space for a geometric random variable is $\mathbb{N}$. A geometric random variable can be thought of as the number of times we have to flip a coin before we see the first head. We begin by developing the basic properties of the geometric distribution.

Theorem 9.4.1 *Suppose that X is a geometric random variable with parameter p. The probability that $X = x$ is given by $p(x) = p(1 - p)^x$. The expected value of X is given by $E(X) = \frac{1-p}{p}$. The variance of X is given by $V(X) = \frac{1-p}{p^2}$.*

Proof In order to have x failures followed by a single success, we must have a fixed string of length $x+1$. The first x elements of the string are 0 and the final element is 1. Each of these x failures occurs with probability $1-p$. Thus the probability that the first x trials are all failures is $(1-p)^x$. The probability that the final trial is a success is p. Thus, we have that

$$p(x) = p(1-p)^x.$$

We now compute the expected value of X as follows:

$$\begin{aligned} E(X) &= \sum_{x\in\mathbb{N}} xp(1-p)^x = p\sum_{x=0}^{\infty} x(1-p)^x \\ &= p\left(\sum_{x=0}^{\infty}(x+1)(1-p)^x - \sum_{x=0}^{\infty}(1-p)^x\right). \end{aligned}$$

Note that

$$\sum_{x=0}^{\infty}(1-p)^x = \frac{1}{1-(1-p)} = \frac{1}{p}$$

as it is a geometric series. Similarly, Example 5.2.2 implies that

$$\sum_{x=0}^{\infty}(x+1)(1-p)^x = \frac{1}{(1-(1-p))^2} = \frac{1}{p^2}.$$

Applying both of these sums to the above equation yields

$$\begin{aligned} E(X) &= p\left(\sum_{x=0}^{\infty}(x+1)(1-p)^x - \sum_{x=0}^{\infty}(1-p)^x\right) \\ &= p\left(\frac{1}{p^2} - \frac{1}{p}\right) = \frac{1-p}{p}. \end{aligned}$$

To compute the variance, we begin by computing $E(X^2)$. This can be done as follows:

$$\begin{aligned} E(X^2) &= \sum_{x=0}^{\infty} x^2 p(1-p)^x \\ &= p\left(\sum_{x=0}^{\infty}(x^2+3x+2)(1-p)^x - 3\sum_{x=0}^{\infty}(x+1)(1-p)^x + \sum_{x=0}^{\infty}(1-p)^x\right) \\ &= p\left(\sum_{x=0}^{\infty}(x+1)(x+2)(1-p)^x - 3\sum_{x=0}^{\infty}(x+1)(1-p)^x + \sum_{x=0}^{\infty}(1-p)^x\right). \end{aligned}$$

We have seen how to deal with the last two sums when we computed $E(X)$. As for the first sum, Exercise 5.2.8 implies that

$$\sum_{x=0}^{\infty}(x+1)(x+2)(1-p)^x = \frac{2}{(1-(1-p))^3} = \frac{2}{p^3}.$$

Applying these results to the above expression yields

$$E(X^2) = p\left(\sum_{x=0}^{\infty}(x+1)(x+2)(1-p)^x - 3\sum_{x=0}^{\infty}(x+1)(1-p)^x + \sum_{x=0}^{\infty}(1-p)^x\right)$$
$$= p\left(\frac{2}{p^3} - \frac{3}{p^2} + \frac{1}{p}\right) = \frac{2-3p+p^2}{p^2}.$$

Thus,

$$V(X) = E(X^2) - \mu^2$$
$$= \frac{2-3p+p^2}{p^2} - \frac{(1-p)^2}{p^2} = \frac{1-p}{p^2}.$$

■

Example 9.4.2 Suppose that X is a geometric random variable with parameter $p = .3$.

(i) Find $E(X)$ and $V(X)$.
(ii) Find the probability that $3 \leq x \leq 5$.
(iii) Find the probability that $x \geq 3$.

Solution We apply Theorem 9.4.1 to obtain the desired results.

(i) Note that $E(X) = \frac{1-.3}{.3} = \frac{7}{3}$ and $V(X) = \frac{1-.3}{.3^2} = \frac{70}{9}$.
(ii) By the Addition Principle, it suffices to compute

$$p(3) + p(4) + p(5) = .3(1-.3)^3 + .3(1-.3)^4 + .3(1-.3)^5 = .225351.$$

(iii) Let A = "at least three failures are needed before the first success." It is easier to compute the probability of A^c = "no more than two failures are needed before the first success." Thus,

$$p(A) = 1 - p(A^c) = 1 - (p(0) + p(1) + p(2))$$
$$= 1 - (.3(1-.3)^0 + .3(1-.3)^1 + .3(1-.3)^2) = .343.$$

□

Our next distribution is very closely related to the geometric distribution. Again, suppose that we are allowed infinitely many independent Bernoulli trials each with parameter p. Here, our random variable counts the number of failures that occur before we observe n successes. Such a random variable is said to have a *negative binomial distribution* with parameters n and p. The Bernoulli trials are independent. Thus, as soon as we encounter a success, we begin counting the failures again until we are successful again. For this reason, if X is negative binomial random variable, then $X = X_1 + \cdots + X_n$, where the X_i are independent geometric random variables. This observation will be instrumental in proving the following theorem.

Theorem 9.4.3 *Suppose that X is a negative binomial random variable with parameters n and p. The probability that $X = x$ is given by $p(x) = \binom{x+n-1}{n-1}p^n(1-p)^x$. The expected value of X is $E(X) = \frac{n(1-p)}{p}$. The variance of X is $V(X) = \frac{n(1-p)}{p^2}$.*

Proof By Stars and Bars, there are $\binom{x+n-1}{n-1}$ ways to arrange x 0's and n 1's such that the last symbol is a 1 (see also Theorem 4.2.6). The probability that the x selected trials are failures is $(1-p)^x$. The probability that the n selected trials are successes is p^n. Thus, the probability that we have x failures before n successes is given by $p(x) = \binom{x+n-1}{n-1}p^n(1-p)^x$.

If X has a negative binomial distribution, then $X = X_1 + \cdots + X_n$, where the X_i are independent geometric random variables. Since $E(X_i) = \frac{1-p}{p}$ and $V(X_i) = \frac{1-p}{p^2}$ it follows from Theorems 9.2.1 and 9.2.3 that $E(X) = \frac{n(1-p)}{p}$ and $V(X) = \frac{n(1-p)}{p^2}$. ■

Example 9.4.4 Suppose that X is a negative binomial distribution with parameters $n = 12$ and $p = .7$.

(i) Find $E(X)$ and $V(X)$.
(ii) Find the probability of exactly 14 failures before 12 successes.
(iii) Find the probability that we have between 9 and 11 failures before we have 12 successes.
(iv) Find the probability that we have at least one failure before we have 12 successes.

Solution

(i) Using Theorem 9.4.3, we have that $E(X) = \frac{n(1-p)}{p} = \frac{12(1-.7)}{.7} = \frac{36}{7}$ and $V(X) = \frac{n(1-p)}{p^2} = \frac{12(1-.7)}{.7^2} = \frac{360}{49}$.

(ii) Again, we apply Theorem 9.4.3 to yield the desired result:

$$p(14) = \binom{14+12-1}{12-1}.7^{12}(1-.7)^{14} \approx .002951.$$

(iii) By the Addition Principle, it suffices to compute $p(9) + p(10) + p(11)$:

$$\binom{20}{11}.7^{12}(1-.7)^9 + \binom{21}{11}.7^{12}(1-.7)^{10} + \binom{22}{11}.7^{12}(1-.7)^{11} \approx .091883.$$

(iv) If A ="at least one failure," then it is easier to compute the probability of A^c ="no failures." Thus,

$$p(A) = 1 - p(A^c) = 1 - p(0) = 1 - \binom{0+12-1}{12-1} .7^{12}(1-.7)^0 \approx .986159.$$

□

Exercise 9.4.5 Prove that $\sum_{x=0}^{\infty} p(1-p)^x = 1$.

Exercise 9.4.6 Suppose that X is a geometric random variable with parameter $p = .45$.

(i) Find $E(X)$ and $V(X)$.
(ii) Find the probability that $4 \leq x \leq 7$.
(iii) Find the probability that $x \geq 2$.

Exercise 9.4.7 Suppose that X is a geometric random variable with parameter $p = .6$.

(i) Find $E(X)$ and $V(X)$.
(ii) Find the probability that $6 \leq x \leq 10$.
(iii) Find the probability that $x \geq 3$.

Exercise 9.4.8 Prove that $\sum_{x=0}^{\infty} \binom{x+n-1}{n-1} p^n(1-p)^x = 1$.

Exercise 9.4.9 Suppose that X is a negative binomial distribution with parameters $n = 15$ and $p = .4$.

(i) Find $E(X)$ and $V(X)$.
(ii) Find the probability of exactly 11 failures before 15 successes.
(iii) Find the probability that we have between 5 and 8 failures before we have 15 successes.
(iv) Find the probability that we have at least 1 failure before we have 15 successes.

Exercise 9.4.10 Suppose that X is a negative binomial distribution with parameters $n = 18$ and $p = .5$.

(i) Find $E(X)$ and $V(X)$.
(ii) Find the probability of exactly 9 failures before 18 successes.
(iii) Find the probability that we have between 7 and 10 failures before we have 18 successes.
(iv) Find the probability that we have at least 1 failure before we have 18 successes.

9.5 The Poisson Distribution

In many applications, such as the service industry, we can expect that a constant number of individuals will enter into the system at a particular time. For example, at a bank we could expect that the same number of customers enter the bank between 2 and 3 P.M. on Tuesdays, regardless of the individual week. Suppose that we take our particular time interval and partition it into n subintervals of equal width. For each of these subintervals, we can assume that there is an equal probability of success (in this case, a success is a customer entering the bank). We assume that the successes are independent. Thus, if X is the random variable that counts the number of successes on this interval, then it is essentially a binomial random variable with parameters n and p. However, when we partition our interval, multiple successes may occur in the same subinterval. Returning to our example with the bank, we might partition the interval into ten minute segments. In each of these ten minute segments, multiple customers may enter into the bank. However, if we partition our interval into small enough subintervals, then we can assume that at most one person can enter into the bank during that time.

Since we want each subinterval to be as small as possible, we will take as many subintervals as possible. In other words, we will assume that $n \to \infty$. Logically, as the length of the subintervals decreases, the probability that a success occurs in that interval will approach zero. However, our expected number of successes in the interval should remain constant. For this reason, we set $\mu = np$ as in the binomial distribution. We then replace $p = \frac{\mu}{n}$ in binomial probability distribution function. Now consider the following limit:

$$\begin{aligned}\lim_{n\to\infty}\binom{n}{x}p^x(1-p)^{n-x} &= \lim_{n\to\infty}\binom{n}{x}\left(\frac{\mu}{n}\right)^x\left(1-\frac{\mu}{n}\right)^{n-x}\\ &= \lim_{n\to\infty}\left(\frac{n(n-1)...(n-x+1)}{x!}\right)\left(\frac{\mu^x}{n^x}\right)\left(1-\frac{\mu}{n}\right)^n\left(1-\frac{\mu}{n}\right)^{-x}\\ &= \lim_{n\to\infty}\left(\frac{(n-1)...(n-x+1)}{n^{x-1}}\right)\left(\frac{\mu^x}{x!}\right)\left(1-\frac{\mu}{n}\right)^n\left(1-\frac{\mu}{n}\right)^{-x}.\end{aligned}$$

We now examine the individual terms in the above limit. Note that:

$$\lim_{n\to\infty}\left(1-\frac{\mu}{n}\right)^{-x} = 1.$$

Similarly,

$$\lim_{n\to\infty}\frac{(n-1)...(n-x+1)}{n^{x-1}} = 1.$$

As shown in an elementary calculus class,

$$\lim_{n\to\infty}\left(1-\frac{\mu}{n}\right)^n = e^{-\mu}.$$

Thus,

$$\lim_{n\to\infty}\binom{n}{x}p^x(1-p)^{n-x} = \frac{\mu^x}{x!}e^{-\mu}.$$

Suppose that X is a random variable that counts the number of successes in an interval, where the expected number of successes in the interval is μ. In this case, X has a *Poisson distribution* with parameter μ.

Theorem 9.5.1 *Suppose that X is a Poisson random variable with parameter μ. The probability that $X = x$ is given by $p(x) = \frac{\mu^x}{x!}e^{-\mu}$. The expected value of X is $E(X) = \mu$. The variance of X is $V(X) = \mu$.*

Proof The probability distribution function follows from the above comments. Based on the above definitions, we anticipate that $E(X) = \mu$. However, we can confirm this with direct computation as follows:

$$E(X) = \sum_{x=0}^{\infty} x\frac{\mu^x}{x!}e^{-\mu} = \sum_{x=1}^{\infty} x\frac{\mu^x}{x!}e^{-\mu}$$

$$= \mu e^{-\mu}\sum_{x=1}^{\infty}\frac{\mu^{x-1}}{(x-1)!} = \mu e^{-\mu}e^{\mu} = \mu.$$

As usual, we will compute the variance by first computing $E(X^2)$:

$$E(X^2) = \sum_{x=0}^{\infty} x^2\frac{\mu^x}{x!}e^{-\mu} = \mu e^{-\mu}\sum_{x=1}^{\infty} x\frac{\mu^{x-1}}{(x-1)!}$$

$$= \mu e^{-\mu}\left(\sum_{x=1}^{\infty}(x-1)\frac{\mu^{x-1}}{(x-1)!} + \sum_{x=1}^{\infty}\frac{\mu^{x-1}}{(x-1)!}\right)$$

$$= \mu e^{-\mu}\left(\sum_{x=2}^{\infty}(x-1)\frac{\mu^{x-1}}{(x-1)!} + e^{\mu}\right)$$

$$= \mu e^{-\mu}\left(\mu\sum_{x=2}^{\infty}\frac{\mu^{x-2}}{(x-2)!} + e^{\mu}\right)$$

$$= \mu e^{-\mu}\left(\mu e^{\mu} + e^{\mu}\right) = \mu^2 + \mu.$$

From this it follows that

$$V(X) = E(X^2) - (E(X))^2 = (\mu^2 + \mu) - \mu^2 = \mu.$$

■

Example 9.5.2 Suppose that X is a Poisson random variable with parameter $\mu = 2.1$.

(i) Find the probability that $2 \leq X \leq 4$.
(ii) Find the probability that $X \geq 1$.
(iii) Find the probability that $X \leq 5$.

Solution

(i) By the Addition Principle, it suffices to compute $p(2) + p(3) + p(4)$:

$$p(2) + p(3) + p(4) = \frac{2.1^2}{2!}e^{-2.1} + \frac{2.1^3}{3!}e^{-2.1} + \frac{2.1^4}{4!}e^{-2.1} \approx .558259.$$

(ii) As usual, if A ="at least one success," then it is easier to compute the probability of A^c. Thus,

$$\begin{aligned} p(A) &= 1 - p(A^c) = 1 - p(0) \\ &= 1 - \frac{2.1^0}{0!}e^{-2.1} \approx .877544. \end{aligned}$$

(iii) We have already computed the probability that $2 \leq X \leq 4$ in (i). Therefore, we need only compute $p(0) + p(1)$:

$$p(0) + p(1) = \frac{2.1^0}{0!}e^{-2.1} + \frac{2.1^1}{1!}e^{-2.1} \approx .379615.$$

Adding this result with (i), we have that the probability that $X \leq 5$ is approximately .937874. □

Exercise 9.5.3 Prove that $\lim_{n\to\infty}\left(1 - \frac{\mu}{n}\right)^n = e^{-\mu}$.

Exercise 9.5.4 Prove that $\sum_{x=0}^{\infty} \frac{\mu^x}{x!}e^{-\mu} = 1$.

Exercise 9.5.5 Suppose that X is a Poisson random variable with parameter $\mu = 3.4$.

(i) Find the probability that $3 \leq X \leq 6$.
(ii) Find the probability that $X \geq 1$.
(iii) Find the probability that $X \leq 7$.

Exercise 9.5.6 Suppose that X is a Poisson random variable with parameter $\mu = 5.2$.

(i) Find the probability that $6 \leq X \leq 10$.
(ii) Find the probability that $X \geq 1$.
(iii) Find the probability that $X \leq 4$.

9.6 The Hypergeometric Distribution

Suppose that we have a box containing n_1 white balls and n_2 black balls. At random, we are to select k balls from this box without replacement. Let X be the random variable that counts the number of white balls that we have selected. We want to know the probability that exactly x white balls are selected. To compute this probability, it is convenient to think of the balls as being labeled. The total number of ways to select k balls from the box is given by $\binom{n_1+n_2}{k}$. We now count the number of ways to select exactly x white balls from the box. This can be done as follows:

(i) Select x of the n_1 white balls. There are $\binom{n_1}{x}$ ways to do this.
(ii) Select $k-x$ of the n_2 black balls. There are $\binom{n_2}{k-x}$ ways to do this.

By the Multiplication Principle, the number of ways of selecting exactly x white balls is $\binom{n_1}{x}\binom{n_2}{k-x}$. From this, it follows that the probability of selecting exactly x white balls is given by

$$p(x) = \frac{\begin{pmatrix} n_1 \\ x \end{pmatrix}\begin{pmatrix} n_2 \\ k-x \end{pmatrix}}{\begin{pmatrix} n_1+n_2 \\ k \end{pmatrix}}.$$

Suppose that X is the random variable that counts the number of white balls selected when picking k balls at random from a box containing n_1 white balls and n_2 black balls. In this case, X has a *hypergeometric distribution* with parameters k, n_1, and n_2. Clearly, we have to select at least zero white balls. Similarly, if $k > n_2$, then the Pigeonhole Principle implies that we must select at least $k - n_2$ white balls. This $X \geq \max\{0, k-n_2\}$. Using an analogous argument, $X \leq \min\{k, n_1\}$. Thus, the random variable X takes on values from the set

$$\mathbb{Z} \cap [\max\{0, k-n_2\}, \min\{k, n_1\}].$$

However, for computational purposes, we may assume that $S = \{0, 1, ..., k\}$ because the other values of x will cause one of the binomial coefficients to be zero. We now develop additional properties of the hypergeometric distribution before proceeding with an example.

Theorem 9.6.1 *Suppose that X is a hypergeometric random variable with parameters k, n_1, and n_2. The probability that $X = x$ is given by*

$$p(x) = \frac{\begin{pmatrix} n_1 \\ x \end{pmatrix}\begin{pmatrix} n_2 \\ k-x \end{pmatrix}}{\begin{pmatrix} n_1+n_2 \\ k \end{pmatrix}}.$$

The expected value of X is $E(X) = \frac{kn_1}{n_1+n_2}$. The variance of X is

$$V(X) = \frac{kn_1n_2(n_1+n_2-k)}{(n_1+n_2)^2(n_1+n_2-1)}.$$

Proof The probability distribution function follows from the above comments. To compute the expected value and the variance, we will make ample use of Vandermonde's Sum. We begin with the expected value:

$$E(X) = \sum_{x=0}^{k} x \frac{\binom{n_1}{x}\binom{n_2}{k-x}}{\binom{n_1+n_2}{k}}$$

$$= \frac{1}{\binom{n_1+n_2}{k}} \sum_{x=1}^{k} x \binom{n_1}{x}\binom{n_2}{k-x}$$

$$= \frac{k}{(n_1+n_2)\binom{n_1+n_2-1}{k-1}} \sum_{x=1}^{k} x \frac{n_1}{x} \binom{n_1-1}{x-1}\binom{n_2}{k-x} \text{(see Exercise 3.4.6)}$$

$$= \frac{kn_1}{(n_1+n_2)\binom{n_1+n_2-1}{k-1}} \sum_{x=1}^{k} \binom{n_1-1}{x-1}\binom{n_2}{k-1-(x-1)}.$$

Let $y = x - 1$. Thus, when $x = 1$, $y = 0$ and when $x = k$, $y = k - 1$. Hence,

$$E(X) = \frac{kn_1}{(n_1+n_2)\binom{n_1+n_2-1}{k-1}} \sum_{y=0}^{k-1} \binom{n_1-1}{y}\binom{n_2}{k-1-y}$$

$$= \frac{kn_1}{(n_1+n_2)\binom{n_1+n_2-1}{k-1}} \binom{n_1+n_2-1}{k-1} \text{(Vandermonde' s Sum)}$$

$$= \frac{kn_1}{n_1+n_2}.$$

As usual, to compute the variance, we will first compute $E(X^2)$.

$$E(X^2) = \sum_{x=0}^{k} x^2 \frac{\binom{n_1}{x}\binom{n_2}{k-x}}{\binom{n_1+n_2}{k}}$$

$$= \sum_{x=1}^{k} x(x-1)\frac{\binom{n_1}{x}\binom{n_2}{k-x}}{\binom{n_1+n_2}{k}} + \sum_{x=1}^{k} x\frac{\binom{n_1}{x}\binom{n_2}{k-x}}{\binom{n_1+n_2}{k}}.$$

Notice that the second summation is simply the expected value. Thus, we simply need to compute the value of the first summation:

$$\sum_{x=1}^{k} x(x-1)\frac{\binom{n_1}{x}\binom{n_2}{k-x}}{\binom{n_1+n_2}{k}}$$

$$= \frac{kn_1(k-1)(n_1-1)}{(n_1+n_2)(n_1+n_2-1)\binom{n_1+n_2-2}{k-2}} \sum_{x=2}^{k}\binom{n_1-2}{x-2}\binom{n_2}{k-x} \text{(see Exercise 3.4.6)}$$

$$= \frac{kn_1(k-1)(n_1-1)}{(n_1+n_2)(n_1+n_2-1)\binom{n_1+n_2-2}{k-2}} \sum_{x=2}^{k}\binom{n_1-2}{x-2}\binom{n_2}{k-2-(x-2)}.$$

Again, we re-index the summation to take advantage of Vandermonde's Sum. Namely, we let $y = x - 2$ which implies that when $x = 2$, $y = 0$ and when $x = k$, $y = k - 2$. This yields

$$\sum_{x=1}^{k} x(x-1)\frac{\binom{n_1}{x}\binom{n_2}{k-x}}{\binom{n_1+n_2}{k}}$$

$$= \frac{kn_1(k-1)(n_1-1)}{(n_1+n_2)(n_1+n_2-1)\binom{n_1+n_2-2}{k-2}} \sum_{y=0}^{k-2} \binom{n_1-2}{x-2}\binom{n_2}{k-2-y}$$

$$= \frac{kn_1(k-1)(n_1-1)}{(n_1+n_2)(n_1+n_2-1)\binom{n_1+n_2-2}{k-2}} \binom{n_1+n_2-2}{k-2} \text{(Vandermonde' s Sum)}$$

$$= \frac{kn_1(k-1)(n_1-1)}{(n_1+n_2)(n_1+n_2-1)}.$$

Hence,

$$E(X^2) = \frac{kn_1}{n_1+n_2} + \frac{kn_1(k-1)(n_1-1)}{(n_1+n_2)(n_1+n_2-1)}.$$

We are now prepared to compute the variance.

$$\begin{aligned}
V(X) &= E(X^2) - (E(X))^2 \\
&= \frac{kn_1}{n_1+n_2} + \frac{kn_1(k-1)(n_1-1)}{(n_1+n_2)(n_1+n_2-1)} - \frac{k^2n_1^2}{(n_1+n_2)^2} \\
&= \frac{kn_1}{n_1+n_2}\left(1 + \frac{(n_1-1)(k-1)}{n_1+n_2-1} - \frac{kn_1}{n_1+n_2}\right) \\
&= \frac{kn_1}{(n_1+n_2)^2(n_1+n_2-1)}\left(n_1n_2 + n_2^2 - kn_2\right) \\
&= \frac{kn_1n_2(n_1+n_2-k)}{(n_1+n_2)^2(n_1+n_2-1)}.
\end{aligned}$$

■

Example 9.6.2 Suppose that we randomly draw 9 balls without replacement from a box containing 10 white balls and 15 black balls. Let X be the number of white balls selected.

(i) Find $E(X)$ and $V(X)$.
(ii) Find the probability that exactly 6 white balls are selected.
(iii) Find the probability that between 2 and 5 white balls are selected.
(iv) Find the probability that at least 1 white ball is selected.

Solution

(i) Here, we have a hypergeometric distribution with parameters $k = 9$, $n_1 = 10$, and $n_2 = 15$. Applying these specific values to Theorem 9.6.1 yields

$$E(X) = \frac{kn_1}{n_1 + n_2} = \frac{9(10)}{10 + 15} = \frac{18}{5} \quad \text{and}$$

$$\begin{aligned} V(X) &= \frac{kn_1 n_2(n_1 + n_2 - k)}{(n_1 + n_2)^2(n_1 + n_2 - 1)} \\ &= \left(\frac{9(10)(15)}{(10 + 25)^2}\right)\left(\frac{10 + 25 - 9}{10 + 25 - 1}\right) = \frac{7020}{833}. \end{aligned}$$

(ii) Again, we apply Theorem 9.6.1 to obtain the desired probability:

$$p(6) = \frac{\binom{10}{6}\binom{15}{3}}{\binom{25}{9}} = \frac{3822}{81719} \approx .04677.$$

(iii) By the Addition Principle, it suffices to compute:

$$p(2) + p(3) + p(4) + p(5)$$

$$\begin{aligned} &= \frac{1}{\binom{25}{9}}\left(\binom{10}{2}\binom{15}{7} + \binom{10}{3}\binom{15}{6} + \binom{10}{4}\binom{15}{5} + \binom{10}{5}\binom{15}{4}\right) \\ &= \frac{1864785}{2042975} \approx .912779. \end{aligned}$$

(iv) As usual, if A = "at least one white ball," then it is easier to compute $p(A^c) = p(0)$:

$$\begin{aligned} p(A) &= 1 - p(A^c) = 1 - p(0) \\ &= 1 - \frac{\binom{10}{0}\binom{15}{9}}{\binom{25}{9}} = 1 - \frac{5005}{2042975} \approx .99755. \end{aligned}$$

□

Exercise 9.6.3 Prove that

$$\sum_{x=0}^{k} \frac{\binom{n_1}{x}\binom{n_2}{k-x}}{\binom{n_1+n_2}{k}} = 1.$$

Exercise 9.6.4 Suppose that we randomly draw 11 balls without replacement from a box containing 15 white balls and 17 black balls. Let X be the number of white balls selected.

(i) Find $E(X)$ and $V(X)$.
(ii) Find the probability that exactly four white balls are selected.
(iii) Find the probability that between five and eight white balls are selected.
(iv) Find the probability that at least one white ball is selected.

Exercise 9.6.5 Suppose that we randomly draw 12 balls without replacement from a box containing 14 white balls and 10 black balls. Let X be the number of white balls selected.

(i) Find $E(X)$ and $V(X)$.
(ii) Find the probability that exactly eight white balls are selected.
(iii) Find the probability that between five and nine white balls are selected.
(iv) Find the probability that at least one white ball is selected.

Chapter 10
Application: Combinatorial Designs

10.1 Introduction

Informally, a *combinatorial design* is a way of selecting subsets such that certain conditions are satisfied. The goal in studying combinatorial designs is not to determine *how many ways* the conditions can be satisfied. Rather, the principal question is "Can the conditions be satisfied?" This being the case, it is often desirable to provide an explicit construction that satisfies the design. In this chapter, we will give only a brief treatment of this area. For more information, the interested reader is referred to *Design Theory* by Hughes and Piper [29] and *Combinatorial Designs* by W. D. Wallis [43].

We consider one of the earliest combinatorial designs in the next example.

Example 10.1.1 (Kirkman's schoolgirl problem) Suppose fifteen schoolgirls are to walk to school each day in a week. The girls are to walk three abreast. Each pair of girls can walk in the same row precisely once during the week How can this be accomplished?

Solution Label the girls with the elements of $\{0, 1, \ldots, 14\}$. One way of accomplishing this is the following:

Monday	0, 1, 2	3, 4, 5	6, 7, 8	9, 10, 11	12, 13, 14
Tuesday	0, 3, 10	1, 4, 9,	2, 7, 12	5, 6, 13	8, 11, 14
Wednesday	0, 4, 8	1, 5, 7	2, 10, 14	3, 11, 13	6, 9, 12
Thursday	0, 5, 12	1, 3, 14	2, 8, 9	4, 6, 11	7, 10, 13
Friday	0, 6, 14	1, 8, 13	2, 5, 11	3, 7, 9	4, 10, 12
Saturday	0, 7, 11	1, 6, 10	2, 4, 13	3, 8, 12	5, 9, 14
Sunday	0, 9, 13	1, 11, 12	2, 3, 6	4, 7, 14	5, 8, 10

□

The girls in this problem are traditionally referred to as *treatments* or *varieties*. The number of treatments in a design is denoted v. Hence there are 15 varieties in the

R. A. Beeler, *How to Count*, DOI 10.1007/978-3-319-13844-2_10

above example. The rows in which we arrange the girls are called *blocks*. The number of blocks in the design is denoted b. Thus $b = 35$ in the above example. Since any two girls walk together exactly once, we say that the *covalency* of any two treatments is one. In general, the covalency of x and y, denoted $\lambda_{x,y}$, is the number of blocks that contain both x and y. The notion of covalency is important if we are interested in the interactions between two treatments. For example, we may be interested in how each treatment compares with every other treatment. In general, $\lambda_{x,y}$ does not have to be a constant for all choices of x and y. If $\lambda_{x,y} = \lambda$ for all choices of x and y, then the design is *balanced*. This is the case with the above example. We note that every girl appears in exactly seven rows. In general, the *replication number* or *frequency* of a treatment is the number of blocks that contain that treatment.

In studying more general designs, some natural questions occur:

(i) Is order important in arranging the blocks? In other words, do we consider the blocks $0, 1, 2$ and $1, 2, 0$ to be the same? In our example, we have assumed that the internal order of the blocks is unimportant. This is the usual assumption when considering sets. This is evident in our listing the elements of each block in ascending numerical order. Further, this is the traditional interpretation of Kirkman's schoolgirl problem. If we do not consider the order to be important, then the design is called a *block design*.
(ii) Do all of the blocks contain the same number of treatments? If all of the blocks have the same size, then we denote this common size k. In Kirkman's schoolgirl problem, $k = 3$. For the rest of this chapter, we will assume that all blocks have the same size.
(iii) Does each treatment appear in the same number of blocks? If each treatment belongs to the same number of blocks, then the design is a *regular design*. The replication number in a regular design is denoted r. In Kirkman's schoolgirl problem, $r = 7$.
(iv) Is the covalency a constant for all pairs of treatments? In Kirkman's schoolgirl problem, the covalency is constant for every pair of girls. In this case, $\lambda = 1$.

We now present additional examples of combinatorial designs.

Example 10.1.2 Suppose that a farmer has four different crops (denoted 1, 2, 3, 4). He has the option of planting each crop in any of his four fields (North, South, East, West). Each field is divided into four identical sections each of which can hold one crop. The farmer wishes to determine their yield when using four different brands of fertilizer (A, B, C, and D). What is an efficient design of experiment for this problem?

In attempting to solve this problem, we might first assign put a single corp in each field and use a different fertilizer on each section. This is illustrated below:

	A	B	C	D
N	1	1	1	1
S	2	2	2	2
E	3	3	3	3
W	4	4	4	4

We would expect that the East field would get different sun exposure than the West field. Unfortunately, this does not account for the fact that different crops may have different requirements for sunlight and water. This would mean that crop 3 would consistently get different sun than the other crops. For this reason, this is not an efficient design.

As a second attempt, we might place a different crop in each section of the four fields. We then assign the same fertilizer to each crop. This is illustrated below.

	A	B	C	D
N	1	2	3	4
S	1	2	3	4
E	1	2	3	4
W	1	2	3	4

While each crop is placed in each of the fields, unfortunately each crop is given a specific fertilizer. For instance, crop 3 is given fertilizer C, while none of the other crops are given this fertilizer. Hence, this design does not account for the fact that different crops may react differently to different fertilizers. Further, this design does not consider the specific interaction between the fields and the fertilizer.

Thus, our goal should be to place each crop in each field and to use each fertilizer on each crop. The result will be a 4×4 matrix that has 1, 2, 3, and 4 appearing exactly once in each row and column. Such a design is referred to as a *Latin square*. To accomplish this, we can simply place the numbers 1, 2, 3, and 4 into the first row in any order. For the subsequent rows, we simply rotate this arrangement. An example of this is given below.

	A	B	C	D
N	1	2	3	4
S	2	3	4	1
E	3	4	1	2
W	4	1	2	3

A popular example of Latin squares is the logic puzzle *Sudoku*. Sudoku is based on a 9×9 Latin square with the additional restriction that each of the numbers

1, ...,9 appears exactly once in each of the 3×3 subgrids. For more information on Latin squares, the interested reader is referred to the survey by Colbourn and Dinitz [16]. The book *Taking Sudoku Seriously* by Rosenhouse and Taalman [36] has more information on the mathematics of Sudoku.

Exercise 10.1.3 Arrange the numbers $\{1, 2, 3, 4, 5\}$ into a 5×5 Latin square.

Exercise 10.1.4 The numbers 0,1, ...,20 are to be arranged into subsets of size three. Every pair of numbers is to be a subset together exactly once. How many blocks will be in this design?

Exercise 10.1.5 The numbers 0,1,...,39 are to be arranged into subsets of size four. Every pair of numbers is to be a subset together exactly once. How many blocks will be in this design?

Exercise 10.1.6 The numbers 0,1,...,9 are to be arranged into subsets of size three. Every pair of numbers is to be a subset together exactly twice. How many blocks will be in this design?

10.2 Block Designs

In this section, we will study *block designs*. In this case, the blocks are unordered sets of treatments. This is different from a Latin square in which the order of the treatments is important. For instance, in the crop example in the previous section, the order the treatments are placed does matter.

We will focus on regular designs. A regular design is a collection of k-sets from S such that every member of S belongs to exactly r of the sets. Hence a regular design can be characterized by four parameters:

(i) The number of treatments, denoted v;
(ii) The number of blocks, denoted b;
(iii) The size of each block, denoted k;
(iv) The number of blocks to which each treatment belongs (in other words, the replication number r).

For example, Kirkman's schoolgirl problem is an example of such a design. In this case $v = 15$, $b = 35$, $k = 3$, and $r = 7$. A natural question is whether these parameters are related. In the case of Kirkman's schoolgirl problem, we have that $35(3) = 15(7) = 105$. In other words, $bk = vr$. This fact holds in general.

Theorem 10.2.1 *In any regular design, $bk = vr$.*

Proof Consider the problem of counting the number of ordered pairs (x, y) such that treatment x belongs to block y. This can be counted in two different ways.

(i) Each treatment x belongs to exactly r distinct blocks. Since there are v such treatments, there are vr valid ordered pairs by the Multiplication Principle.

(ii) Each block y contains exactly k distinct varieties. There are b such blocks. Hence there are bk valid pairs by the Multiplication Principle.

Thus $bk = vr$.

■

This allows us to eliminate many possibilities for the parameters of a regular design. For example, there is no regular design with $v = 10$, $r = 1$, $b = 5$, and $k = 3$. Suppose that we have v, r, b, and k that satisfy $bk = vr$. The question of whether there exists a regular design satisfying these parameters is more difficult. In most cases, establishing the existence of such a design relies on direct construction. This is the case with our next example.

Example 10.2.2 Construct a regular design with the following parameters:

(i) $v = 10, r = 1, b = 5$, and $k = 2$;
(ii) $v = 5, r = 6, b = 10$, and $k = 3$.

Solution *(i)* With a replication number of one, each element is in exactly one block. Further, every block is size two. Therefore, a simple (and natural) way to construct this design is given below:

$$\{1,2\},\{3,4\},\{5,6\},\{7,8\},\{9,10\}.$$

(ii) Each of the five treatments is in exactly six blocks. Further, each of the ten blocks is of size three. One possibility for an acceptable design is:

$$\{1,2,3\},\{1,4,5\},\{1,2,4\},\{1,3,5\},\{1,2,5\},$$
$$\{1,3,4\},\{2,3,4\},\{2,3,5\},\{2,4,5\},\{3,4,5\}.$$

□

There are several special types designs. If all v treatments occur in a block, then that block is called *complete*. In the case of a regular designs, we have that $k = v$ and thus $b = r$ by Theorem 10.2.1. Such designs are called *complete designs*. Further, the construction of such a design is trivial because every treatment appears in every block. For this reason, we will concentrate our efforts on *incomplete designs*. In an incomplete design, there is at least one block that is not complete. If there are as many treatments as blocks (in other words, $b = v$), then the design is *symmetric*.

Suppose that we are designing an experiment to test the effect of various drugs. In this case, we are concerned with how the different drugs interact with each other in the system. Hence, it is desirable that every pair of treatments occur in the same number of blocks. Recall that such a design is called a *balanced design*. In such a design, we denote the common covalency by λ. This is sometimes referred to as the *index* of the design.

If a regular incomplete is also balanced, then we refer to it as a *balanced incomplete block design* or *BIBD* for short. Such a design is characterized by five parameters:

(i) The number of treatments, denoted v;

(ii) The number of blocks, denoted b;
(iii) The size of each block, denoted k;
(iv) The replication number r);
(v) The covalency of the design, λ.

It is common to characterize balanced incomplete block design by these five parameters. Hence a (v, b, r, k, λ)-BIBD is a balanced incomplete design with v treatments, b blocks, replication number r, block size k, and covalency λ. For example, the second design in Example 10.2.2 is a $(5, 10, 6, 3, 3)$-BIBD. Like with regular designs, these parameters are not independent.

Theorem 10.2.3 *In a (v, b, r, k, λ)-BIBD, $bk = vr$ and $r(k-1) = \lambda(v-1)$.*

Proof Since a balanced incomplete block design is a regular design, we have that $bk = vr$ by Theorem 10.2.1.

Consider the set of all blocks that contain a given treatment x. There are r blocks in this set. Because the design is balanced, every treatment other than x appears in exactly λ of these blocks. If we list all the entries of these blocks, then x is listed r times while every other entry is listed λ times. Hence the list contains $r + \lambda(v-1)$ entries. Similarly, each treatments occur in r of the blocks and each block will contain k treatments. Thus, the list will contain rk entries by the Multiplication Principle. From this it follows that:

$$\begin{aligned} rk &= r + \lambda(v-1) \\ \Rightarrow rk - r &= \lambda(v-1) \\ r(k-1) &= \lambda(v-1). \end{aligned}$$

■

Again, Theorem 10.2.3 allows us to show that certain sets of parameters are not admissible for a BIBD. For example, there is no $(12, 6, 3, 4, \lambda)$-BIBD for any value of λ, despite the fact that there is a regular design with $v = 12$, $b = 6$, $r = 2$, and $k = 4$. For more information on balanced incomplete block designs, the interested reader is referred to the article by Abel and Greig [1].

Exercise 10.2.4 Determine whether there is a regular design with each of the following sets of parameters. If there is an acceptable design, then give the explicit construction.

(i) $v = 13$, $r = 2$, $b = 2$, and $k = 7$;
(ii) $v = 5$, $r = 1$, $b = 3$, and $k = 4$;
(iii) $v = 14$, $r = 1$, $b = 2$, and $k = 7$;
(iv) $v = 6$, $r = 2$, $b = 3$, and $k = 4$.

Exercise 10.2.5

(i) Show that there is no $(15, 5, 4, 12, 3)$-BIBD.
(ii) Show that there is no $(10, 5, 2, 4, 2)$-BIBD.

Exercise 10.2.6 Construct a $(6, 10, 5, 3, 2)$-BIBD.

Exercise 10.2.7 Construct a $(7, 7, 3, 3, 1)$-BIBD.

10.3 Steiner Triple Systems

In this section, we consider a variation of Kirkman's schoolgirl problem. Suppose that we have v treatments that are to be arranged into blocks of size three such that each pair of treatments is in exactly one block. In other words, we have a design with covalency $\lambda = 1$ and $k = 3$. Such a design is called a *Steiner triple system.* These are systems are named in honor of Jacob Steiner [41] who solved the problem unaware of the earlier work by Kirkman [32]. A Steiner triple system with v treatments is denoted $STS(v)$.

A natural question is to classify the values of v for which a Steiner triple system $STS(v)$ exists. Without loss of generality, we assume that the treatments are the elements $0, 1, \ldots, v-1$. A natural first step is to determine the necessary conditions for such a design to exist.

Theorem 10.3.1 *An $STS(v)$ is a balanced regular design with $b = \frac{v(v-1)}{6}$ and $r = \frac{v-1}{2}$. For this reason, $v \equiv 1, 3 \pmod 6$ is a necessary condition for such a design to exist.*

Proof Note that there are $\binom{v}{2} = \frac{v(v-1)}{2}$ pairs of treatments. Each pair is in exactly one block, whereas there are three pairs in each block. Thus, there are $b = \frac{1}{3}\binom{v}{2} = \frac{v(v-1)}{6}$ blocks. Since the number of blocks is an integer, $v(v-1)$ must be divisible by six. Equivalently, $v \equiv 0, 1, 3 \pmod 6$ is necessary for the existence of a $STS(v)$.

Similarly, suppose that x is a treatment. The treatment x must share a block with each of the other $v-1$ treatments exactly once. Each block containing x will contain exactly two other treatments. Thus, x is in $\frac{v-1}{2}$ blocks. As x was chosen arbitrarily, it follows that $r = \frac{v-1}{2}$. It follows that $v \equiv 1 \pmod 2$ is necessary for the existence of a $STS(v)$. Note that if $v \equiv 1 \pmod 2$, then $v \not\equiv 0 \pmod 6$. Thus, $v \equiv 1, 3 \pmod 6$ is necessary.

■

In Theorem 10.3.1, necessary conditions for the existence of a $STS(v)$ are established. However, this does not answer the question of sufficiency. Namely, if v satisfies the conditions of Theorem 10.3.1, does there exist a $STS(v)$? Our treatment of this problem will follow that of "A Direct Method to Construct Triple Systems" by Hwang and Lin [30].

Our basic method will be to consider the treatments as the elements $0, 1, \ldots, v-1$. Each pair of elements must share a common block. Thus, if we think of treatments in a block as the vertices of a triangle, then the edges of this triangle are labeled with the *differences.* These differences are computed modulo v. Further, they are taken from the set of "positive residues" modulo v. In other words, they are taken from the set $\{1, \ldots, \lfloor v/2 \rfloor\}$. Our goal will be to represent each one of the differences in the set

$\{1, \ldots, \lfloor v/2 \rfloor\}$ exactly once in a "base block." These differences d_1, d_2, and d_3 must satisfy one of the following:

(i) $d_1 + d_2 + d_3 \equiv 0 \pmod{v}$ or
(ii) $d_1 + d_2 \equiv d_3 \pmod{v}$.

It is easy to verify that choosing a differences that satisfy one of the above conditions will guarantee that we can form a triangle with our base block. This base block is then "rotated" to give the complete triple system.

As a concrete example, consider forming a $STS(7)$. The set of positive differences modulo 7 is $\{1, 2, 3\}$. Conveniently, $1 + 2 = 3$ as in (ii). This gives us the required differences $d_1 = 1, d_2 = 2$, and $d_3 = 3$. Thus, we choose our base block to represent these differences. For example, suppose that we take the block with treatments 0, 1, and 3. We will represent this block as $[0, 1, 3]$. In general, $[x, y, z]$ will represent the block containing the treatments x, y, and z. The block $[0, 1, 3]$ will contain the differences $1 - 0 = 1 = d_1$, $3 - 1 = 2 = d_2$, and $3 - 0 = 3 = d_3$. When we "rotate" the block $[0, 1, 3]$, we add i (for $i = 0, 1, \ldots, 6$) to each component of the block, reducing modulo 7. This results in the blocks $[0, 1, 3]$, $[1, 2, 4]$, $[2, 3, 5]$, $[3, 4, 6]$, $[4, 5, 0]$, $[5, 6, 1]$, and $[6, 1, 2]$. This can be more compactly represented as $[i, i + 1, i + 3]$ for $i = 0, 1, \ldots, 6$. In this case, it is understood that the computations are being done modulo 7.

As an illustration of this method, suppose that we want to find the block in which 3 and 6 appear together. The difference between 3 and 6 is $3 = d_3$. This difference is between the third and first components of our block. Thus, if we let $i = 3$, then the required block is $[3, 4, 6]$. We now provide a second example of this method to construct $STS(13)$.

Example 10.3.2 Construct a $STS(13)$.

Solution Our treatments will be the numbers $0, 1, \ldots, 12$. The set of positive differences modulo 13 is $\{1, \ldots, 6\}$. Our goal is to partition this set into two parts of size three such that each of the parts satisfies the conditions above. One possible way of partitioning the elements is to use the difference sets $\{1, 3, 4\}$ (as $1 + 3 = 4$) and $\{2, 5, 6\}$ (as $2+5+6 = 13$). The first difference set generates the blocks $[i, i+1, i+4]$ for $i = 0, 1, \ldots, 12$. The second set of differences generates the blocks $[i, i+2, i+7]$ for $i = 0, 1, \ldots, 12$. Direct inspection confirms that these blocks give the required Steiner triple system. □

The case for $v \equiv 3 \pmod{6}$ is a bit different. For example, the set of positive differences modulo 9 is $\{1, 2, 3, 4\}$. This set cannot be partitioned into sets of size three. Hence, slightly different techniques are required. These technique are illustrated in Examples 10.3.3 and 10.3.4.

Example 10.3.3 Construct a $STS(9)$.

Solution To accomplish this, suppose that the treatments are $0, 1, \ldots, 7$, and ∞. For the most part, we will be working modulo 8. However, if x is a treatment and $x \neq \infty$, then the difference between x and ∞ is ∞. Thus, the set of positive differences is $\{1, 2, 3, 4, \infty\}$. It is important to note that with the difference 4, we can only rotate the corresponding block half way through the treatments. The reason for this is that the difference 4 is its own additive inverse modulo 8. Thus, if we do a complete rotation of a block, say $[0, 1, 4]$, then this would result in the block $[4, 5, 0]$. Hence, the pair of treatments 0 and 4 would be repeated. Thus, we partition our differences into the sets $\{1, 2, 3\}$ (similar to the previous examples) and $\{4, \infty\}$. The first partition generates the blocks $[i, i + 1, i + 3]$ for $i = 0, 1, \ldots, 7$. The second partition generates the blocks $[j, j + 4, \infty]$ for $j = 0, 1, 2, 3$. Direct inspection confirms that these blocks give the required Steiner triple system. □

The method in Example 10.3.3 (pulling off a point "at infinity") is more common in designs where v is even. This is because the difference $v/2$ can create problems, as we discussed in Example 10.3.3. In fact, this technique for constructing Steiner triple systems is unique to $v = 9$. Thus, we will use a slightly different strategy for the case where $v = 15$. The method used in Example 10.3.4 will be generalized for all $v \equiv 3 \pmod 6$ such that $v \neq 9$.

Example 10.3.4 Construct a $STS(15)$.

Solution Take the set of treatments to be $\{0, 1, \ldots, 14\}$. The set of positive differences is $\{1, 2, 3, 4, 5, 6, 7\}$. The difference 5 will be treated special. Namely, each block that has the difference 5, will have that difference three times. However, we only rotate that block one third of the way through the treatments. This results in the blocks $[0, 5, 10]$, $[1, 6, 11]$, $[2, 7, 12]$, $[3, 8, 13]$, and $[4, 9, 14]$. The remaining differences can be partitioned as $\{1, 3, 4\}$ and $\{2, 6, 7\}$ (visual inspection shows that these sets satisfy the conditions above). The first set of differences generates the blocks $[i, i+1, i+4]$ for $i = 0, 1, \ldots, 14$. The second set of differences generates the blocks $[i, i+2, i+8]$ for $i = 0, 1, \ldots, 14$. These blocks together with the blocks above give the required Steiner triple system. □

We now proceed with a general construction.

Theorem 10.3.5 *For $v \geq 3$ and $v \equiv 1, 3 \pmod 6$, there exists a $STS(v)$.*

Proof Note that the case where $v \equiv 1 \pmod 6$, can further be divided into the subcases $v \equiv 1 \pmod{24}$, $v \equiv 7 \pmod{24}$, $v \equiv 13 \pmod{24}$, and $v \equiv 19 \pmod{24}$. Similarly, the case where $v \equiv 3 \pmod 6$ can be further divided into the subcases $v \equiv 3 \pmod{24}$, $v \equiv 9 \pmod{24}$, $v \equiv 15 \pmod{24}$, and $v \equiv 21 \pmod{24}$. We will consider each of these case in turn. In all cases, computations on the components are assumed to be done modulo v.

Suppose that $v \equiv 1 \pmod{24}$. Thus, there exists $k \in \mathbb{N}$ such that $v = 24k + 1$. Since $v = 1$ is not possible, we can assume that $k \geq 1$. For $j = 0, 1, \ldots, k - 1$ and $i = 0, 1, \ldots, 24k$ we use the blocks $[i, i+2j+1, i+j+11k+1]$, $[i, i+2j+3k+1, i+j+9k+1]$, $[i, i+2j+3k+2, i+j+6k+2]$, and $[i, i+2k, i+8k+1]$. If $k \geq 2$, then we also use the blocks $[i, i+2\ell+2, i+8k+\ell+2]$ for $\ell = 0, 1, \ldots, k-2$ and $i = 0, 1, \ldots, 24k$.

Suppose that $v \equiv 3 \pmod{24}$. Thus, there exists $k \in \mathbb{N}$ such that $v = 24k + 3$. Since the case where $v = 3$ is trivial, we can assume that $k \geq 1$. For $\ell = 0, 1, \ldots, 8k$, we use the blocks $[\ell, \ell + 8k + 1, \ell + 16k + 2]$. In addition, we use the blocks $[i, i +$

$2j+1, i+j+11k+2]$, $[i, i+2j+2, i+j+8k+2]$, $[i, i+2j+3k+2, i+j+9k+2]$, and $[i, i+2j+3k+1, i+j+6k+1]$ for $j = 0, 1, \ldots, k-1$ and $i = 0, 1, \ldots, 24k+2$.

Suppose that $v \equiv 7 \pmod{24}$. Thus, there exists $k \in \mathbb{N}$ such that $v = 24k+7$. We use the blocks $[i, i+2k+1, i+8k+3]$ for $i = 0, 1, \ldots, 24k+6$. If $k \geq 1$, then we use the additional blocks $[i, i+2j+1, i+j+11k+4]$, $[i, i+2j+2, i+j+8k+4]$, $[i, i+2j+3k+3, i+j+9k+4]$, and $[i, i+2j+3k+2, i+j+6k+3]$ for $j = 0, 1, \ldots, k-1$ and $i = 0, 1, \ldots, 24+6$.

Suppose that $v \equiv 9 \pmod{24}$. Thus, there exists $k \in \mathbb{N}$ such that $v = 24k+9$. The case where $v = 9$ is done in Example 10.3.3. Hence, we can assume that $k \geq 1$. For $\ell = 0, 1, \ldots, 8k+2$, use the blocks $[\ell, \ell+8k+3, \ell+16k+6]$. For $k \geq 1$, we use the additional blocks $[i, i+2k-1, i+5k+2]$, $[i, i+3k, i+12k+3]$, $[i, i+3k+1, i+12k+5]$, $[i, i+2j+3k+2, i+j+9k+5]$, and $[i, i+2j+3k+5, i+j+6k+4]$ for $j = 0, 1, \ldots, k-1$ and $i = 0, 1, \ldots, 24k+8$. If $k \geq 2$, then we additionally use the blocks $[i, i+2j+1, i+j+11k+4]$ and $[i, i+2j+2, i+j+8k+4]$ for $j = 0, 1, \ldots, k-2$ and $i = 0, 1, \ldots, 24k+8$.

Suppose that $v \equiv 13 \pmod{24}$. Thus, there exists $k \in \mathbb{N}$ such that $v = 24k+13$. If $k \geq 0$, then we use the blocks $[i, i+2k+1, i+8k+4]$ and $[i, i+3k+2, i+12k+7]$ for $i = 0, 1, \ldots, 24k+12$. If $k \geq 1$, then we use the additional blocks $[i, i+2j+1, i+j+11k+6]$, $[i, i+2j+2, i+j+8k+5]$, $[i, i+2j+3k+4, i+j+9k+6]$, and $[i, i+2j+3k+3, i+j+6k+4]$ for $j = 0, 1, \ldots, k-1$ and $i = 0, 1, \ldots, 24k+12$.

Suppose that $v \equiv 15 \pmod{24}$. Thus, there exists $k \in \mathbb{N}$ such that $v = 24k+15$. For $\ell = 0, 1, \ldots, 8k+4$, use the blocks $[\ell, \ell+8k+5, \ell+16k+10]$. Additionally, we use the blocks $[i, i+3k+2, i+12k+8]$ and $[i, i+2j+3k+3, i+j+6k+4]$ for $j = 0, 1, \ldots, k$ and $i = 0, 1, \ldots, 24k+14$. If $k \geq 1$, then we use the additional blocks $[i, i+2j+1, i+j+11k+7]$, $[i, i+2j+2, i+j+8k+6]$, and $[i, i+2j+3k+4, i+j+9k+7]$ for $j = 0, 1, \ldots, k-1$ and $i = 0, 1, \ldots, 24k+14$.

Suppose that $v \equiv 19 \pmod{24}$. Thus, there exists $k \in \mathbb{N}$ such that $v = 24k+19$. If $k \geq 0$, then we use the blocks $[i, i+2k+1, i+8k+5]$, $[i, i+3k+2, i+12k+8]$, and $[i, i+3k+3, i+12k+10]$ for $i = 0, 1, \ldots, 24k+18$. If $k \geq 1$, then we use the additional blocks $[i, i+2j+1, i+j+11k+8]$, $[i, i+2j+2, i+j+8k+6]$, $[i, i+2j+3k+5, i+j+9k+8]$, and $[i, i+2j+3k+4, i+j+6k+5]$ for $j = 0, 1, \ldots, k-1$ and $i = 0, 1, \ldots, 24k+18$.

Suppose that $v \equiv 21 \pmod{24}$. Thus, there exists $k \in \mathbb{N}$ such that $v = 24k+21$. For $\ell = 0, 1, \ldots, 8k+6$, use the blocks $[\ell, \ell+8k+7, \ell+16k+14]$. Additionally, use the blocks $[i, i+2j+1, i+j+11k+10]$, $[i, i+2j+3k+3, i+j+9k+8]$, and $[i, i+2j+3k+4, i+j+6k+6]$ for $j = 0, 1, \ldots, k$ and $i = 0, 1, \ldots, 24k+20$. If $k \geq 1$, then use the additional blocks $[i, i+2j+2, i+j+8k+8]$ for $j = 0, 1, \ldots, k-1$ and $i = 0, 1, \ldots, 24k+20$.

■

To confirm that the above construction works, we need only confirm that all possible differences are accounted for in all eight of the above cases. We will illustrate this with the case where $v \equiv 1 \pmod{24}$. Confirming the remaining cases will be left as exercises to the reader. If $v = 24k+1$, where $k \geq 1$, then the possible differences are $1, \ldots, 12k$. The block $[i, i+2j+1, i+j+11k+1]$ contains the differences $2j+1$, $11k-j$, and $j+11k+1$ for $j = 0, 1, \ldots, k-1$. This accounts for all differences $10k+1, 10k+2, \ldots, 12k$ as well as the differences $1, 3, \ldots, 2k-1$. The block $[i, i+2j+3k+1, i+j+9k+1]$ contains differences $2j+3k+1, 6k-j$, and $j+9k+1$ for $j=0, 1, \ldots, k-1$. This accounts for the differences $5k+1, 5k+2, \ldots, 6k$, the differences $9k+1, 9k+2, \ldots, 10k$, and the differences $3k+1, 3k+3, \ldots, 5k-1$. The block $[i, i+2j+3k+2, i+j+6k+2]$ contains the differences $2j+3k+2$, $3k-j$, and $j+6k+2$ for $j=0, 1, \ldots, k-1$. This accounts for the differences $2k+1, 2k+2, \ldots, 3k$, the differences $6k+2, 6k+3, \ldots, 7k+1$, and the differences $3k+2, 3k+4, \ldots, 5k$. Further, the block $[i, i+2k, i+8k+1]$ accounts for the differences $2k$, $6k+1$, and $8k+1$. So far, we have accounted for the differences $2k-1, 2k, \ldots, 7k+1$, the differences $9k+1, 9k+2, \ldots, 12k$, and the differences $1, 3, \ldots, 2k-3$, and $8k+1$. If $k=1$, then we have accounted for all the differences. If $k \geq 2$, then the remaining differences are $2, 4, \ldots, 2k-2$, $7k+2, 7k+3, \ldots, 8k$, and $8k+2, 8k+2, \ldots, 9k$. The remaining block for this case is $[i, i+2j+2, i+j+6k+2]$, where $j=0, 1, \ldots, k-2$. This accounts for differences $2j+2$, $8k-j$, and $j+8k+2$ for $j=0, 1, \ldots, k-2$. This gives us the remaining differences.

For more information on Steiner triple systems, the interested reader is referred to the article by Colbourn and Mathon [17].

Exercise 10.3.6 Consider the problem of arranging v treatments into blocks of size three such that every pair of treatments appears in exactly two blocks. Find necessary and sufficient conditions for the existence of such a design.

Exercise 10.3.7 Consider the problem of arranging v treatments into blocks of size four such that every pair of treatments appears in exactly one block. Find necessary conditions for the existence of such a design.

Exercise 10.3.8 Constructing a Steiner triple system has been described as partitioning the treatments so that each partition gives the vertices of a triangle. Consider a design where we are required to partition the treatments such that each partition gives the vertices of a square.

(i) Is this a block design?
(ii) What are the necessary conditions for the existence of such a design?
(iii) Suppose that we wish to use a difference method to construct such a design. What conditions must each partition of the difference set satisfy?

Exercise 10.3.9 Confirm that the construction in Theorem 10.3.5 works for the case where $v \equiv 3 \pmod{24}$.

Exercise 10.3.10 Confirm that the construction in Theorem 10.3.5 works for the case where $v \equiv 7 \pmod{24}$.

Exercise 10.3.11 Confirm that the construction in Theorem 10.3.5 works for the case where $v \equiv 9 \pmod{24}$.

Exercise 10.3.12 Confirm that the construction in Theorem 10.3.5 works for the case where $v \equiv 13 \pmod{24}$.

Exercise 10.3.13 Confirm that the construction in Theorem 10.3.5 works for the case where $v \equiv 15 \pmod{24}$.

Exercise 10.3.14 Confirm that the construction in Theorem 10.3.5 works for the case where $v \equiv 19 \pmod{24}$.

Exercise 10.3.15 Confirm that the construction in Theorem 10.3.5 works for the case where $v \equiv 21 \pmod{24}$.

10.4 Finite Projective Planes

The *Euclidean plane* consists of a set of *points* P and a set of *lines* L. In the Euclidean plane, two lines are *parallel* if they do not intersect. The points and the lines of the Euclidean plane satisfy the following properties:

(i) Between any two distinct points there is a unique line passing through those points;
(ii) Any two non-parallel lines intersect at a unique point;
(iii) Given a line ℓ and a point p not on ℓ, there is a unique line parallel to ℓ passing through p.

Traditionally, L consists of all lines in the plane. Similarly, the set of points P consists of all possible points in the plane. In this case, we can conclude that there are an infinite number of lines passing through each point. Also, there will be an infinite number of points along any line.

During the Renaissance, artists began considering the problem of *perspective*. This became important when considering parallel lines which are moving away from the observer. For instance, suppose that you are standing directly above train tracks that are in a straight line. To draw these parallel lines in perspective, you would draw the tracks gradually growing closer together until they meet in the horizon (in other words, the point "at infinity"). This notion of perspective led mathematicians to consider *projective geometries*.

We define the *projective plane* in a similar way as the Euclidean plane was defined above. Again, the projective plane consists of a set of points P and a set of lines L. These objects satisfy the following properties:

(i) Between any two distinct points there is a unique line passing through those points;
(ii) Any two distinct lines intersect at a unique point.

Notice that the second property implies that that there are no parallel lines in the projective plane. This corresponds to the artists' intuition that parallel lines will intersect at a point, even if that point is "at infinity."

For our purposes, we will restrict our attention to *finite projective planes*. In this case, there are only finitely many points in the projective plane. Since any two lines must intersect at a point, this implies that the set of lines must also be finite. Thus we

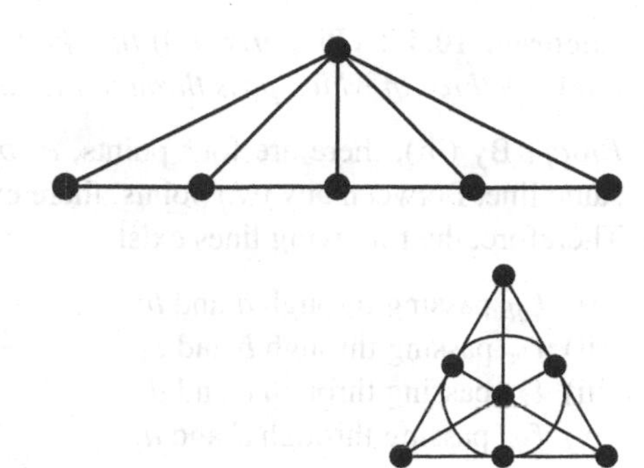

Fig. 10.1 A degenerate plane

Fig. 10.2 Projective plane of order 2

will assume that $|P| = n$. One of the principal questions will be: For what values of n is there a projective plane with n points?

Note that if $n = 2$, then the properties are trivially satisfied by having a single line between the two points. In general, the two properties are trivially satisfied by placing all the points on a single line. A third trivial example arises by placing three points on the corners of a triangle. These trivial examples can be avoided by introducing a third property:

(iii) There are four distinct points, no three of which lie on the same line.

A projective plane that satisfies this third property is called *non-degenerate*. An example of a degenerate plane (sometimes called a "pencil") is given in Fig. 10.1 We will assume that all finite projective planes are non-degenerate. The third property implies that the smallest non-degenerate plane has four points.

Proposition 10.4.1 *There is no non-degenerate projective plane with four points.*

Proof Assume to the contrary that such a plane exists. Then there are four points a, b, c, and d, no three of which lie on a line. Thus there is a line ℓ_{ab} that passes through a and b. Since no three points lie on the same line, c and d are not on ℓ_{ab}. Similarly, there is a line ℓ_{cd} passing through c and d that does not contain the points a and b. Thus, ℓ_{ab} and ℓ_{cd} contain no common points. This contradicts the second property. ■

As we will later see, there are no projective planes with five or six points either. An example of a projective plane with seven points (known as the *Fano plane*) is given in Fig. 10.2. Notice that one of the "lines" is represented by a circle passing through three of the points.

In the first two properties, the roles "point" and "line" can be interchanged resulting in equivalent statements. This is a notion known as *duality*. Two statements regarding projective planes, Q and R, are *dual statements* if R can be obtained from Q by reversing the roles of "points" and "lines." Hence (i) and (ii) are dual statements. We can also obtain a dual statement of the third property.

Theorem 10.4.2 *(Property (iv)) In a finite projective plane, there are four distinct lines, no three of which pass through the same point.*

Proof By (iii), there are four points, a, b, c, and d such that no three lie on the same line. Between any two points, there exists a line passing through these points. Therefore, the following lines exist:

(i) ℓ_{ab} passing through a and b;
(ii) ℓ_{bc} passing through b and c;
(iii) ℓ_{cd} passing through c and d;
(iv) ℓ_{da} passing through d and a.

Since no three points are on the same line, these lines are all distinct. It suffices to show that no three of these lines pass through a common point. Assume to the contrary that there is a point x such that at least three of the above lines pass through x. Without loss of generality, assume that ℓ_{ab}, ℓ_{bc}, and ℓ_{cd} all have point x in common. Thus, $x \neq b$ as $b \notin \ell_{cd}$. This implies that ℓ_{ab} and ℓ_{bc} have two points in common, namely x and b. This contradicts the fact that two lines share at most one common point. ■

The above theorem which reverses the roles of points and lines is far from unique. In fact, given any result about projective planes, we can obtain another result by reversing the roles of points and lines. This theorem, referred to as the *duality principle* is given below.

Theorem 10.4.3 *(Duality principle) If Q is a true statement about projective planes, then the dual statement R is also a true statement.*

Proof Suppose that Q is a true statement about projective planes. First, we reverse the roles of "point" and 'line' in the proof. Next, we perform a similar function regarding the properties of projective planes. Namely, we replace every instance of property (i) with property (ii). Analogously, we replace every instance of property (ii) with property (i). Similarly, we replace every instance of property (iii) with property (iv), and vice versa. ■

Finite projective planes are an example of a combinatorial design. In this case, the treatments are the points. The blocks are the lines in the projective plane. The duality principle implies that there are as many points as lines in the projective plane. Thus the finite projective plane is an example of a symmetric design. We will now show that this design is regular.

Theorem 10.4.4 *In a finite projective plane, every point lies on the same number of lines. Further, every line passes through the same number of points.*

Proof Take two lines in the projective plane, say ℓ_1 and ℓ_2. We begin by showing there is a point on neither line. By property (iii), there are four points a, b, c, and d, no three of which lie on the same line. Further, suppose all of these are on ℓ_1 or ℓ_2. Since no three of these are on the same line, we must have two of the points on

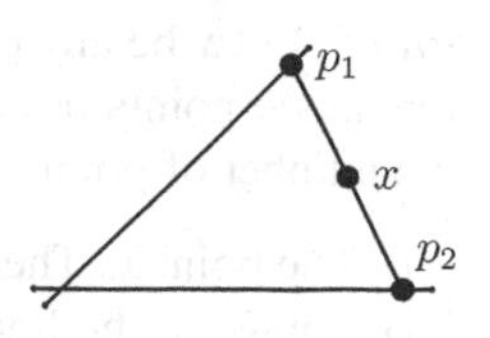

Fig. 10.3 A projection of a point onto a line

each line, say $a, b \in \ell_1$ and $c, d \in \ell_2$. By property (ii), there is a line ℓ_{ac} through a and c and a line ℓ_{bd} through b and d. Since any two lines share a common point, ℓ_{ac} and ℓ_{bd} share a common point, x. If $x \in \ell_1$, then x and b share two common lines, namely ℓ_{bd} and ℓ_1. This is contrary to the fact that two points lie on a unique line. If $x \in \ell_2$, then a similar argument holds for x and c. In either case, x is the required point.

To show that ℓ_1 and ℓ_2 have the same number of points, it suffices to establish a bijection between the points of ℓ_1 and the points of ℓ_2. Given a point $p_1 \in \ell_1$, there is a line through p_1 and x. This line intersects ℓ_2 at a point, p_2. We define p_2 to be the *projection* of p_1 through x onto ℓ_2 (see Fig. 10.3).

We will show that this projection is the required bijection. To show that this is an injective map, take $q_1 \in \ell_1$ and suppose that q_2 is the projection of q_1 through x onto ℓ_2. Note that q_1, q_2, and x are on the same line. Further, p_1 and q_1 are both on ℓ_1. Since any two points lie on a unique line, it follows that if $p_1 \neq q_1$, then $p_2 \neq q_2$. Hence, the projection is an injective function. To show that the projection is surjective, take any point $r_2 \in \ell_2$. We can obtain $r_1 \in \ell_1$ by projecting r_2 through x onto ℓ_1. Thus the projection function is a bijection. Hence ℓ_1 and ℓ_2 have the same number of points.

The duality principle shows that every point lies on the same number of lines.

■

Theorem 10.4.5 *In a finite projective plane, the number of lines through each point is the same as the number of points on each line.*

Proof It suffices to establish the required bijection. Let ℓ be a line in the projective plane. By property (iii), there is a point $x \notin \ell$. Let y be any point on ℓ. By property (i), there is a unique line, $\ell(y)$ passing through both x and y. Similarly, if $\ell(z)$ is any line passing through x, then $\ell(z)$ intersects ℓ at a unique point z. Further, since ℓ does not pass through x and $\ell(z)$ passes through x, then $\ell \neq \ell(z)$. Thus, $\ell(y)$ is a bijection between the points of ℓ and the lines through x. Hence the number of lines through each point is the same as the number of points on each line.

■

Finite projective planes are typically parameterized by their *order*. If q is the order of the projective plane, then there are $q + 1$ points on each line and $q + 1$ lines through each point.

Theorem 10.4.6 *A projective plane of order q has $q^2 + q + 1$ points and $q^2 + q + 1$ lines.*

Proof Let x be any point. By definition, there are $q+1$ lines through x. Each such line has q points besides x. Every point lies on one and only one line with x. Thus the number of points in the plane can be counted in two disjoint, exhaustive sets:

(i) The point x. There is one such point.
(ii) Points on the lines through x, not including x. There are $q+1$ lines that pass through x. Each of these lines contain q points other than x. So there are $q(q+1)$ points in this set by the Multiplication Principle.

By the Addition Principle, there are $q(q+1)+1 = q^2+q+1$ such points.

■

In abstract algebra, it is well known that there are finite fields of order q if and only if q is a prime power. The interested reader is referred to Fraleigh [22] for a description of how to construct such a field. We will denote the finite field of order q as $\mathbb{F}_q$.

Theorem 10.4.7 *If q is a prime power, then there is a projective plane of order q.*

Proof Consider the set of non-zero vectors where the entries come from $\mathbb{F}_q$. In other words, we look at vectors of the form (x, y, z), where $x, y, z \in \mathbb{F}_q$ and x, y, and z are all not zero. We say that two vectors, $\overrightarrow{v_1}$ and $\overrightarrow{v_2}$ are equivalent if there is a non-zero scalar $\lambda \in \mathbb{F}_q$ such that $\lambda \overrightarrow{v_1} = \overrightarrow{v_2}$. If z is not zero, then it has a multiplicative inverse in $\mathbb{F}_q$. Thus, the vector (x, y, z) is equivalent to the vector $(xz^{-1}, yz^{-1}, 1)$. Hence, we can assume without loss of generality that all such vectors are of the form $(x, y, 1)$, where $x, y \in \mathbb{F}_q$.

If $z = 0$, then either x or y is non-zero. If y is non-zero, then by a similar argument we can assume that all such vectors are of the form $(x, 1, 0)$, where $x \in \mathbb{F}_q$. Finally, if $y = z = 0$, then x must be non-zero. For this reason, we can assume that all such vectors are of the form $(1, 0, 0)$. Thus, we have the following sets of vectors (up to equivalence):

(i) Vectors of the form $(x, y, 1)$, where $x, y \in \mathbb{F}_q$. There are q^2 such vectors.
(ii) Vectors of the form $(x, 1, 0)$, where $x \in \mathbb{F}_q$. There are q such vectors.
(iii) The unique vector of the form $(1, 0, 0)$.

By the Addition Principle, there are $q^2 + q + 1$ such vectors. Define these to be the points of our projective plane.

We now construct our lines. Suppose $\ell_{\vec{a}, \vec{b}}$ is the (unique) line passing through the points $\vec{a}$ and $\vec{b}$. A point $\vec{c}$ is on $\ell_{\vec{a}, \vec{b}}$ if and only if $\vec{c}$ is equivalent to a vector of the form $\lambda_1 \vec{a} + \lambda_2 \vec{b}$, where $\lambda_1, \lambda_2 \in \mathbb{F}_q$.

We leave it as an exercise to the reader to prove that this construction satisfies the properties of the projective plane.

■

Example 10.4.8 Construct a projective plane of order 2.

Solution Using Theorem 10.4.7, the required points are $\overrightarrow{p_1} = (0, 0, 1)$, $\overrightarrow{p_2} = (0, 1, 1)$, $\overrightarrow{p_3} = (1, 0, 1)$, $\overrightarrow{p_4} = (1, 1, 1)$, $\overrightarrow{p_5} = (0, 1, 0)$, $\overrightarrow{p_6} = (1, 1, 0)$, and $\overrightarrow{p_7} = (1, 0, 0)$. The set of lines can be determined by direct computation:

$$\begin{aligned}
\overrightarrow{p_1} + \overrightarrow{p_2} &= \overrightarrow{p_5} \Rightarrow \ell_1 = \{\overrightarrow{p_1}, \overrightarrow{p_2}, \overrightarrow{p_5}\},\\
\overrightarrow{p_1} + \overrightarrow{p_3} &= \overrightarrow{p_7} \Rightarrow \ell_2 = \{\overrightarrow{p_1}, \overrightarrow{p_3}, \overrightarrow{p_7}\},\\
\overrightarrow{p_1} + \overrightarrow{p_4} &= \overrightarrow{p_6} \Rightarrow \ell_3 = \{\overrightarrow{p_1}, \overrightarrow{p_4}, \overrightarrow{p_6}\},\\
\overrightarrow{p_2} + \overrightarrow{p_3} &= \overrightarrow{p_6} \Rightarrow \ell_4 = \{\overrightarrow{p_2}, \overrightarrow{p_3}, \overrightarrow{p_6}\},\\
\overrightarrow{p_2} + \overrightarrow{p_4} &= \overrightarrow{p_7} \Rightarrow \ell_5 = \{\overrightarrow{p_2}, \overrightarrow{p_4}, \overrightarrow{p_7}\},\\
\overrightarrow{p_3} + \overrightarrow{p_4} &= \overrightarrow{p_5} \Rightarrow \ell_6 = \{\overrightarrow{p_3}, \overrightarrow{p_4}, \overrightarrow{p_5}\},\\
\overrightarrow{p_5} + \overrightarrow{p_6} &= \overrightarrow{p_7} \Rightarrow \ell_7 = \{\overrightarrow{p_5}, \overrightarrow{p_6}, \overrightarrow{p_7}\}.
\end{aligned}$$

□

Theorem 10.4.7 shows that there is a projective plane of order q whenever q is a prime power. A natural, and unsolved, question is to whether all finite projective planes have prime power order. However, it is conjectured that this is the case, as stated below.

Conjecture 10.4.9 (The Prime Power Conjecture) *If q is not a prime power, then there is no projective plane of order q.*

So far, there has been little progress proving the Prime Power Conjecture. Perhaps the most important general result providing partial progress towards the Prime Power Conjecture is given below.

Theorem 10.4.10 *(Bruck-Ryser-Chowla Theorem [12, 15]) Suppose that there exists a projective plane of order q, where $q \equiv 1 \pmod 4$ or $q \equiv 2 \pmod 4$. It is necessary that either q is a prime power or $q = s^2 + t^2$, where $s, t \in \mathbb{Z}^+$.*

Example 10.4.11 Show that there is no projective plane of order 6 or 14.

Solution Note that neither number is a prime power. Further, both numbers are congruent to 2 (mod 4). We begin with 6. Note that the only possibilities that are smaller than 6 are $1^2 + 1^2$ or $1^2 + 2^2$, both of which are too small. Hence there is no projective plane of order 6 by the Bruck-Ryser-Chowla Theorem.

For 14, if we use $s = 1$, then $t = \sqrt{13} \notin \mathbb{Z}$. If $s = 2$, then $t = \sqrt{10} \notin \mathbb{Z}$. If $s = 3$, then $t = \sqrt{5} \notin \mathbb{Z}$. Any other value of s will exceed 14. Since 14 cannot be written as the sum of two squares, there is no projective plane of order 14 by the Bruck-Ryser-Chowla Theorem. □

Using an exhaustive computer search, it has been shown that there is no projective plane of order 10. However, it is not known if there is a projective plane of order 12.

Exercise 10.4.12 Construct the projective plane of order three.

Exercise 10.4.13 Construct the projective plane of order five.

Exercise 10.4.14 Prove that the construction in Theorem 10.4.7 satisfies the properties of a projective plane.

Exercise 10.4.15 Show that there is no projective plane of order 21.

Exercise 10.4.16 Show that there is no projective plane of order 22.

Chapter 11
Application: Graph Theory

11.1 What is a Graph?

Graph theory is a branch of combinatorics that specializes in modeling relationships between objects. For instance, the objects may be individuals and they are related if they are acquainted. As a second example, suppose that our objects are airports. We would say that two airports are related if there is a direct flight from one to the other. As a final example, consider the shipping network of a company. The objects in question may be the warehouses in the network. Two warehouses are related if they ship from to one another. As we see from these examples, graph theory has numerous applications in communications, transportation, and computer science.

The objects that we are considering are referred to as *vertices*. Visually, these are represented by points in the plane. The relationships are represented by *edges* between the vertices. These are typically represented by (not necessarily straight) lines between pairs of vertices. A *graph* G is a set of vertices $V(G)$ and a set of edges $E(G)$. If the vertices u and v are connected by an edge, then we denote this edge as uv. In this case, u and v are *adjacent*. Analogously, the edge uv is *incident* with both u and v.

As an example, consider the graph G given in Fig. 11.1. Here, $V(G) = \{0, 1, ..., 5\}$ and $E(G) = \{12, 14, 15, 23, 24, 45\}$. So 1 is adjacent to 2, 4, and 5. The edge 24 is incident with its endpoints 2 and 4.

Unless otherwise noted, we assume that the "lengths" of the edges in the visual representation of the graph do not indicate the actual distance between vertices. Similarly, the relative positions of the vertices is considered irrelevant. For instance, suppose that our graph represents the set of airports in the United States. In our representation, we may place the vertices representing O'Hare, LAX, and La Guardia on a straight line, despite the fact that these airports do not lie along a straight line in reality. Similarly, we traditionally represent all vertices as the same size, despite the relative "sizes" of the objects in questions. In other words, larger airports would not warrant larger vertices. For this reason, graph theory was initially described as "geometry without position."

Before proceeding further, we list several additional terms that are useful when discussing graphs. The *order* of a graph G, denoted $n(G)$, is the cardinality of the

R. A. Beeler, *How to Count*, DOI 10.1007/978-3-319-13844-2_11

Fig. 11.1 A graph

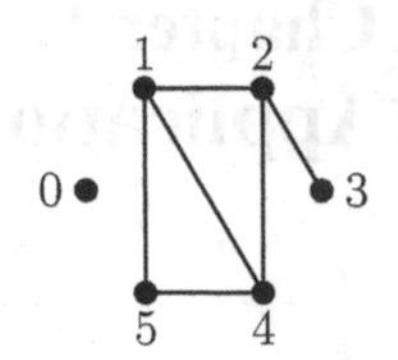

vertex set of G. The *size* of a graph G, denoted $e(G)$, is the cardinality of the edge set of G. In other words, $n(G) = |V(G)|$ and $e(G) = |E(G)|$. The *degree* of a vertex v, denoted $deg(v)$ is the number of edges incident to v.

Example 11.1.1 Find the order and size of the graph in Fig. 11.1. For each vertex in this graph, give the degree of the vertex.

Solution Since $V(G) = \{0, 1, ..., 5\}$, the order of G is 6. The edge set of G is $\{12, 14, 15, 23, 24, 45\}$. Hence the size of G is also 6. As for the degrees, we have $deg(0) = 0$, $deg(1) = 3$, $deg(2) = 3$, $deg(3) = 1$, $deg(4) = 3$, and $deg(5) = 2$. ■

Notice that in Example 11.1.1, the sum of the degrees is $0+3+3+1+3+2 = 12$ which is twice the size of G. This actually holds in general, as shown in the next theorem.

Theorem 11.1.2 *(Handshaking Lemma) For any graph G,*

$$\sum_{v \in V(G)} deg(v) = 2e(G).$$

Proof It suffices to show that each edge in $E(G)$ is counted exactly twice by the left side of the equation. Let $uv \in E(G)$. This edge is counted once by $deg(u)$ and once by $deg(v)$. ■

We now present the most common interpretation of Theorem 11.1.2. Suppose that we have n people at a party. These will be represented by vertices in our graph. If two people shake hands while at the party, then there is an edge between their respective vertices. Thus, $e(G)$ counts the number of (initial) handshakes that occur at the party. Similarly, $\sum_{v \in V(G)} deg(v)$ counts the number of hands that have been shaken. Therefore, the number of hands that have been shaken should be twice the number of handshakes because there are two hands involved in each shake. Because of this analogy, Theorem 11.1.2 is often referred to as the *Handshaking Lemma*.

One of the most important considerations for many graph theory applications is whether there is a sequence of edges between two given vertices. For example, suppose that you are planning air travel from Los Angeles to London. It is not necessarily the case that there will be a direct flight, let alone one near your planned departure time. For this reason, it may be necessary to change planes at some intermediate airport. In fact, several connecting flights may be necessary depending on your departure and destination. However, it is reasonable to expect you can get from any airport to any other airport using some sequence of connecting flights. The notions of connectivity and distance in graphs is generalized below.

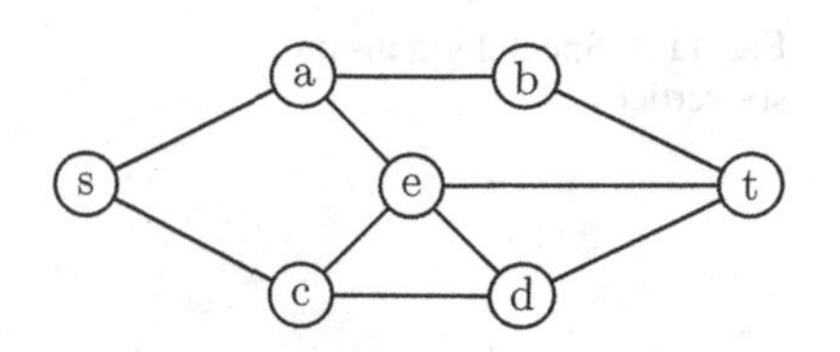

Fig. 11.2 An example of distance in graphs

A *walk* is an ordered list of vertices such that if two vertices appear consecutively on the list, then they are adjacent. Given $u, v \in G$, a *(u, v)-path* is a walk between u and v such that no vertex appears twice on the list. A graph is *connected* if given any two vertices $u, v \in G$, there is a (u, v)-path. The *distance* between u and v is the length of the shortest (u, v)-path. The distance between u and v is denoted $d(u, v)$. If no such path exists, then $d(u, v) = \infty$.

Example 11.1.3 Find the distance between every pair of vertices in the graph in Fig. 11.2.

Solution Like all distance functions, the distance between two vertices is symmetric. That is, $d(u, v) = d(v, u)$ for all $u, v \in V(G)$. For this reason, it is sufficient to list the distance once for each distinct pair of vertices. The edge set of this graph is

$$E(G) = \{sa, sc, ab, ae, bt, et, ec, ed, cd, dt\}.$$

Thus,

$$\begin{aligned} d(s,a) &= d(s,c) = d(a,b) = d(a,e) = d(b,t) \\ &= d(e,t) = d(e,c) = d(e,d) = d(c,d) = d(d,t) = 1. \end{aligned}$$

There is a path of length two between s and each of the vertices e, b, and d. Hence,

$$d(s,e) = d(s,b) = d(s,d) = 2.$$

Similarly, the shortest path between s and t is of length three. Thus $d(s,t) = 3$.

The shortest path between a and each of c, d, and t is two. Therefore,

$$d(a,c) = d(a,d) = d(a,t) = 2.$$

By a similar argument,

$$d(b,e) = d(b,d) = d(c,t) = 2$$

and $d(b,c) = 3$. ■

In general, it is more difficult to determine the shortest distance between two vertices. Fortunately, the graph in Fig. 11.2 is relatively small. Hence, the distances involved can easily be confirmed.

In graph theory, there are several important classes of graphs that appear repeatedly. A *path* is a walk such that no vertex appears twice on the list. A path on n vertices is denoted P_n. A *cycle* is a walk such that the first and last vertex on the list

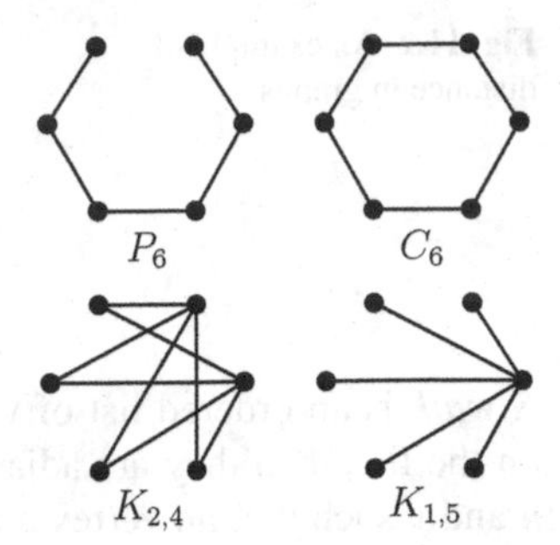

Fig. 11.3 Special graphs on six vertices

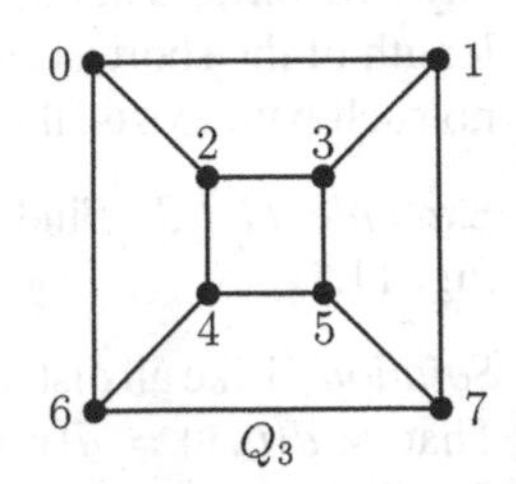

Fig. 11.4 The 3-dimensional hypercube, Q_3

are the same and no other vertex is repeated. A cycle on n vertices is denoted C_n. A *complete bipartite graph* is a graph whose vertex set can be partitioned into two parts such that two vertices are adjacent if and only if they are in different parts. When the parts contain n and m vertices, we denote this graph by $K_{n,m}$. In particular, $K_{1,n}$ is called a *star*. Figure 11.3 shows P_6, C_6, $K_{2,4}$, and $K_{1,5}$.

A *complete graph* is a graph where any two distinct vertices are adjacent (see Fig. 8.4.6). The complete graph on n vertices is denoted K_n.

An *n-dimensional hypercube* is a graph where every vertex corresponds to an n-tuple with entries in $\{0, 1\}$. There is an edge between two vertices if and only if their corresponding n-tuples differ in exactly one position. The n-dimensional hypercube is denoted Q_n. The particular example of Q_3 is found in Fig. 11.4.

A k-dimensional *face* of an n-dimensional hypercube is a portion (often called a *subgraph*) of the n-dimensional hypercube that is a connected k-dimensional hypercube. A natural combinatorial question is to determine the number of k-dimensional faces in an n-dimensional hypercube.

Proposition 11.1.4 *The number of k-dimensional faces in an n-dimensional hypercube is given by* $\binom{n}{k}2^{n-k}$.

Proof Each vertex in Q_n can be represented as a binary string of length n. A k-dimensional face in this cube can be thought of a k-subset of this string. This strings varies over all 2^k possibilities, while the remaining $n - k$ elements of the string remain fixed. Thus, we can find the number of faces by:

(i) Selecting k positions within the string to vary. There are $\binom{n}{k}$ ways to do this.
(ii) For each of the remaining $n - k$ elements of the string, choose either 0 or 1. Therefore, there are 2^{n-k} possibilities.

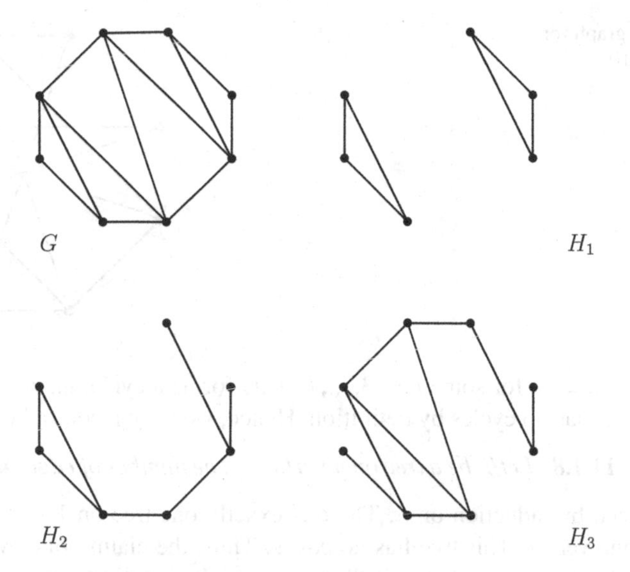

Fig. 11.5 A graph and three subgraphs

Thus, there are $\binom{n}{k}2^{n-k}$ faces by the Multiplication Principle. ■

We now formalize the definition of a subgraph.

Definition 11.1.5 A graph H is a *subgraph* of G if $V(H) \subseteq V(G)$ and $E(H) \subseteq E(G)$. Further, if $V(H) = V(G)$, then H is a *spanning subgraph* of G.

In Fig. 11.5, we give a graph G as well as three subgraphs of G. The subgraph H_1 is not connected and it is not a spanning subgraph. The subgraph H_2 is connected, but it is not a spanning subgraph. Of these, only H_3 is a spanning subgraph and it is also connected.

One of the most important families of graphs is the family of *trees*. Examples of trees include paths, stars, and the graph in Fig. 8.3.

Definition 11.1.6 A *tree* is a connected graph such that no subgraph is a cycle. A *forest* is collection of trees.

Trees are often studied in computer science, because many search algorithms can be modeled as trees. A list of trees of small order can be found in the appendix to Harary [25] and the small graph database [24]. We now prove two results regarding trees.

Proposition 11.1.7 *If T is a tree with at least two vertices, then it has at least two vertices of degree one.*

Proof Let $v_1, \ldots, v_\ell$ be a path of maximum length in T. If v_1 and v_ℓ are of degree one, then we have the required vertices. If not, then there exists $v_0 \in V(T)$ such that $v_0v_1 \in E(T)$ and $v_0 \neq v_2$. If $v_0 \neq v_i$ for all $i = 3, ..., \ell$, then $v_0, v_1, \ldots, v_\ell$ is a path of length $\ell + 1$. This contradicts the assumption that $v_1, \ldots, v_\ell$ is a path of maximum

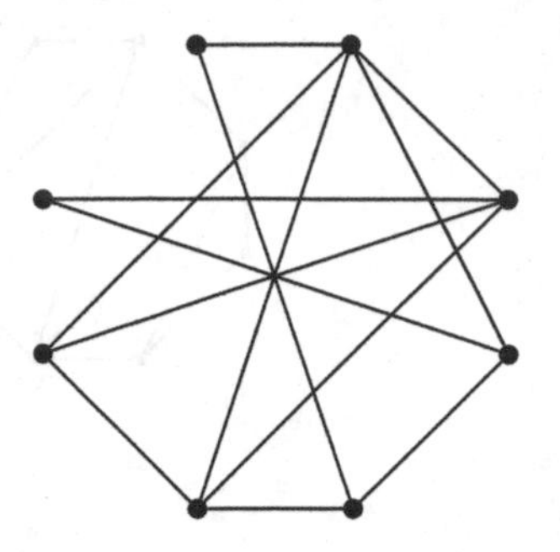

Fig. 11.6 The graph for Exercise 11.1.10

length. Thus, $v_0 = v_i$ for some $i = 3, ..., \ell$. This forms a cycle, namely $v_1, \ldots, v_i, v_1$. However, T has no cycles by definition. Hence, we have a contradiction. ■

Proposition 11.1.8 *Let T be a tree on n vertices. The number of edges in T is $n-1$.*

Proof Proceed by induction on n. There is exactly one tree on 1 vertex. Namely, the tree on one vertex. This tree has no edges. Thus, the claim holds when $n = 1$. Assume that for some $n \geq 1$ that if T is a tree with n vertices, then T has $n - 1$ edges.

Let T be a tree with $n+1$ vertices. Since T has at least two vertices, T must have at least two vertices of degree one. Call one of these vertices v. Let T' be obtained from T by deleting v and its incident edge. Since, v has degree one in T, its deletion does not result in a disconnected graph. Further, since T contains no cycles, it follows that any of its subgraphs, including T', contain no cycles. Since T' is a connected graph with no cyclic subgraphs, it follows that T' is a tree by definition. Moreover, T' is a tree with n vertices. Therefore, it has $n - 1$ edges by hypothesis. Since v is a vertex is of degree one, it follows that it contributes one vertex and one edge to T. Thus, the number of edges in T is $n - 1 + 1 = n$. The claim holds by the Principle of Mathematical Induction. ■

Proposition 11.1.8 shows that trees represent a class of minimally connected graphs. That is, of all connected graphs on n vertices, trees have the fewest number of edges.

Exercise 11.1.9 Let k be the number of vertices in a graph G that have odd degree. Prove that for every graph G, k must be even.

Exercise 11.1.10 Let G be the graph in Fig. 11.6. Find the order and size of G. For every pair of vertices, $u, v \in V(G)$, find $d(u, v)$.

Exercise 11.1.11 Prove that if G is a graph with at least two vertices, then G has at least two vertices of equal degree.

Exercise 11.1.12 Find the distance between every pair of vertices in the graph from Fig. 11.5.

Exercise 11.1.13 Let $u, v \in V(Q_n)$. Define $s(u, v)$ to be the number of positions in which the binary representations of u and v differ. Prove that $d(u, v) = s(u, v)$.

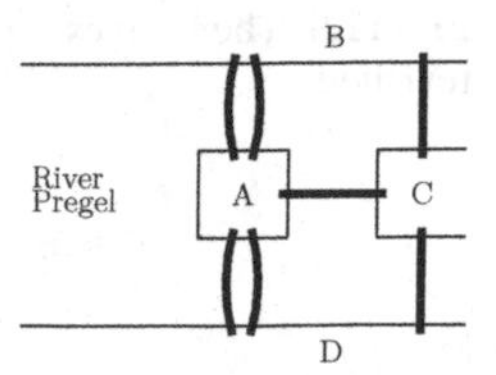

Fig. 11.7 The bridges of Königsberg

Exercise 11.1.14 Let G be the graph in Fig. 11.2. Find subgraphs H_1, H_2, H_3, and H_4 of G such that:

(i) H_1 is not connected and does not span;
(ii) H_2 is connected but does not span;
(iii) H_3 is not connected, but spans;
(iv) H_4 is a spanning tree.

Exercise 11.1.15 Let G be the graph in Fig. 11.6. Find subgraphs H_1, H_2, H_3, and H_4 of G such that:

(i) H_1 is not connected and does not span;
(ii) H_2 is connected but does not span;
(iii) H_3 is not connected, but spans;
(iv) H_4 is a spanning tree.

Exercise 11.1.16 Prove that any connected graph has a spanning tree.

Exercise 11.1.17 A *d-regular* graph is a graph G such that every vertex has degree d. Prove that every d-regular graph on n vertices has $\frac{dn}{2}$ edges.

Exercise 11.1.18 Suppose that T is a tree. Prove that if $u, v \in V(T)$, then there is a unique (u, v)-path.

Exercise 11.1.19 Suppose that G is a connected graph with order n, where $n \geq 3$. If $deg(v) = 2$ for all $v \in V(G)$, then prove that G is the cycle on n vertices.

11.2 Cycles Within Graphs

Most math historians agree that the study of graph theory began in 1735 with a paper by Leonard Euler. Euler's paper "*Solutio problematis ad geometriam situs pertinentis,*" [9, 20] was concerned with determining the solution to a particular diversion of the citizens of Königsberg. The city of Königsberg is divided into four land masses by the river Pregel. Between these land masses, there are seven bridges, as illustrated in Fig. 11.7. The people of Königsberg wondered whether it was possible to start at any land mass, cross each bridge exactly once, and return to your starting position. This is often referred to as a *walking tour*.

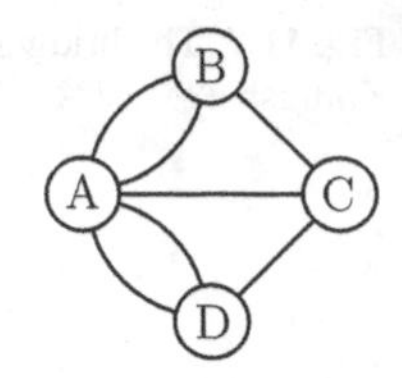

Fig. 11.8 The bridges revisited

Euler realized that the size and relative position of the four land masses was irrelevant. He further realized that the lengths of the seven bridges was irrelevant. Thus, the bridges of Königsberg can be represented as a *multigraph*, as shown in Fig. 11.8. In general, a *multigraph* is a graph in which we allow multiple edges between the same pair of vertices. In other words, the edge set of a multigraph is a multiset.

However, representing the bridges of Königsberg as a graph does not answer the question of whether one could take a walking tour of the bridges. Fortunately, this concern is addressed by Euler's final observation. Namely, Euler observed that in a walking tour, each time you enter a land mass via a bridge, you must leave that land mass along a different bridge. Hence, a simple necessary condition for the existence of a walking tour is that each land mass must have an even number of bridges. To translate this into the vocabulary of the previous section, each vertex must have even degree. For this reason, it is impossible to complete the walking tour of the bridges of Königsberg.

Because of Euler's contributions to the bridges of Königsberg problem (and the foundations of graph theory in general), cycles of the type described are named in his honor. An *euler cycle* in a graph G is a walk beginning at any vertex v that contains every edge of G exactly once and returns to v. A graph is *eulerian* if it contains an euler cycle.

While Euler's observations provide a simple necessary condition for the existence of an euler cycle, they do not answer the question of sufficiency. Namely, if G is a graph in which every vertex has even degree, then does G have an euler cycle? Euler stated that his simple necessary condition was also sufficient. However, he did not provide a proof. The proof was finally provided in 1873 by Carl Hierholzer [28]. The results are provided in the following theorem.

Theorem 11.2.1 *A connected graph G is eulerian if and only if every vertex of G has even degree.*

Proof Suppose that G has an euler cycle. Once you enter a vertex v via an edge on the cycle, it must be possible to leave that vertex along a different edge. Thus, each time the euler cycle passes through v it contributes 2 to $deg(v)$. For this reason, the degree of v must be even. Since v was chosen arbitrarily, this holds for all vertices of G.

Conversely, suppose that every vertex of G has even degree. Begin a walk beginning at v_1 along one of incident edges, say $v_1 v_1'$. Since the degree of v_1' is also even, there is an unused edge that is incident with v_1'. Thus we leave along this edge. This is true for any vertex along the walk, except perhaps for when the walk eventually returns to v. Let C_1 denote the constructed cycle. If C_1 contains all the edges of G, then we have constructed the required euler cycle. If not, then there is a first vertex along this cycle, say v_2, which has incident edges that are not in C_1. Repeat this process beginning at v_2. Since the degree of every vertex in G is even, then we will eventually return to v_2. Denote this cycle C_2. If $C_1 \cup C_2$ contains all the edges of G, then this is the required euler cycle. If not, then repeat this process with a vertex v_3 on $C_1 \cup C_2$. Repeat this process until we have a sequence of cycles $C_1, ..., C_k$ such that every edge in G is in $C_1 \cup ... \cup C_k$.

To construct the required cycle, begin at v_1 and traverse the edges of C_1 until v_2 is encountered for the first time. At this point, traverse the edges of C_2 until either we return to v_2 or we have encounter v_3 for the first time. At this point we traverse the edges of C_3. In general, we traverse the edges of $C_1 \cup ... \cup C_i$ in this manner until we encounter v_{i+1} for the first time. At this point, we traverse the edges of C_{i+1} as described above. Continue this process until we have encountered $v_1, \ldots ,v_k$. By hypothesis, $C_1 \cup ... \cup C_k$ contains all the edges of G. Thus, we have constructed the required cycle. ■

While we have shown that there is no euler cycle for the bridges of Königsberg, we might then ask if we can cross each bridge exactly once without the restriction of returning to our original position. An *euler path* is a walk beginning at a vertex v that traverses each edge exactly once. However, an euler path need not return to vertex v. The following theorem provides necessary and sufficient conditions for the existence of an euler path. Since its proof is very similar to the proof of Theorem 11.2.1, we leave the proof as an exercise to the reader.

Theorem 11.2.2 *A connected graph G has an euler path if and only if there are at most two vertices in G with odd degree.*

Since all four vertices in the bridges of Königsberg have odd degree, it follows that it does not have an euler path. The next natural question regarding cycles within graphs would be to determine if there exists a cycle within a graph G that visits each vertex exactly once. Such cycles are named *hamilton cycles* in honor of William Rowan Hamilton, who in 1857 invented the *icosian game*. In the icosian game, the player is given a dodecahedral board, as shown in Fig. 11.9. The goal is to trace out a cycle that passes through each vertex exactly once. A solution of the icosian game is given by the dark edges in Fig. 11.9. If there is a hamilton cycle in a graph, then we say that the graph is *hamiltonian*.

Unlike eulerian graphs, there is no (known) simple necessary and sufficient condition for the existence of hamilton cycles. Of course, the existence, or lack thereof, is known for many graphs and many families of graphs. For instance, it is trivial to show that C_n and K_n are hamiltonian for $n \geq 3$. We can show a similar result for the n-dimensional hypercube.

Theorem 11.2.3 *The graph Q_n is hamiltonian for $n \geq 2$.*

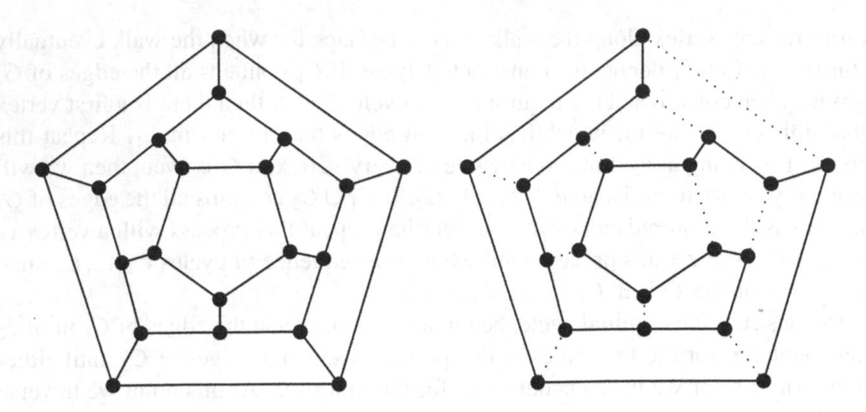

Fig. 11.9 The icosian game

Proof Proceed by induction on n. If $n = 2$, then the graph is identical to C_4, which is trivially hamiltonian. Assume that Q_n is hamiltonian for some $n \geq 2$ and consider Q_{n+1}.

Partition the vertices of Q_{n+1} into two sets. The first set, denoted A, will be the set of all vertices whose binary representation begins with a 0. Analogously, the second set, denoted B, will be the set of all vertices whose binary representation begins with a 1. Let $G[A]$ be the subgraph formed by taking all the vertices in A and all edges of the form uv, where $u, v \in A$. Since all the vertices in this graph have a binary representation that begins with 0, we can simply ignore this first 0. Hence, $G[A]$ is a Q_n. Analogously, we can define $G[B]$ which is also a Q_n by the same argument.

By inductive hypothesis, both $G[A]$ and $G[B]$ have a hamilton cycle. Let C_A and C_B be these cycles. Without loss of generality, we may assume that these cycles use corresponding edges in both $G[A]$ and $G[B]$. In other words, suppose that C_A uses the edge between $0s_1$ and $0s_2$, where s_1 and s_2 are binary strings of length n. We will assume that C_B uses the corresponding edge between $1s_1$ and $1s_2$.

Take any edge u_0v_0 in C_A and the corresponding edge u_1v_1 in C_B. Note that u_0u_1 and v_0v_1 are edges in Q_{n+1}. To construct our hamilton cycle in Q_{n+1}, begin at u_0 and go to u_1. Follow C_B in such a way that v_1 is the last vertex reached. At this point, go to v_0 and follow C_A in such a way that u_0 is the last vertex reached. This gives the required hamilton cycle. ■

An illustration of Theorem 11.2.3 is given in Fig. 11.10. In this case, dashed lines indicate the common edge that was "cut" from both C_A and C_B. Solid lines indicate the constructed cycle. Dotted lines are edges that are not used in the constructed cycle. For simplicity, we omit all of the "transitional" edges between $G[A]$ and $G[B]$ except those used in the constructed cycle.

As mentioned earlier, there is no known condition on graphs that is necessary and sufficient for the existence of a hamilton cycle on G. A trivial necessary condition for the existence of a hamilton cycle is that ever vertex has degree at least two. However, this condition is not sufficient. As a counterexample, consider the graph in Fig. 11.11. Every vertex has degree at least two. However, any cycle that includes e must pass

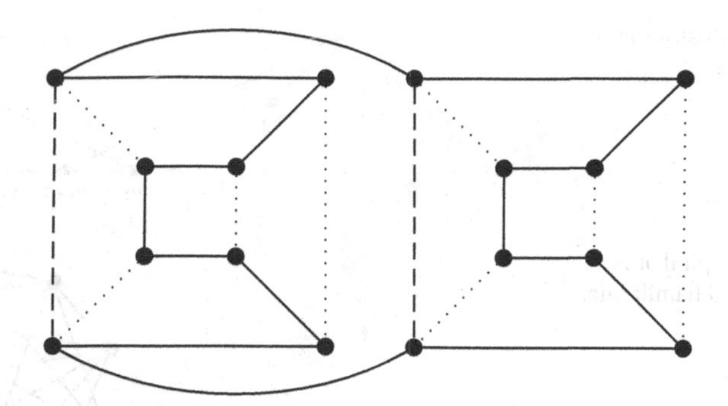

Fig. 11.10 An illustration of Theorem 11.2.3 using Q_4

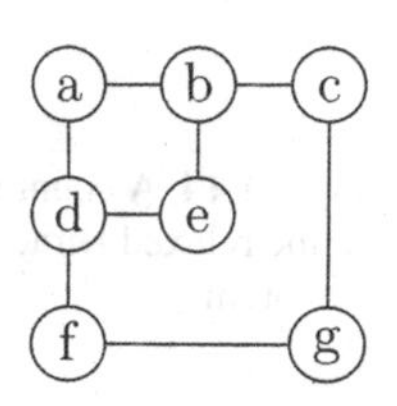

Fig. 11.11 The necessary condition for hamiltonicity is not sufficient

through both b and d. Similarly, any cycle that includes a must pass through both b and d. However any hamilton cycle may use at most two edges incident with a given vertex. Further, $\{a,b,d,e\} \neq V(G)$. Hence, the graph is not hamiltonian.

Sufficient conditions are more difficult to obtain. A quick observation is that a graph with a large number of edges relative to its order is more likely to be hamiltonian. This observation was made rigorous by Gabriel Dirac in 1952 [19].

Theorem 11.2.4 *(Dirac's Theorem) If G is a graph with $n(G) \geq 3$ such that $deg(v) \geq n/2$ for all $v \in V(G)$, then G is hamiltonian.*

Proof Our proof will consist of two major steps. First, we will show that G contains a cycle. Second, we will show that every vertex in G is on this cycle.

First, suppose that $v_1, \ldots, v_\ell$ is a path of maximum length of G. If v_1 is adjacent to a vertex v_0 that is not on this path, then we can construct a longer path, namely $v_0, v_1, \ldots, v_\ell$. Thus, if v_1 is adjacent to a vertex u, then $u \in \{v_2, ..., v_\ell\}$. By the same argument, if v_ℓ is adjacent to u, then $u \in \{v_1, ..., v_\ell\}$. Let S be the set of vertices that are adjacent to v_1. Note that if $v_\ell \in S$, then we have the required cycle. So, we can assume without loss of generality that $v_\ell \notin S$. Define $S^- = \{v_j : v_{j+1} \in S\}$. Since $deg(v_1) \geq n/2$, it follows that $|S| = |S^-| \geq n/2$. Further, $deg(v_\ell) \geq n/2$. It follows from the Pigeonhole Principle that v_ℓ is adjacent to at least one vertex in S^-, say v_i. By definition of these sets, v_1 is adjacent to the corresponding vertex in S, namely v_{i+1}, as shown in Fig. 11.12.

We now construct a cycle among the vertices of $v_1, \ldots, v_\ell$. Begin with v_1 and proceed along the path of maximum length until v_i is reached. At which point, proceed directly to v_ℓ. Now, proceed along the path of maximum length (in reverse)

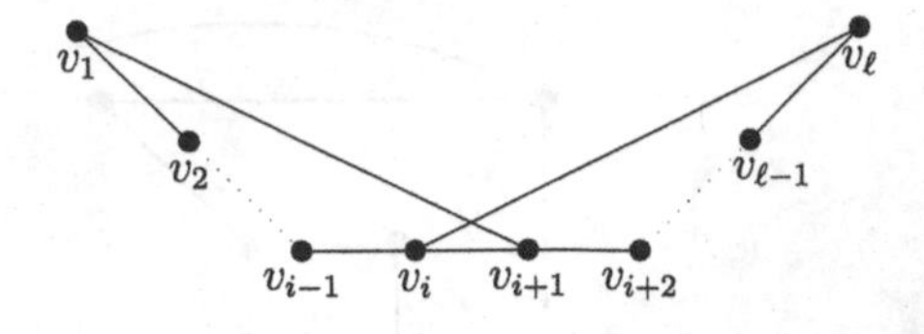

Fig. 11.12 An illustration of Dirac's Theorem

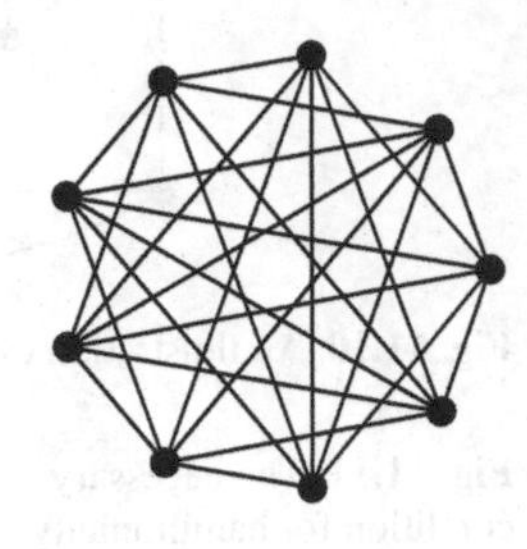

Fig. 11.13 A graph that is both eulerian and hamiltonian

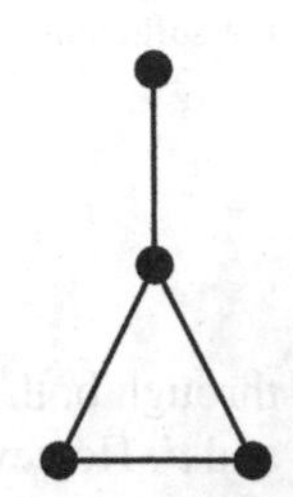

Fig. 11.14 A counterexample to the relaxed Dirac's Theorem

until v_{i+1} is reached. Finally, return to v_1. Thus, the desired cycle is $v_1, v_2, \ldots, v_i, v_\ell, v_{\ell-1}, \ldots, v_{i+1}, v_1$.

Finally, we must show that $V(G) = \{v_1, ..., v_\ell\}$. Suppose to the contrary that $u \in V(G)$ and $u \notin \{v_1, ..., v_\ell\}$. Since $S \subset \{v_1, ..., v_\ell\}$ and $|S| \geq n/2$, it follows that $|\{v_1, ..., v_\ell\}| \geq n/2$. Since $deg(u) \geq n/2$, it follows from the Pigeonhole Principle that u is adjacent to at least one vertex in $\{v_1, ..., v_\ell\}$, say v_j. Relabel the elements of this set so that $v_j = v_1'$ and $v_1', \ldots, v_\ell'$ form a cycle. We can construct a path of length $\ell + 1$ by taking the path $u, v_1', \ldots, v_\ell'$. However, this is contrary to $v_1, \ldots, v_\ell$ being a path of maximum length. Thus, $V(G) = \{v_1, ..., v_\ell\}$. Ergo, G is hamiltonian by definition. ■

We now consider an example of both of the major results in this section. Consider the graph in Fig. 11.13. This is a graph of order nine in which every vertex is of degree six. Since every vertex is of even degree, it follows from Theorem 11.2.1 that this graph is eulerian. Similarly, since every vertex has degree at least 9/2, it follows from Dirac's Theorem that the graph is hamiltonian.

When learning theorems, it is advisable to determine what happens if the hypotheses of the theorem are relaxed. For instance, suppose that we relax the condition on Dirac's Theorem so that there is a single vertex of degree less than $n/2$. Now, we are no longer guaranteed that the graph is hamiltonian. As an example, consider the graph in Fig. 11.14. This graph has four vertices. Three of the vertices have degree at least 2. However, as the fourth vertex has degree one, this graph cannot have a hamilton cycle.

Two hamilton cycles on a graph G are *distinct* if they traverse different edges of G. Thus, a natural, and difficult, combinatorial question is to determine the number of distinct hamilton cycles in a given graph. Essentially, this is equivalent to a table setting problem. The edges of the graph dictate the various restrictions of the table setting. However, unlike traditional table setting problems, we do not consider rotations and reflections of the same hamilton cycle to be distinct.

Example 11.2.5 Find the number of distinct hamilton cycles in K_n and $K_{n,n}$.

Solution Suppose that $V(K_n) = \{v_1, ..., v_n\}$. Since K_n has all possible edges, it follow that this is equivalent to a table setting in which there are no restrictions. Without loss of generality, select v_1 as the initial vertex. There are then $(n-1)!$ ways to arrange the remaining vertices in the cycle. Since reflections are not considered distinct, we have counted each cycle twice. For example, if $n = 5$, then v_1, v_2, v_3, v_4, v_5 and v_1, v_5, v_4, v_3, v_2 are the same cycle. To compensate for this, we divide by 2. Hence, the number of distinct hamilton cycles in K_n is given by $\frac{(n-1)!}{2}$.

Suppose that $V(K_{n,n}) = X \cup Y$, where $X = \{x_1, ..., x_n\}$ and $Y = \{y_1, ..., y_n\}$, where $xy \in E(K_{n,n})$ if and only if $x \in X$ and $y \in Y$. Since any hamilton cycle must alternate between X and Y, it follows that this is equivalent to a sex alternating seating with n men and n women. The number of ways to seat n men and n women around a circular dinner table (with labeled seats) in such a way that sexes alternate. The number of such settings is given by $2(n!)^2$ (see Example 2.3.2) . However, when considering the number of hamilton cycles, we must compensate for the number of rotations and reflections. The number of reflections is 2 and the number of rotations is $2n$. Thus, the number of distinct hamilton cycles is given by:

$$\frac{2(n!)^2}{2 * 2n} = \frac{n!(n-1)!}{2}.$$ ■

Exercise 11.2.6 Does the graph in Fig. 11.11 have an euler cycle? Does it have an euler path?

Exercise 11.2.7 Under what conditions (on n and m) does $K_{n,m}$ have an euler cycle? Under what conditions does it have an euler path?

Exercise 11.2.8 Prove Theorem 11.2.2.

Exercise 11.2.9 Find an euler cycle for the graph in Fig. 11.15. Is the graph hamiltonian?

Exercise 11.2.10 Prove that $K_{n,m}$ is hamiltonian if and only if $n = m$ and $m \geq 2$.

Exercise 11.2.11 Construct a connected graph G with $n \geq 3$ vertices such that the following are satisfied:

(i) Every vertex of G has degree at least two;
(ii) Every vertex but one has degree at least $n/2$;
(iii) G is not hamiltonian.

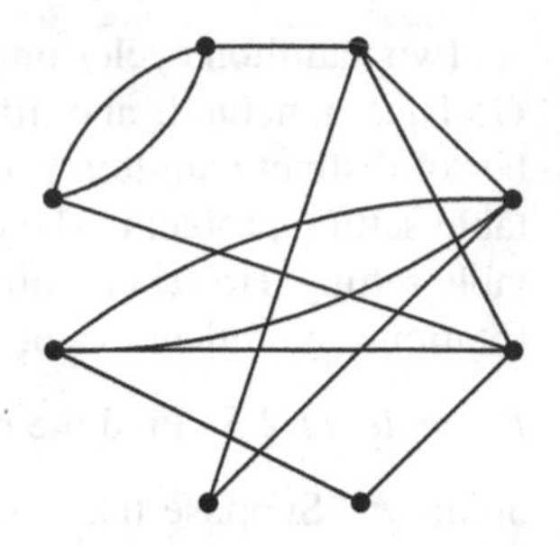

Fig. 11.15 The graph for Exercise 11.2.9

Exercise 11.2.12 Consider the graph obtained from K_n by deleting a single edge. Give an explicit example of a table setting problem that is equivalent to determining the number of distinct hamilton cycles in this graph. Find the number of distinct hamilton cycles in this graph.

Exercise 11.2.13 A *matching*, M, is a set of edges in a graph such that every vertex is the endpoint of exactly one edge in M. Suppose that G is the graph obtained from K_{2n} by deleting a matching. Give an explicit example of a table setting problem that is equivalent to determining the number of distinct hamilton cycles in this graph. Find the number of distinct hamilton cycles in this graph.

Exercise 11.2.14 Suppose that G is the graph obtained from $K_{n,n}$ by deleting a matching. Give an explicit example of a table setting problem that is equivalent to determining the number of distinct hamilton cycles in this graph. Find the number of distinct hamilton cycles in this graph.

11.3 Planar Graphs

We begin this section with a simple puzzle. Suppose that we want to connect three houses to each of three utilities, or wells. No two of the structures can share the same space (in other words, a house cannot be inside of another house). In such a problem, it is desirable that no two of the connecting lines cross, otherwise we might be digging a water main over a gas line. Is it possible to position the three houses and three utilities in such a way that when each house is connected to each utility, no two of the connecting lines cross?

First, we rewrite this problem into the language of graph theory. Three houses that are adjacent to each of three utilities can be thought of as a $K_{3,3}$. Hence, we must determine if $K_{3,3}$ can be drawn in such a way that no edges cross.

Note that just because we may draw a graph in such a way that edges cross does not mean that there is *no* way to draw the graph in such a way that no edges cross. For example, in Fig. 11.16, the graph K_4 is drawn in two different ways. The first way is one in which edges cross. The second is one in which no edges cross.

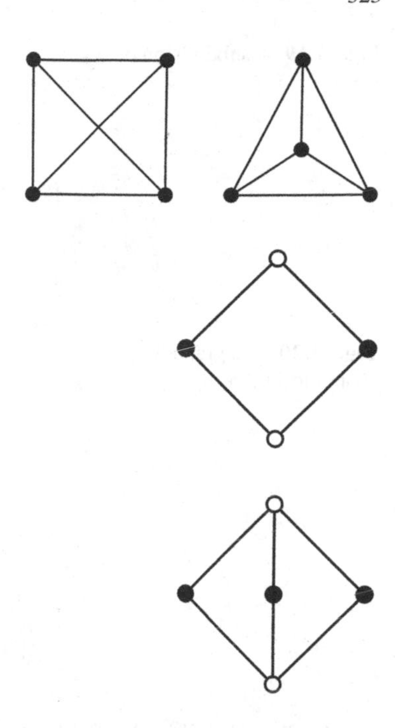

Fig. 11.16 The graph K_4 drawn in two different ways

Fig. 11.17 Two houses and two utilities

Fig. 11.18 Three houses and two utilities

Definition 11.3.1 A graph is *planar* if it can be drawn in the plane in such a way that no edges cross.

The goal of this section is to give various results regarding planar graphs. We begin by answering the question that motivated this section.

Proposition 11.3.2 *The graph $K_{3,3}$ is not planar.*

Proof Begin with the easier problem of drawing $K_{2,2}$ in the plane. This graph is the same as C_4. Thus, there is essentially one way to draw this graph in the plane. This is shown in Fig. 11.17. In this figure, hollow vertices represent utilities and solid vertices represent houses.

Notice that this graph separates the plane into two regions. Namely, the region inside of the cycle and the region outside of the cycle. Hence, there are essentially two places to place the next vertex, which represents a "house." This "house" must be inside of the cycle or outside of the cycle. If we place this "house" outside of the cycle and connect it to each of the "utilities," then one of the original houses is now inside of the cycle. Therefore, we may assume without loss of generality that the third "house" is placed inside of the cycle. This is illustrated in Fig. 11.18.

Notice that this graph separates the plane into three regions. Thus, our third "utility" must be placed into one of these three regions. If the "utility" is placed in the outside region, then it cannot be connected to the inside "house" without crossing an edge. Similarly, if the third "utility" is placed in the left interior region, then it cannot be connected to the right house without crossing an edge. Finally, if the third

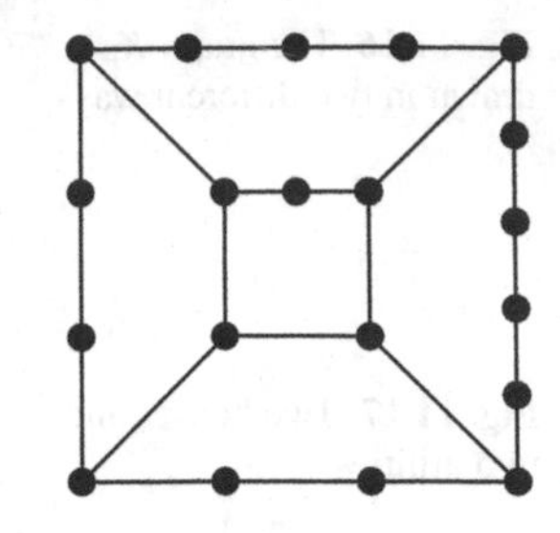

Fig. 11.19 A subdivision of Q_3

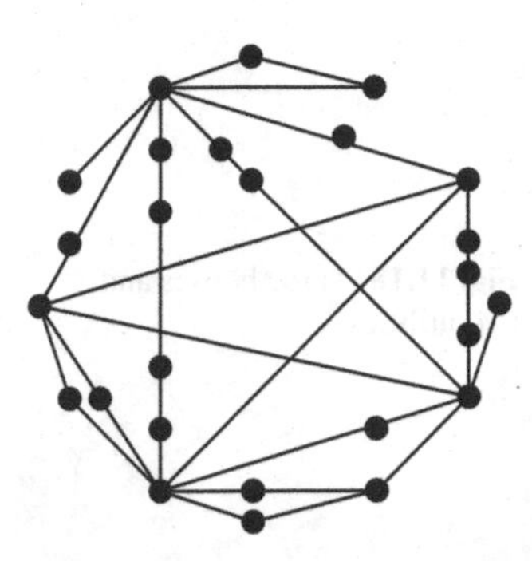

Fig. 11.20 The graph for Example 11.3.5

"utility" is placed in the right interior region, then it cannot be connected to the left house without crossing an edge. Hence, $K_{3,3}$ is not planar. ■

The graph $K_{3,3}$ is far from the only graph that is not planar. Clearly, any graph that has $K_{3,3}$ as a subgraph cannot be planar. Further, if a graph is obtained from $K_{3,3}$ by placing additional vertices along the edges of $K_{3,3}$, then that graph is also not planar. This notion is generalized in the definition below.

Definition 11.3.3 A graph H is a *subdivision* of the graph G if H is obtained from G by inserting at least one vertex of degree two into at least one edge of G.

An example of a subdivision of a graph is given in Fig. 11.19.

Thus, $K_{3,3}$ provides a basis for many non-planar graphs. A second graph that performs this function is K_5. The proof that K_5 is non-planar will be left as an exercise to the reader. Are there other graphs that form the basis for non-planar graphs? This was answered in 1930 by Kazimierz Kuratowski [34]. The proof of Kuratowski's Theorem can be found in any graph theory text.

Theorem 11.3.4 *(Kuratowski's Theorem) A graph G is planar if and only if it does not have a subdivision of K_5 or $K_{3,3}$ as a subgraph.*

Example 11.3.5 Show that the graph in Fig. 11.20 is not planar.

Solution By Kuratowski's Theorem, we must find a subgraph of this graph that is a subdivision of K_5 or $K_{3,3}$. We notice the star pattern in the center of the graph and thus try to show that it has a subdivision of K_5. Ignoring the vertices that are not part of this star pattern yields the graph on the left of Fig. 11.21. This graph is a subdivision of K_5, as shown in the right of Fig. 11.21. ■

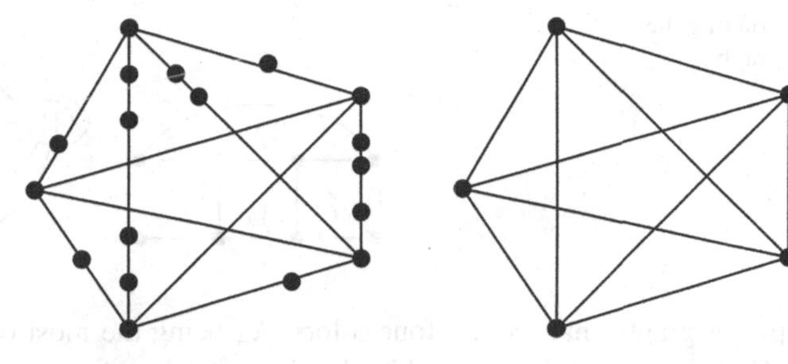

Fig. 11.21 The solution for Example 11.3.5

Recall that in the proof of Proposition 11.3.2, the result was dependent on partitioning the plane into regions. These regions are usually referred to as the *faces* of the planar graph. One of Leonard Euler's more striking results relates the number of vertices, edges, and faces of a planar graph.

Theorem 11.3.6 *(Euler's Criterion for Planar Graphs) Suppose that G is a connected planar graph with V vertices, E edges, and F faces. The graph G satisfies $V - E + F = 2$.*

Proof Since G is connected, it has a spanning tree by Exercise 11.1.16. Let T be a spanning tree of G. Note that T has V vertices, $V - 1$ edges (by Proposition 11.1.8), and 1 face, namely the outside face. Thus, T satisfies

$$V - (V - 1) + 1 = 2.$$

Hence, the result holds for trees. Each time a edge is added from G to T, it encloses a new face. Thus, the number of edges increases by one and the number of faces increases by one. Hence, the net change in $V - E + F$ is zero. Therefore, the result holds. ■

For more information on Euler's Criterion, the interested reader is referred to *Euler's Gem* by David Richeson [35]. We now consider several examples of Euler's Criterion.

Example 11.3.7 Confirm that Euler's Criterion holds for Q_3, K_4, and $K_{2,3}$.

Solution The cube Q_3 is a planar graph as illustrated in Fig. 11.4. This graph has 8 vertices, 12 edges, and 6 faces. As expected by Euler's Criterion, we have $8 - 12 + 6 = 2$.

The graph K_4 is a planar graph as illustrated by Fig. 11.16. This graph has 4 vertices, 6 edges, and 4 faces. As expected, we have that $4 - 6 + 4 = 2$.

The graph $K_{2,3}$ is a planar graph as shown in Fig. 11.18. This graph has 5 vertices, 6 edges, and 3 faces. Again, we have $5 - 6 + 3 = 2$.

One of the most famous (or infamous) results involving planar maps is the *Four Color Theorem*. Suppose that we wish to color the faces (or equivalently, the vertices) of a planar graph so that if two faces share a common edge, then they must receive different colors. What is the minimum number of colors that are required?

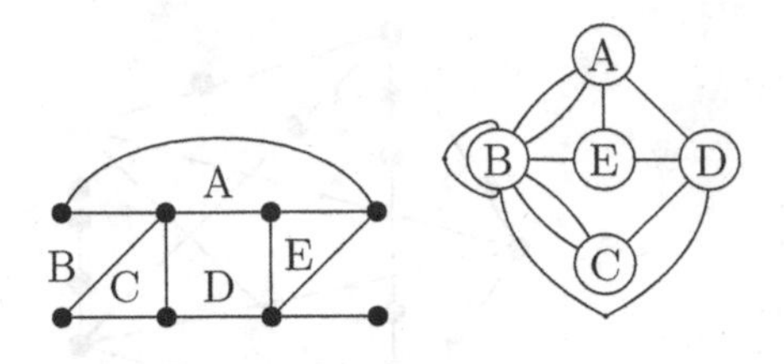

Fig. 11.22 Constructing the dual of a planar graph

There are planar graphs that require four colors, K_4 being the most obvious. In 1852, Francis Guthrie noticed that he could color the countries of any map using at most four colors. He conjectured that any planar map requires at most four colors. While this seems to to be a simple enough idea, these are sometimes the most difficult to prove in mathematics. In 1890, Percy Heawood [26] gave an elementary proof that every planar map can be colored with at most five colors. It was not until 1977 that Appel and Haken [3] proved the Four Color Theorem making extensive use of computers. For more information on the Four Color Theorem, the reader is referred to Robin Wilson's book, *Four Colors Suffice* [46].

We end this section with a discussion of the dual of a planar graph. Informally, the *dual* of a planar graph is obtained by reversing the roles of faces and vertices in a planar graph. This is formalized in the following definition.

Definition 11.3.8 Let G be a planar graph. The *dual* of G, denoted G', has a vertex for each face in G. Two vertices in G' are adjacent if and only if their corresponding faces share a common edge in G.

Before proceeding with an example, we give some reminders about constructing the dual:

(i) If G is a multigraph, then any enclosure caused by a multiple edge counts as a face of G.
(ii) If G has *loops*, that is an edge that goes from a vertex to itself, then the enclosure created by the loop counts as a face of G.
(iii) If G has no multiple edges, then the dual may or may not have multiple edges.
(iv) If G has no loops, then the dual may or may not have loops.

As an example, consider the graph on the left side of Fig. 11.22. Let A denote the face created by the edge going from the top left and top right vertex. Let B be the outside face. Let C be the left interior face, D be the middle interior face, and E be the right interior face. The face A shares two edges with B, one with D, and one with E. The outer face B shares two edges with C, one with D, one with E, and one edge with itself (namely, the edge at the lower right of the graph). Finally, the face D shares an edge with C and an edge with E. The resulting dual is given in the right of Fig. 11.22.

Exercise 11.3.9 Show that K_5 is not planar.

Exercise 11.3.10 Use Kuratowski's Theorem to show that the graph in Fig. 11.13 is not planar.

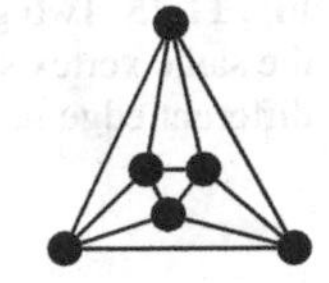
Fig. 11.23 The octahedron

Fig. 11.24 The graph for Exercise 11.3.15

Exercise 11.3.11 Use Kuratowski's Theorem to show that Q_n is not planar for $n \geq 4$.

Exercise 11.3.12 Confirm that Euler's Criterion holds for the dodecahedron (see Fig. 11.9).

Exercise 11.3.13 Confirm that Euler's Criterion holds for the octahedron (see Fig. 11.23).

Exercise 11.3.14 Draw the graph in Fig. 11.15 as a planar graph. Confirm that Euler's Criterion holds for this graph. Draw the dual of this graph.

Exercise 11.3.15 Construct the dual of the graph in Fig. 11.24.

Exercise 11.3.16 Suppose that G is a planar graph. Prove that the dual of G is also a planar graph.

Exercise 11.3.17 Draw the dual of K_4, Q_3, the dodecahedron (see Fig. 11.9), and the octahedron (see Fig. 11.23).

11.4 Counting Labeled Graphs

In this section and the one that follows, we consider the problem of counting graphs. In order to do this, we will need to assume something about the vertex set. As a simple example, we consider two graphs on three vertices. The graph G_1 has $V(G_1) = \{a, b, c\}$ and $E(G_1) = \{ab, bc\}$. The graph G_2 has $V(G_2) = \{1, 2, 3\}$ and $E(G_2) = \{12, 23\}$. Technically, these graphs are different, as they have different vertex sets. So, if we assumed that two graphs are different if they have different vertex sets, then the problem of counting graphs becomes trivial. Namely, the number of different graphs on n vertices is equal to the number of different n-sets. Hence, if we assumed that two graphs with different vertex sets are different, then there are infinitely many graphs on n vertices.

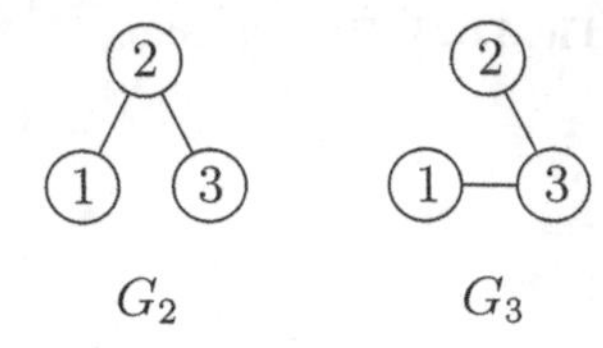

Fig. 11.25 Two graphs with the same vertex set but different edge sets

For this reason, we will assume in this section that all graphs on n vertices have their vertices labeled with the same n-set. Without loss of generality, we will assume that this n-set is $[n]$. Thus, if we were to relabel the vertices of G_1 by replacing the label of a with 1, the label of b with 2, and c with 3, then the two graphs would have the same vertex set and the same edge set. Therefore, we would count these as the same graph. However, if we were to consider a third graph, G_3, with $V(G_3) = [3]$ and $E(G_3) = \{13, 23\}$. This graph has the same vertex set as G_2 and the same shape (see Fig. 11.25). But, it has a different edge set. Thus, we will consider it to be a different graph. The next definition will formalize this observation.

Definition 11.4.1 A graph on n vertices is *labeled* if its vertex set is the set $[n]$. Two labeled graphs are *different* if they have different edge sets or a different number of vertices.

Our goal in this section is to determine the number of labeled graphs on n vertices. However, some of these graphs will have the same shape, as in Fig. 11.25. The problem of counting only those graphs which have different shapes will be dealt with in the next section. We begin with determining the number of labeled graphs on n vertices.

Proposition 11.4.2 *For any labeled graph G on n vertices, $|E(G)| \leq \binom{n}{2}$. Further, equality is achieved by the complete graph.*

Proof Let G be a labeled graph on n vertices. Each edge of G can be thought of as a 2-element subset of $[n]$. There are $\binom{n}{2}$ such subsets. Thus, there are at most $\binom{n}{2}$ edges in G.

Each vertex in K_n is adjacent to every other vertex. So, for every $v \in V(K_n)$, $deg(v) = n - 1$. By the Handshaking Lemma, the number of edges in K_n is given by

$$|E(K_n)| = \frac{n(n-1)}{2} = \binom{n}{2}.$$

■

We can use the above proposition to determine the number of labeled graphs on n vertices.

Proposition 11.4.3 *There are $2^{\binom{n}{2}}$ labeled graphs on n vertices.*

Proof By Proposition 11.4.2, there are $\binom{n}{2}$ "potential" edges of a labeled graph G on n vertices. Each of these potential edges may be in the edge set or not. Thus, there are $2^{\binom{n}{2}}$ distinct possibilities for the edge set of G. Hence, there are $2^{\binom{n}{2}}$ distinct labeled graphs on n vertices.

■

Proposition 11.4.3 gives the total number of labeled graphs on n vertices. However, in reality we are much more interested in the number of labeled graphs on a fixed number of edges. This is covered in the next theorem.

Theorem 11.4.4 *The number of labeled graphs on n vertices and m edges is given by*

$$\binom{\binom{n}{2}}{m}.$$

Proof The fact that the n vertices are labeled implies that the $\binom{n}{2}$ potential edges are also labeled. Thus, we think of each of these $\binom{n}{2}$ potential edges as a labeled urn. We are to place m edges on this graph. These m edges are unlabeled until they are placed between a pair of vertices. For this reason, these m edges can be thought of as unlabeled balls. Further, each of the $\binom{n}{2}$ "urns" can receive at most one of these "balls." By Proposition 4.2.2, the number of ways to distribute m unlabeled balls into $\binom{n}{2}$ labeled urns such that no urn receives more than one ball is given by

$$\binom{\binom{n}{2}}{m}.$$

■

Suppose that we wanted to use Theorem 11.4.4 to determine the number of labeled trees on n vertices. A tree on n vertices will have $n-1$ edges by Proposition 11.1.8. However, Theorem 11.4.4 makes no insurance that the graph will be connected, whereas we require trees to be connected. Further, the method employed by Theorem 11.4.4 may result in a cyclic subgraph, which again violates the definition of a tree.

To aid in determining the number of labeled trees, we need to introduce the concept of a *rooted tree*. A *rooted tree* is simply a labeled tree where one vertex has been designated as the *root*. There are n choices for the root of tree on n vertices. Thus, there are n times as many rooted trees as labeled trees. Similarly, a *rooted forest* is a forest in which every tree in the forest is a rooted tree. With this concept in mind, we are now prepared to prove Cayley's Tree Counting Formula [13].

Theorem 11.4.5 *(Cayley's Tree Counting Formula) The number of labeled trees on n vertices is given by* n^{n-2}.

Proof We instead count the number of rooted trees on n vertices. To do this, we consider a sequence of rooted forests $F_1, \ldots, F_n$. Suppose that this sequence satisfies the following:

(i) Each F_i has n vertices.
(ii) The forest F_i consists of i rooted trees.
(iii) The forest F_{i+1} is contained in F_i for all i. That is, every tree in the forest F_{i+1} is a spanning subgraph of F_i (see Fig. 11.26).

This definition implies that F_1 is a rooted tree. Further, F_{i+1} has one less edge than F_i. For this reason, F_n consists of n vertices, each of degree zero.

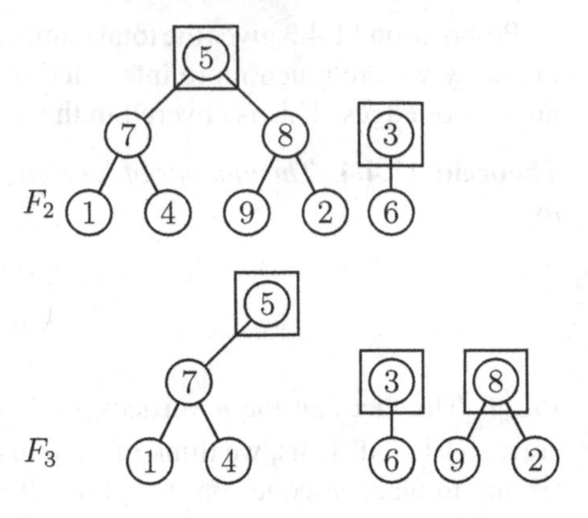

Fig. 11.26 A forest contained in another

Let N_n be the number of such sequences ending in F_n. Further, let T_n be the number of rooted trees with n vertices. Our goal is to determine N_n in two different ways. This will give a formula for T_n. This in turn will give the number of labeled trees on n vertices.

We compute N_n as follows:

(i) Select a rooted tree on n vertices. There are T_n ways to do this.
(ii) Let F_1 be the selected tree. The $n-1$ edges of F_1 can be deleted in any order to obtain F_n. Thus, there are $(n-1)!$ ways to delete the edges.

The Multiplication Principle implies that $N_n = T_n(n-1)!$.

To obtain our formula, we need to count N_n in a second way, one that does not involve T_n. To do this, we count the number of ways to obtain a possible F_i from a given F_{i+1}. This can be done as follows:

(i) Select one vertex of F_{i+1}. There are n ways to do this.
(ii) Suppose that v is the vertex chosen in (i). Select any of the i rooted trees in F_{i+1} that do not contain v. Add an edge from v to the root of the selected tree.

By the Multiplication Principle, there are ni ways to obtain a F_i from a given F_{i+1}. Ergo, the number of sequence that end in F_n is given by

$$N_n = \prod_{i=1}^{n-1} ni = n^{n-1}(n-1)!.$$

Equating these two formulas for N_n, we have that:

$$T_n(n-1)! = n^{n-1}(n-1)!$$
$$\Rightarrow T_n = n^{n-1}.$$

The number of rooted trees on n vertices is exactly n times the number of labeled trees on n vertices. Thus, the number of labeled trees on n vertices is $T_n/n = n^{n-2}$. ■

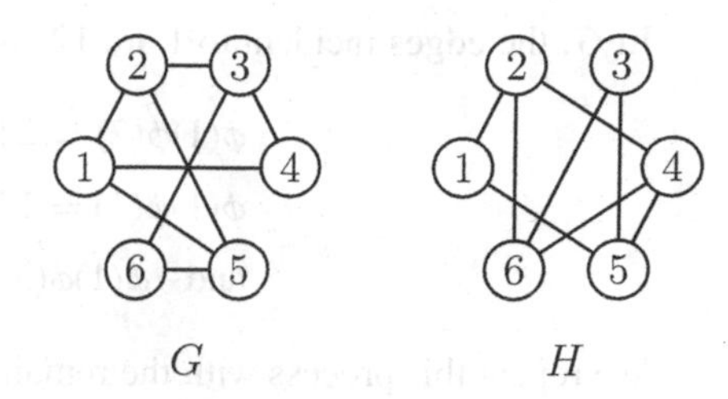

Fig. 11.27 Two isomorphic graphs

Exercise 11.4.6 Find the number of labeled graphs on n vertices for $n \leq 10$.

Exercise 11.4.7 In a *directed graph*, the edges are ordered pairs, rather than subsets. Find the number of directed labeled graphs on n vertices.

Exercise 11.4.8 Let $n \geq 3$. Determine the number of labeled graphs on n vertices such that neither 12 nor 13 is an edge in the graph.

Exercise 11.4.9 For $n \leq 6$ and $m \leq 6$, determine the number of labeled graphs on n vertices and m edges.

Exercise 11.4.10 Find the number of directed labeled graphs on n vertices and m directed edges (see Exercise 11.4.7).

Exercise 11.4.11 Determine the number of labeled multigraphs on n vertices and m edges.

Exercise 11.4.12 Let $n \geq 3$ and $m \geq 2$. Determine the number of labeled graphs on n vertices such that neither 12 nor 13 is an edge in the graph.

Exercise 11.4.13 Find the number of labeled trees on n vertices for $n \leq 10$.

11.5 Pólya Theory Revisited

In the previous section, we discussed counting labeled graphs. However, many of the labeled graphs will actually have the same shape. This is the case in Fig. 11.25. Often times, we only want to count each distinct shape once. To do this, we first formally define what we mean if two graphs have the same "shape."

Definition 11.5.1 Let G and H be graphs. An *isomorphism* ϕ is a bijection $\phi : V(G) \to V(H)$ such that $uv \in E(G)$ if and only if $\phi(u)\phi(v) \in E(H)$. If such a function exists, then we say that G and H are *isomorphic*. This situation is denoted $G \approx H$.

Example 11.5.2 Show that the two graphs in Fig. 11.27 are isomorphic.

Solution Let $\phi : V(G) \to V(H)$ be defined by $\phi(1) = 2$, $\phi(2) = 4$, $\phi(3) = 5$, $\phi(4) = 1$, $\phi(5) = 6$, and $\phi(6) = 3$. Clearly, ϕ is a bijection. Thus, it suffices to show that it preserves adjacency.

In G, the edges incident to 1 are 12, 14, and 15. Note that

$$\begin{aligned}\phi(1)\phi(2) &= 24 \in E(H),\\ \phi(1)\phi(4) &= 12 \in E(H),\\ \text{and}\quad \phi(1)\phi(5) &= 26 \in E(H).\end{aligned}$$

We repeat this process with the remaining edges of G:

$$\begin{aligned}23 \in E(G) \quad &\text{and} \quad \phi(2)\phi(3) = 45 \in E(H);\\ 25 \in E(G) \quad &\text{and} \quad \phi(2)\phi(5) = 46 \in E(H);\\ 34 \in E(G) \quad &\text{and} \quad \phi(3)\phi(4) = 15 \in E(H);\\ 36 \in E(G) \quad &\text{and} \quad \phi(3)\phi(6) = 35 \in E(H);\\ 56 \in E(G) \quad &\text{and} \quad \phi(5)\phi(6) = 36 \in E(H).\end{aligned}$$

Note that $|E(G)| = |E(H)| = 9$. Thus, $xy \in E(G)$ if and only if $\phi(x)\phi(y) \in E(H)$. Thus, we have the require bijection.

We now show that graph isomorphism is an equivalence relation. In this case, the equivalence classes are precisely those labeled graphs that are isomorphic.

Theorem 11.5.3 *Graph isomorphism is an equivalence relation.*

Proof It suffices to show that graph isomorphism is reflexive, symmetric, and transitive.

(i) **(Reflexive)**—In this case, the required bijection is the identity map. Hence, $G \approx G$.
(ii) **(Symmetric)**—Suppose that $G \approx H$. Then, there exists a bijection $\phi : V(G) \to V(H)$ such that $uv \in E(G)$ if and only if $\phi(u)\phi(v) \in E(H)$. Since ϕ is a bijection, it has an inverse, ϕ^{-1}. Since $G \approx H$, if $wz \in E(H)$, then there exists $xy \in E(G)$ such that $\phi(x) = w$ and $\phi(y) = z$. Thus, $\phi^{-1}(w) = x$ and $\phi^{-1}(z) = y$. It follows that $\phi^{-1}(w)\phi^{-1}(z) = xy \in E(G)$. So, ϕ^{-1} is an isomorphism. Therefore, $H \approx G$.
(iii) **(Transitive)**—Suppose that $G \approx H$ and $H \approx K$. Thus, there exists a bijection $\phi : V(G) \to V(H)$ such that $uv \in E(G)$ if and only if $\phi(u)\phi(v) \in E(H)$. Similarly, there exists a bijection $\psi : V(H) \to V(K)$ such that $wz \in E(H)$ if and only if $\psi(w)\psi(z) \in E(K)$. Note that the composition of bijections is likewise a bijection (see Exercise 1.4.8). We claim that $\psi \circ \phi$ is the required bijection. Suppose that $uv \in E(G)$. Thus, $\phi(u)\phi(v) \in E(H)$. This implies that $\psi(\phi(u))\psi(\phi(v)) \in E(K)$. A similar argument holds if $uv \notin E(G)$. Ergo, $G \approx K$. ■

To determine the number of non-isomorphic graphs on n vertices, we need to express the language of graphs into that of Pólya Theory. As usual, we will assume that the vertex set is $[n]$. Each edge, or potential edge, of a graph can be thought of as a

2-element subset of $[n]$. We denote the collection of all 2-element subsets of $[n]$ as $[n]^{(2)}$. The edge set of a graph G can be thought of as a 2-coloring of $[n]^{(2)}$. In other words, there is a function $C : [n]^{(2)} \to \{0, 1\}$ such that:

$$C(\{x, y\}) = \begin{cases} 0, & \text{if } xy \notin E(G) \\ 1, & \text{if } xy \in E(G). \end{cases}$$

To determine the number of non-isomorphic graphs on n vertices, we need to determine the number of distinguishable 2-colorings under the group action. Unfortunately, we have yet to determine the appropriate group action. Any vertex of a graph on n vertices can be relabeled with any element on $[n]$. Thus, the group acting on the vertices is the nth symmetric group, S_n. However, this is not necessarily the group acting on the edges. Each permutation in S_n will induce a permutation on the set of potential edges, $[n]^{(2)}$. We denote this group of induced permutations as $S_n^{(2)}$. This is sometimes referred to as the *pair group*. Our goal is to determine the structure of this group. In particular, to implement Pólya's Enumeration Theorem, we will need the cycle index polynomial associated with this group.

We now examine how a permutation on $[n]$ induces a permutation on $[n]^{(2)}$. Consider the permutation $(1, 2)(3) \in S_3$. This permutation flips 1 and 2 and leaves 3 fixed. The elements of $[3]^{(2)}$ are $\{1, 2\}$, $\{1, 3\}$, and $\{2, 3\}$. The permutation on $[3]^{(2)}$ induced by $(1, 2)(3)$ will leave $\{1, 2\}$ fixed. Since 1 and 2 are flipped by our original permutation, the 2-sets $\{1, 3\}$ and $\{2, 3\}$ will be swapped in the induced permutation. Thus, the permutation in $[3]^{(2)}$ induced by $(1, 2)(3)$ is $(\{1, 2\})(\{1, 3\}, \{2, 3\})$. Note that if we are working with small n, then it is much more common to represent the 2-sets without brackets or commas. Thus, $\{1, 2\}$ can be more compactly represented as 12. This will be our convention for the rest of this section.

Example 11.5.4 For each permutation in S_3, find the corresponding permutation in $S_3^{(2)}$.

Solution As discussed above, the permutation $(1, 2)(3) \in S_3$ induces the permutation $(12)(13, 23)$ in $S_3^{(2)}$. By a similar argument, the permutation $(1, 3)(2) \in S_3$ induces $(12, 23)(13)$ in $S_3^{(2)}$ and $(1)(2, 3) \in S_3$ induces $(12, 13)(23)$ in $S_3^{(2)}$. Obviously, the identity element $(1)(2)(3) \in S_3$ induces the corresponding element $(12)(13)(23) \in S_3^{(2)}$.

Finally, $(1, 2, 3) \in S_3$ rotates the three elements. Therefore in $[3]^{(2)}$, 12 is rotated into the position originally occupied by 23, 23 is rotated into the position occupied by 13, and 13 is rotated into the position of 12. This corresponds to the permutation $(12, 23, 13) \in S_3^{(2)}$. By a similar argument, $(1, 3, 2) \in S_3$ induces the permutation $(12, 13, 23) \in S_3^{(2)}$. ■

It turns out that $S_3^{(2)}$ has the same cycle index polynomial as S_3. However, this is not the case in general. Now, we will use the observations of Example 11.5.4 to find the number of non-isomorphic graphs with three vertices.

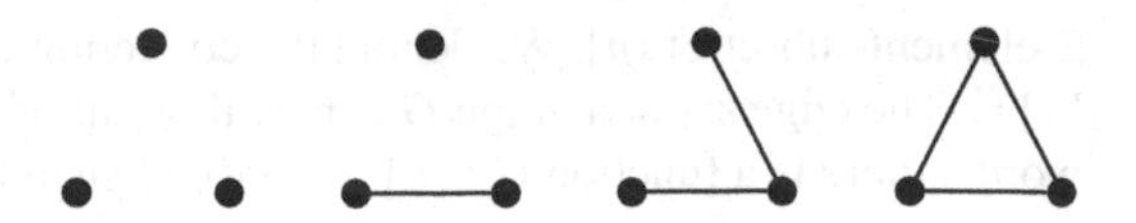

Fig. 11.28 Non-isomorphic graphs on three vertices

Example 11.5.5 Find the number of non-isomorphic graphs with three vertices.

Solution By Example 11.5.4, the cycle index polynomial associated with $S_3^{(2)}$ is

$$Z_{S_3^{(2)}}(x_1, x_2, x_3) = \frac{1}{6}\left(x_1^3 + 3x_1x_2 + 2x_3\right).$$

To apply Burnside's Lemma, we let $x_i = 2$ in the above equation. This gives

$$\frac{1}{6}\left(2^3 + 3(2^2) + 2(2)\right) = 4.$$

These four non-isomorphic graphs are given in Fig. 11.28. ■

Note that in many cases, we simply want the number of non-isomorphic graphs with n vertices and m edges. In this case, we use Pólya's Enumeration Theorem. This is only a simple modification of Example 11.5.4. In the above example, we let $x_i = 2$ to apply Burnside's Lemma. For Pólya's Enumeration Theorem, we instead let $x_i = 1+x^i$. We then find the coefficient of x^m in the resulting generating function. This will be examined further in our later examples.

We now generalize Example 11.5.4. Suppose that $n \geq 2$. If $\pi \in S_n$ and $x, y \in [n]$, then π induces a permutation g_π on $[n]^{(2)}$ defined by

$$g_\pi(\{x, y\}) = \{\pi(x), \pi(y)\}.$$

The first and most logical question is whether g_π is indeed a permutation for all $\pi \in S_n$. This is left as an exercise for the reader. Another question is how to efficiently examine the structure of $S_n^{(2)}$. Obviously, we do not want to examine all $n!$ elements of S_n to find their corresponding element in $S_n^{(2)}$. However, Example 11.5.4 suggests that we need only look at one permutation for each cycle type in S_n. Note that each cycle type in S_n is a partition of n (see Sect. 4.3). Further note that we can determine the number of permutations in S_n using the methods discussed in Sect. 2.7. Thus, it is beneficial to establish a formal bijection between the elements of S_n and $S_n^{(2)}$.

Theorem 11.5.6 *The map $g : S_n \to S_n^{(2)}$ defined by*

$$g(\pi) = g_\pi(\{x, y\}) = \{\pi(x), \pi(y)\}$$

is a bijection. Moreover, $g(\pi \circ \sigma) = g(\pi) \circ g(\sigma)$ for all $\pi, \sigma \in S_n$.

Proof It suffices to show that g is one-to-one, onto $S_n^{(2)}$, and that it preserves multiplication.

(i) (g is injective)—Suppose that $\pi \neq \sigma$ and choose $x \in [n]$ such that $\pi(x) \neq \sigma(x)$. Choose $y \in [n]$ such that $y \neq x$ and $\sigma(y) \neq \pi(x)$. The above definitions imply that $\pi(x) \notin \{\sigma(x), \sigma(y)\}$. It then follows that

$$g_\pi(\{x, y\}) = \{\pi(x), \pi(y)\} \quad \text{and} \quad g_\sigma(\{x, y\}) = \{\sigma(x), \sigma(y)\}$$

are different sets because $\pi(x) \notin g_\sigma(\{x, y\})$. Hence, $g_\pi \neq g_\sigma$. Therefore, g is injective.

(ii) (g is onto $S_n^{(2)}$)—This follows immediately from the definition of $S_n^{(2)}$.

(iii) (g preserves multiplication)—Note that

$$\begin{aligned} g(\pi \circ \sigma) &= g_{\pi\circ\sigma}(\{x, y\}) = \{\pi(\sigma(x)), \pi(\sigma(y))\} \\ &= g_\pi(\{\sigma(x), \sigma(y)\}) = g_\pi \circ g_\sigma. \end{aligned}$$

Hence, g preserves multiplication. ■

The mapping in Theorem 11.5.6 is a special case of a *group isomorphism*. For groups G and H, an isomorphism is a bijection $\phi : G \to H$ such that $\phi(x * y) = \phi(x) * \phi(y)$ for all $x, y \in G$. Again, the notion of isomorphism implies that the two groups have the same structure. In a sense, the two groups are algebraically the same. However, this does not mean that the two groups are the same group. For instance, the subgroup generated by the identity element in S_n is isomorphic to the subgroup generated by the identity element in S_k. However, the cycle index polynomial associated with the first subgroup is x_1^n, whereas the cycle index polynomial associated with the second subgroup is x_1^k. We will see additional instances of this in the coming examples.

We now proceed to find the number of non-isomorphic graphs on four vertices.

Example 11.5.7 Determine the number of non-isomorphic graphs on four vertices. In particular, for all $m \in \{0, 1, ..., 6\}$, determine the number of non-isomorphic graphs on four vertices and m edges.

Solution Note that the group acting on the vertices is S_4. Using the techniques of Sect. 2.7 (or a computer algebra system), the cycle index polynomial associated with S_4 is

$$Z_{S_4}(x_1, x_2, x_3, x_4) = \frac{1}{24}\left(x_1^4 + 6x_1^2x_2 + 3x_1^2x_2^2 + 8x_1x_3 + 6x_4\right).$$

Our goal is to determine the cycle index polynomial associated with $S_4^{(2)}$. To do this, we examine one permutation of each cycle type in S_4 and determine the corresponding permutation in $S_4^{(2)}$. Obviously, the identity element in S_4 leaves the elements of $[4]^{(2)}$ unchanged. Therefore, the permutation $(1)(2)(3)(4) \in S_4$ induces the permutation $(12)(13)(14)(23)(24)(34) \in S_4^{(2)}$. The cycle index monomial of this permutation is x_1^6.

Next, we consider the cycle type $[2, 1, 1]$ in S_4. We take $(1, 2)(3)(4)$ as an example of such a permutation. Obviously, this leaves the edges 12 and 34 fixed. This also swaps the edges 13 and 23 as well as the edges 14 and 24. Hence, the corresponding permutation in $S_4^{(2)}$ is $(12)(13, 23)(14, 24)(34)$. This has cycle index monomial $x_1^2x_2^2$.

Our next cycle type $[2, 2]$ will proceed similarly to the previous case. An example of cycle type $[2, 2]$ is $(1, 2)(3, 4)$. Clearly, this will leave the edges 12 and 34 fixed. This will swap the edges 13 and 24 as well as 14 and 23. Therefore, the induced permutation is $(12)(13, 24)(14, 23)(34)$ in $S_4^{(2)}$. This has cycle index monomial $x_1^2x_2^2$.

Table 11.1 Elements in S_4 and their corresponding element in $S_4^{(2)}$

Cycle type	No. in type	Example, π	g_π	$cim(g_\pi)$
[1, 1, 1, 1]	1	(1)(2)(3)(4)	(12)(13)(14)(23)(24)(34)	x_1^6
[2, 1, 1]	6	(1, 2)(3)(4)	(12)(13, 23)(14, 24)(34)	$x_1^2x_2^2$
[2, 2]	3	(1, 2)(3, 4)	(12)(13, 24)(14, 23)(34)	$x_1^2x_2^2$
[3, 1]	8	(1, 2, 3)(4)	(12, 23, 13)(14, 34, 24)	x_3^2
[4]	6	(1, 2, 3, 4)	(12, 23, 34, 14)(13, 24)	x_2x_4

Consider the cycle type [3, 1] in S_4. An example of such a permutation is (1, 2, 3)(4). This maps the edge 12 to the position originally occupied by 23. The edge 23 is in turn moved to position 13, which is moved to position 12. The edge 14 is moved to 34, which is in turn moved to 24, which is moved to 14. Thus, the permutation $(1,2,3)(4) \in S_4$ induces the permutation in $(12,23,13)(14,24,24) \in S_4^{(2)}$. The cycle index monomial for this permutation is x_3^2.

Finally, the cycle type [4] has example $(1,2,3,4) \in S_4$. This maps 12 to 23 which in turn is mapped to 34. Similarly, 34 is mapped to 14 which in turn is mapped to 12. This also swaps edges 13 and 24. Ergo, the associated permutation in $S_4^{(2)}$ is (12, 23, 34, 14)(13, 24). Our cycle index monomial is x_2x_4. A summary of these induced permutations is included in Table 11.1.

By Theorem 11.5.6, there is a bijection between the permutations in S_4 and $S_4^{(2)}$. Thus the cycle index polynomial associated with $S_4^{(2)}$ is

$$Z_{S_4^{(2)}}(x_1,x_2,x_3,x_4) = \frac{1}{24}\left(x_1^6 + 9x_1^2x_2^2 + 8x_3^2 + 6x_2x_4\right). \tag{11.1}$$

To determine the number of non-isomorphic graphs on four vertices, we let $x_i = 2$ for all i in Eq. (11.1). Hence, Burnside's Lemma yields that there are 11 non-isomorphic graphs on four vertices.

To obtain the number of non-isomorphic graphs with a given number of edges, we instead use Pólya's Enumeration Theorem. Here, we let $x_i = 1 + x^i$. This yields the pattern inventory

$$x^6 + x^5 + 2x^4 + 3x^3 + 2x^2 + x + 1.$$

The coefficient of x^m in this pattern inventory yields the number of non-isomorphic graphs on four vertices and m edges. Thus, there is one acceptable graph with six edges, one with five edges, two with four edges, three with three edges, two with two edges, one with one edge, and one with no edges. These graphs are shown in Fig. 11.29. ■

Using a similar technique as above, we can find the induced permutation for every permutation in S_5. These results are summarized in Table 11.2. Confirmation of these permutations is left as an exercise to the reader.

This will allow us to construct the cycle index polynomial as shown below:

Table 11.2 Elements in S_5 and their corresponding element in $S_5^{(2)}$

Cycle type	No. in type	Example, π	g_π	$cim(g_\pi)$
$[1,1,1,1,1]$	1	(1)(2)(3)(4)(5)	(12)(13)(14)(15)(23) (24)(25)(34)(35)(45)	x_1^{10}
$[2,1,1,1]$	10	(1, 2)(3)(4)(5)	(12)(13, 23)(14, 24) (15, 25)(34)(35)(45)	$x_1^4 x_2^3$
$[2,2,1]$	15	(1, 2)(3, 4)(5)	(12)(13, 24)(14, 23) (15, 25)(34)(35, 45)	$x_1^2 x_2^4$
$[3,1,1]$	20	(1, 2, 3)(4)(5)	(12, 23, 13)(14, 24, 34) (15, 25, 35)(45)	$x_1 x_3^3$
$[3,2]$	20	(1, 2, 3)(4, 5)	(12, 23, 13) (14, 25, 34, 15, 24, 35)(45)	$x_1 x_3 x_6$
$[4,1]$	30	(1, 2, 3, 4)(5)	(12, 23, 34, 14)(13, 24) (15, 25, 35, 45)	$x_2 x_4^2$
$[5]$	24	(1, 2, 3, 4, 5)	(12, 23, 34, 45, 15) (13, 24, 35, 14, 25)	x_5^2

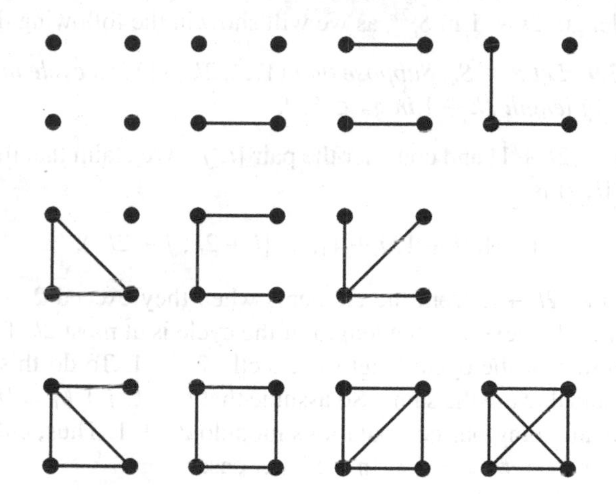

Fig. 11.29 Non-isomorphic graphs on four vertices

$$Z_{S_5^{(2)}}(x_1, x_2, x_3, x_4, x_5, x_6) =$$

$$\frac{1}{120}\left(x_1^{10} + 10x_1^4x_2^3 + 15x_1^2x_2^4 + 20x_1x_3^3 + 20x_1x_3x_6 + 24x_5^2 + 30x_2x_4^2\right).$$

While these methods will allow us to determine $S_n^{(2)}$ for small n, it is not practical for large n. For this, we will need to examine how elements in $[n]^{(2)}$ are affected by a permutation in S_n. Any permutation S_n can be written as a product of disjoint cycles. Thus, it suffices to determine how these cycles affect the edges. In particular, we must determine the cycle type for the corresponding element in $S_n^{(2)}$. In this case, if we take two elements, say 1 and 2, then they are either in the same cycle or on different cycles. We will use this observation to motivate our next three examples. Each of these examples will be used to motivate a more general statement. In each of these examples, we omit the fixed points for simplicity.

Example 11.5.8 Let $(1,2,3,4,5,6,7) \in S_n$. Find the corresponding element in $S_n^{(2)}$.

Solution Let $\pi = (1,2,3,4,5,6,7)$. The edge 12 is moved to position 23. Similarly, 23 is moved to position 34, which in turn is moved to position 45. We also have that 45 is moved to position 56, which is mapped to position 67. Finally, 67 is moved to position 17 which is moved to 12. This gives rise to the cycle $(12,23,34,45,56,67,17)$.

By a similar argument, the corresponding permutation in $S_n^{(2)}$ will have the cycles $(13,24,35,46,57,16,27)$ and $(15,26,37,14,25,36,47)$. Thus, the corresponding permutation in $S_n^{(2)}$ is $(12, 23, 34, 45, 56, 67, 17)(13, 24, 35, 46, 57, 16, 27)(15, 26, 37, 14, 25, 36, 47)$.

Notice that in the above example, our cycle of length seven in S_n gave rise to three cycles of length seven in $S_n^{(2)}$. In general, a cycle of length $2k+1$ in S_n will give rise to k cycles of length $2k+1$ in $S_n^{(2)}$, as we will show in the following theorem.

Theorem 11.5.9 *Let $\pi \in S_n$. Suppose that $(1, ..., 2k+1)$ is a cycle in π. This gives rise to k cycles of length $2k+1$ in $g_\pi \in S_n^{(2)}$.*

Proof Let $i, j \in [2k+1]$ and consider the pair $\{i, j\}$. We claim that the the cycle of g_π containing $\{i, j\}$ is

$$(\{i, j\}, \{i+1, j+1\}, ..., \{i+2k, j+2k\}),$$

where we subtract $2k+1$ from the elements when they exceed $2k+1$. Since g_π maps each pair to the next one, the length of the cycle is at most $2k+1$.

We must show that the cycle length is exactly $2k+1$. To do this, assume that there are two pairs that are the same. So assume that $\{i+\ell, j+\ell\} = \{i+\ell', j+\ell'\}$. Effectively, we are doing our computations modulo $2k+1$. Thus, either $\ell = \ell'$, or $i+\ell = j+\ell'$ and $i+\ell' = j+\ell$. In the latter case,

$$i - j = j - i \text{ (mod 2k + 1)}$$
$$\Rightarrow 2(i-j) = 0 \text{ (mod 2k + 1)}.$$

Since 2 does not divide $2k+1$, we must have that $i = j$, a contradiction. Thus, each induced cycle of g_π has length $2k+1$. These cycles partition the set $[2k+1]^{(2)}$, which has $\binom{2k+1}{2}$ elements. Therefore, there are $\binom{2k+1}{2}/(2m+1) = m$ such cycles. ■

This theorem also holds for cycles of length one. The case when our cycle is of even length has a similar structure. As an example, consider $(1,2,3,4) \in S_n$. This behaves identically to the corresponding cycle in the particular case shown in Example 11.5.7. Thus, the corresponding permutation in $S_n^{(2)}$ is $(12,23,34,14)(13,24)$. So, it induces a cycle of length four and a cycle of length two. We will see a similar result for the 8-cycle in the next example.

Example 11.5.10 Let $(1, 2, 3, 4, 5, 6, 7, 8) \in S_n$. Find the corresponding element in $S_n^{(2)}$.

Solution Finding the individual cycles follows in a similar manner to the above examples. Therefore, we simply list those cycles:

$$(12, 23, 34, 45, 56, 67, 78, 18),$$
$$(13, 24, 35, 46, 57, 68, 17, 28),$$
$$(14, 25, 36, 47, 58, 16, 27, 38),$$
$$\text{and } (15, 26, 37, 48).$$ ■

Again, this 8-cycle induces three cycles of length eight and one cycle of length four. In general, a cycle of length $2k$ gives rise to $k-1$ cycles of length $2k$ and one cycle of length k.

Theorem 11.5.11 *Let $\pi \in S_n$. Suppose that $(1, ..., 2k)$ is a cycle in π. This gives rise to $k-1$ cycles of length $2k$ and one cycle of length k in $g_\pi \in S_n^{(2)}$.*

Proof Let $i, j \in [2k]$ and consider the pair $\{i, j\}$. We have two different cases.

In the first case, suppose that $|i-j| = k$. Without loss of generality, we may assume that $i = 1$ and $j = k+1$. The cycle in g_π containing $\{1, k+1\}$ is

$$(\{1, k+1\}, \{2, k+2\}, .., \{k, 2k\}).$$

This cycle is unique and has length k.

In the second case, we assume that $|i-j| \neq k$. We claim that the cycle in g_π containing $\{i, j\}$ is

$$(\{i, j\}, \{i+1, j+1\}, ..., \{i+2k-1, j+2k-1\}),$$

where we subtract $2k$ from the elements when they exceed $2k$. As with the proof of Theorem 11.5.9, this cycle is of length $2k$. There are $\binom{2k}{2}$ elements in $[2k]^{(2)}$. The first case covers k of these pairs, leaving $2k^2 - 2k$ pairs. Each of the cycles in the second case covers $2k$ pairs. Thus, there are $k-1$ such cycles. ■

Previously, we have dealt with the case where both elements are in the same cycle. Our final case will consider the possibility that the elements are in disjoint cycles.

Example 11.5.12 Consider $\pi = (1, 2, 3, 4, 5)(6)$, $\sigma = (1, 2, 3)(4, 5, 6)$, and $\tau = (1, 2)(3, 4, 5, 6)$. Find the cycles of g_π, g_σ, and g_τ whose members come from disjoint cycles of π, σ, and τ, respectively.

Solution The derivation of the appropriate cycles follows in a similar manner to Example 11.5.7, Theorems 11.5.9, and 11.5.11. Thus, we simply list the required cycles in g_π, g_σ, and g_τ. In g_π, the unique cycle whose members come from disjoint cycles of $\pi = (1, 2, 3, 4, 5)(6)$ is $(16, 26, 36, 46, 56)$. Notice that the cycles of π were of length one and five. Further, this gives rise to $gcd(1, 5) = 1$ cycle of length $lcm(1, 5) = 5$.

Similarly, the cycles in g_σ whose elements come from two different cycles of σ are (14, 25, 36), (15, 26, 34), and (16, 24, 35). Notice that the cycles of σ were both of length three. Further, this gives rise to $gcd(3,3) = 3$ cycles each of length $lcm(3,3) = 3$.

Finally, the cycles in g_τ whose elements come from two different cycles of τ are (13, 24, 15, 26) and (14, 25, 16, 23). Notice that the cycles of τ were of length two and four. Further, this gives rise to $gcd(2,4) = 2$ cycles of length $lcm(2,4) = 4$. ■

We will generalize the above example in the next theorem.

Theorem 11.5.13 *Each distinct pair of disjoint cycles* $(1, ..., k_1)$ *and* $(k_1+1, ..., k_1+k_2)$ *in* π *gives rise to* $gcd(k_1, k_2)$ *disjoint cycles each of length* $lcm(k_1, k_2)$ *in* g_π.

Proof Consider the pair $\{1, k_1 + 1\}$. The cycle in g_π containing this pair is

$$(\{1, k_1 + 1\}, \{2, k_1 + 2\}, ..., \{lcm(k_1, k_2), k_1 + lcm(k_1, k_2)\}),$$

where we subtract k_1 from the first component when it exceeds k_1 and we subtract k_2 from the second component when it exceeds $k_1 + k_2$. Using a similar argument to Theorem 11.5.9, this cycle is of length $lcm(k_1, k_2)$.

By the Multiplication Principle, there are k_1k_2 pairs of elements where one element comes form $(1, ..., k_1)$ and the second element comes from $(k_1 + 1, ..., k_1 + k_2)$. Each of the cycles described above has $lcm(k_1, k_2)$ of these pairs. Ergo, there are $k_1k_2/lcm(k_1, k_2) = gcd(k_1, k_2)$ such cycles. ■

We end this section by giving several examples of how Theorems 11.5.9, 11.5.11, and 11.5.13 interact.

Example 11.5.14 Given the following permutations in S_{14}, find the cycle index monomial for the induced permutation in $S_{14}^{(2)}$:

(i) $\pi = (1, ..., 7)(8, ..., 14)$;
(ii) $\sigma = (1, ..., 6)(7, ..., 14)$;
(iii) $\tau = (1, 2, 3)(4, 5, 6)(7, 8, 9, 10)(11, 12, 13, 14)$.

Solution *(i)* We begin by determining the cycle lengths in g_π that arise from two elements from the same cycle in π. The first cycle of length seven gives rise to three cycles of length seven in g_π by Theorem 11.5.9. Thus, we have a factor of x_7^3 in $cim(g_\pi)$. The same is true for the second 7-cycle. Hence, we have an additional factor of x_7^3 in $cim(g_\pi)$.

We also have terms arising from elements in different cycles. Both cycles are of length seven. So, there are $gcd(7,7) = 7$ cycles of length $lcm(7,7) = 7$ in g_π that contain one element from each cycle by Theorem 11.5.13. Thus, we have a factor of x_7^7 in $cim(g_\pi)$. Therefore,

$$cim(g_\pi) = (x_7^3)(x_7^3)(x_7^7) = x_7^{13}.$$

(ii) Again, we begin by determining the cycle lengths in g_σ that arise from two elements of the same cycle in σ. By Theorem 11.5.11, the 6-cycle gives rise to two cycles of length six and one cycle of length three. This gives a factor of $x_3x_6^2$ in

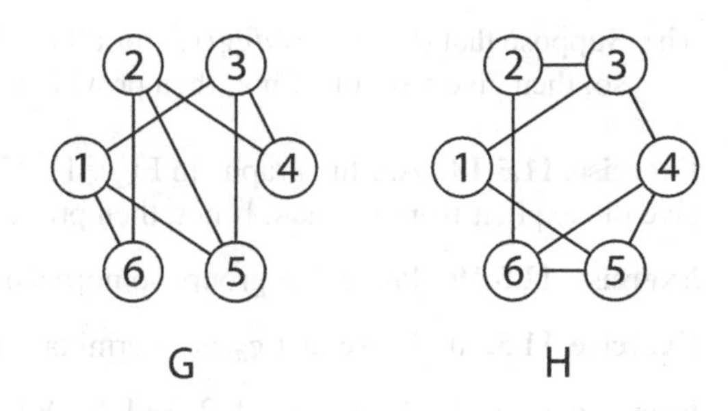

Fig. 11.30 The graphs for Exercise 11.5.15

$cim(g_\sigma)$. Similarly, the 8-cycle gives rise to three cycles of length eight and one cycle of length four by Theorem 11.5.11. Ergo, there is a corresponding factor of $x_4x_8^3$ in $cim(g_\sigma)$.

We now deal with the terms arising from elements in different cycles. The cycles are of length four and six. By Theorem 11.5.13, there are $gcd(4,6) = 2$ cycles of length $lcm(4,6) = 12$ in g_σ. It follows that associated factor in $cim(g_\sigma)$ is x_{12}^2. Therefore,

$$cim(g_\sigma) = (x_3x_6^2)(x_4x_8^3)(x_{12}^2) = x_3x_4x_6^2x_8^3x_{12}^2.$$

(iii) By Theorem 11.5.9 there is one cycle arising from the first 3-cycle of length three. This contributes a factor of x_3 to the cycle index monomial. Likewise for the second 3-cycle. Similarly, Theorem 11.5.11 shows that there is one cycle of length four and one cycle of length two arising from each 4-cycle in τ. Therefore, each contributes a factor of x_2x_4 to $cim(g_\tau)$.

Finally we deal with the term arising from elements of different cycles. In each case we will apply Theorem 11.5.13. The two 3-cycles yield $gcd(3,3) = 3$ cycles of length $lcm(3,3) = 3$ in g_τ. This contributes a factor of x_3^3 to the cycle index monomial. The two 4-cycles yield $gcd(4,4) = 4$ cycles each of length $lcm(4,4) = 4$ in g_τ. This contributes a factor of x_4^4 to $cim(g_\tau)$. A 3-cycle and a 4-cycle yield $gcd(3,4) = 1$ cycle of length $lcm(3,4) = 12$. The associated factor in $cim(g_\tau)$ is x_{12}. There are four ways to select one 3-cycle and one 4-cycle. Thus,

$$cim(g_\tau) = (x_3)^2(x_2x_4)^2(x_3^3)(x_4^4)(x_{12})^4 = x_2^2x_3^5x_4^6x_{12}^4. \quad ■$$

Exercise 11.5.15 Show that the two graphs in Fig. 11.30 are isomorphic.

Exercise 11.5.16 Suppose that G and H are graphs with $|V(G)| = |V(H)|$ and $|E(G)| > |E(H)|$. Show that G and H cannot be isomorphic.

Exercise 11.5.17 Suppose that G and H are graphs with $V(G) = \{u_1, ..., u_n\}$ and $V(H) = \{v_1, ..., v_n\}$. Without loss of generality, we can assume that degrees of the vertices are non-increasing order, in other words, $deg(u_1) \geq ... \geq deg(u_n)$ and $deg(v_1) \geq ... \geq deg(v_n)$.

(i) Suppose that there exists $i \in [n]$ such that $deg(u_i) \neq deg(v_i)$. Show that G and H cannot be isomorphic.

(ii) Suppose that $deg(u_i) = deg(v_i)$ for all $i \in [n]$. Does it follow that $G \approx H$? If so, then give a proof. If not, then provide a counterexample.

Exercise 11.5.18 Are the graphs in Figs. 11.27 and 11.30 isomorphic? If so, then give the explicit isomorphism. If not, then prove that they are not isomorphic.

Exercise 11.5.19 Prove that group isomorphism is an equivalence relation.

Exercise 11.5.20 Prove that g_π is a permutation in $S_n^{(2)}$ for all $\pi \in S_n$.

Exercise 11.5.21 Consider $(1,2,3)(4,5,6)(7) \in S_7$. Find the corresponding permutation in $S_7^{(2)}$.

Exercise 11.5.22 Consider $(1,2)(3,4,5,6,7) \in S_7$. Find the corresponding permutation in $S_7^{(2)}$.

Exercise 11.5.23 Consider $(1,2,3,4)(5,6)(7,8) \in S_8$. Find the corresponding permutation in $S_8^{(2)}$.

Exercise 11.5.24 Confirm the entries in Table 11.2. Use this to confirm the formula for the cycle index polynomial for $S_5^{(2)}$. How many non-isomorphic graphs are there on five vertices? In particular, how many non-isomorphic graphs have five vertices and five edges?

Exercise 11.5.25 Construct a table for $S_6^{(2)}$ using Tables 11.1 and 11.2 as a model. What is the cycle index polynomial associated with $S_6^{(2)}$? How many non-isomorphic graphs are there on six vertices? In particular, how many non-isomorphic graphs have six vertices and ten edges?

Exercise 11.5.26 Let $S_n^{(3)}$ be the permutation group induced on the 3-sets of $[n]$ by S_n. Construct a table for $S_4^{(3)}$ using Tables 11.1 and 11.2 as a model. What is the cycle index polynomial associated with $S_4^{(3)}$?

Exercise 11.5.27 Let $S_n^{(3)}$ be the permutation group induced on the 3-sets of $[n]$ by S_n. Construct a table for $S_5^{(3)}$ using Tables 11.1 and 11.2 as a model. What is the cycle index polynomial associated with $S_5^{(3)}$?

Exercise 11.5.28 Let $S_n^{(4)}$ be the permutation group induced on the 4-sets of $[n]$ by S_n. Construct a table for $S_5^{(4)}$ using Tables 11.1 and 11.2 as a model. What is the cycle index polynomial associated with $S_5^{(4)}$?

Exercise 11.5.29 Let $S_n^{(4)}$ be the permutation group induced on the 4-sets of $[n]$ by S_n. Construct a table for $S_6^{(4)}$ using Tables 11.1 and 11.2 as a model. What is the cycle index polynomial associated with $S_6^{(4)}$?

Exercise 11.5.30 Give a formal proof that the cycles described in Theorem 11.5.11 are of length $2k$.

Exercise 11.5.31 Give a formal proof that the cycles described in Theorem 11.5.13 are of length $lcm(k_1, k_2)$.

Exercise 11.5.32 Given the following permutations in S_{16}, find the cycle index monomial for the induced permutation in $S_{16}^{(2)}$:

(i) $\pi = (1, ..., 9)(10, ..., 16)$;
(ii) $\sigma = (1, ..., 10)(11, ..., 16)$;
(iii) $\tau = (1, ..., 5)(6, ..., 12)(13, ..., 16)$.

Exercise 11.5.33 Given the following permutations in S_{17}, find the cycle index monomial for the induced permutation in $S_{17}^{(2)}$:

(i) $\pi = (1, ..., 7)(8, ..., 17)$;
(ii) $\sigma = (1, ..., 5)(6, ..., 10)(11, ..., 17)$;
(iii) $\tau = (1, 2)(3, 4, 5, 6)(7, 8, 9, 10, 11)(12, 13, 14, 15, 16, 17)$.

Appendices

A.1 On-Line Resources

Over the last two decades, the internet has grown in its scope and its usage. This scope has grown to include resources that would benefit students studying mathematics, including combinatorics. *The author assumes no responsibility for the validity or content of sites other than his own.*

(i) **Author's webpage**
faculty.etsu.edu/beelerr
Contains (reasonably) up to date links to the sites below as well as other potentially useful information.

(ii) **Sloan's On-line Encyclopedia of Integer Sequence**
oeis.org
Allows you to look up specific sequences that you may encounter by both the numbers in the sequence and the name of the sequence. Also gives useful information such as the generating function of the sequence and alternate combinatorial interpretations.

(iii) **The Sage Notebook**
www.sagemath.org
An on-line computer algebra system. Among other things, it allows you to expand large polynomials such as in Chap. 5, factor polynomials such as in Chap. 6, and program.

(iv) ***generatingfunctionology* by Herbert Wilf**
www.math.upenn.edu/~wilf/DownldGF.html
A more advanced book on generating functions generously made available free on-line by the author and the publisher.

(v) **Free Math Books -**
people.math.gatech.edu/~cain/textbooks/onlinebooks.html
A site that contains links to many free math books on a variety of subjects.

(vi) **Wolfram Mathworld -**
mathworld.wolfram.org
An on-line encyclopedia of mathematics. Contains useful information on the

R. A. Beeler, *How to Count*, DOI 10.1007/978-3-319-13844-2

topics covered in this book along with information on virtually any other mathematical topic.

(vii) **The Small Graph Database -**
http://sage.cs.drake.edu/grout/graphs/
A database that includes all graphs on eight vertices or less. The database allows you to restrict your search on criteria such as connectivity, hamiltonicity, and number of edges.

A.2 A List of Notation

Here, we provide a list of notation for easy reference.

$|A|$ the cardinality of a set A

$A \cup B$ the union of sets A and B

$A \cap B$ the intersection of sets A and B

$A - B$ the difference of sets A and B

$A \times B$ the Cartesian product of sets A and B

A^c the complement of a set A

$P(A)$ the power set of A

[n] the set $\{1, ..., n\}$

$\mathbb{Z}$ the set of integers

$\mathbb{Z}^+$ the set of positive integers

$\mathbb{N}$ the set of non-negative integers

$\mathbb{Q}$ the set of quotients

$\mathbb{R}$ the set of real numbers

$\mathbb{C}$ the set of complex numbers

$cyc(\pi)$ the cycle index of a permutation π

F_n The nth Fibonacci number

n^k the number of k-tuples of $[n]$ (replacement allowed)

$n! - n$ factorial, the number of permutations on $[n]$

$P(n,k)$ the number of ordered k-sets of $[n]$

$\binom{n}{k}$ the number of k-element subsets of $[n]$

$s(n,k)$ the number of permutations on $[n]$ with cycle index k

$S(n,k)$ the number of distributions of n labeled balls into k unlabeled urns with no empty urns

$p(n,k)$ the number of partitions of n into k non-negative integers

$p(n)$ the number of partitions of n into non-negative integers

D_n the number of derangements on $[n]$

$\lfloor x \rfloor$ the floor of x

M_n the nth Ménage number

m_n the nth relaxed Ménage number

$cim(\pi)$ the cycle index monomial of a permutation π

Z_G the cycle index polynomial of a group G

C_n	the nth cyclic group
D_n	the nth dihedral group
S_n	the nth symmetric group
$\phi(n)$	the Euler ϕ-function of n
$gcd(n,m)$	the greatest common divisor of n and m
$lcm(n,m)$	the least common multiple of n and m

A.3 A Combinatorialist's Guide to Sage

Sage is an online computer algebra system found at www.sagemath.org. There are several advantages to using Sage:

(i) It is powerful;
(ii) It is free;
(iii) It is available on any computer with Internet access.

In this appendix, we give a brief introduction to the features in Sage that would be of interest to the combinatorialist. For a more comprehensive introduction, the reader is referred to the *Sage Beginner's Guide* by Craig Finch [21] or the resources listed below. Particular interest will be shown to those features referred to in this text.

First of all, help in Sage can be accessed by typing in a key word, followed by the question mark, then hitting the Tab key. So, if we wanted to see help on the Taylor series, we would type

```
taylor?
```

Followed by the Tab key. There is a Tour of Sage located at:

http://www.sagemath.org/pdf/a_tour_of_sage.pdf

and a more in-depth Sage Tutorial located at:

http://www.sagemath.org/pdf/SageTutorial.pdf.

A two page Quick Reference guide is located at:

http://wiki.sagemath.org/quickref

Basic arithmetic can be handled in the way we would expect. For instance, addition can be accomplished with

```
4+3
```

Once this command is entered, you may click on "evaluate" link to calculate the result. Similarly, subtraction is accomplished with '−', multiplication is accomplished with '*', division is accomplished with '/', and exponentiation is accomplished with '^'. Again, once these commands are entered into Sage, you must click on the "evaluate" link to calculate the result.

In algebra, it is typically best to define the variables that will be used. While Sage will give us the variable x for free, we must define any additional variables we wish to use. For instance, if we wish to use the variable y, then we would define it with the command

```
y=var('y')
```

On separate lines (or the same line separated by semicolons), we could them define any additional variables that we might need.

In this text, much time was devoted to solving and factoring algebraic expressions. To find the roots of a polynomial, say $x^2 - 5x + 6$, one would type

```
solve(x^2 - 5*x+6,x)
```

After clicking on the "evaluate" button, we would see the output

```
[x==3, x==2]
```

Similarly, to factor a polynomial, we would type

```
factor(x^2 - 5*x+6,x)
```

Clicking the "evaluate" button will yield

```
(x - 3)*(x - 2)
```

We note that by default Sage will only factor into polynomials with integer coefficients. In order to achieve complex coefficients, we simply specify that we want to work in $\mathbb{C}$. So for instance, inputting

```
realpoly.<x> = PolynomialRing(CC)
factor(x^2+1,x)
```

and clicking "evaluate" will yield

```
(x - I) * (x + I)
```

To differentiate a function, we can use the diff command. For instance, to compute the derivative of x^3 we would type

```
diff(x^3, x)
```

To multiply out a polynomial, such as the generating functions in Chapter 5, we would type

```
expand((x+1)*(x - 1))
```

Again, clicking on the "evaluate" button will yield

```
x^2 - 1
```

Perhaps a more efficient way of dealing with generating functions involves using Taylor series. For instance, the command

```
taylor(1/(1 - x),x,0,4)
```

will return the fourth degree Taylor polynomial for $\frac{1}{1-x}$. Namely,

```
x^4+x^3+x^2+x+1
```

In practice, we need only consider a particular coefficient of the resulting series. This can be done with the following command:

```
taylor(1/(1 - x),x,0,10).coefficient(x^10)
```

This command will return the coefficient of x^{10} in the Taylor series for $\frac{1}{1-x}$.

Sage is also capable of dealing with multivariate Taylor series such as in Sect. 5.4. Suppose that we wish to use Sage to solve the first example in that section. Namely, we want to find the number of non-negative integer solutions to:

$$x_1 + x_2 + x_3 + x_4 = 10;$$
$$x_1 + 2x_2 + 3x_3 + 4x_4 = 20.$$

As discussed in that section, this is given by the coefficient of $x^{10}y^{20}$ in

$$\left(\frac{1}{1-xy}\right)\left(\frac{1}{1-xy^2}\right)\left(\frac{1}{1-xy^3}\right)\left(\frac{1}{1-xy^4}\right).$$

As usual, we will have to specify that y is a variable. Sage will give the thirtieth degree Taylor polynomial with the commands

```
y=var('y')
f(x,y)=(1/(1-x*y))*(1/(1-x*y^2))*(1/(1-x*y^3))*(1/(1-x*y^4))
taylor(f(x,y),(x,0),(y,0),30)
```

In this case, the 30 specifies that the sum of the exponents on x and y will never exceed 30. As usual, we want individual coefficients. If we replace the last line of this code with

```
taylor(f(x,y),(x,0),(y,0),30).coefficient(x^10)
```

then clicking "evaluate" will yield:

```
14*y^20 + 12*y^19 + 10*y^18 + 8*y^17 + 7*y^16 + 5*y^15 + 4*y^14 +
3*y^13 + 2*y^12 + y^11 + y^10
```

Namely, this will consider y as a constant, and return a polynomial in terms of y. In order to return the coefficient of $x^{10}y^{20}$, we would replace the last line of code with

```
taylor(f(x,y),(x,0),(y,0),30).coefficient(x^10).coefficient(y^20)
```

Clicking "evaluate" will yield 14.

If we instead want all terms in the Taylor series such that the exponent on x is at most 10 and the exponent on y is at most 20, then we would input

```
T=taylor(f(x,y), (x,0), (y,0), 30).polynomial(QQ)
T.mod(x^11).mod(y^21)
```

Clicking evaluate will yield the required polynomial. If we wish to evaluate this polynomial at a specific value of x and y (say, $x = y = 1$), then we replace the last line with:

```
g(x,y)=T.mod(x^11).mod(y^21)
```

g(1,1)

Clicking evaluate will yield 528.

Sage will also do partial fraction decompositions. For instance, suppose we want to find the partial fraction decomposition of

$$f(x) = \frac{1}{x^2 - 5x + 6}.$$

In Sage, we would type

f = 1/(x^2 − 5*x + 6)

to define the function $f(x)$. On the next line, we would type

f.partial_fraction(x)

Clicking on "evaluate" will yield

1/(x − 3)−1/(x − 2)

Again, Sage will do partial fractions into terms with linear coefficients unless otherwise specified.

We now turn our attention to linear algebra and solving a system of linear equations, such as what was necessary in Chap. 6. Matrices are defined in an intuitive way. For instance,

A = Matrix([[2,3,4], [1,2,5], [1,2,3]])

will define the matrix,

$$A = \begin{bmatrix} 2 & 3 & 4 \\ 1 & 2 & 5 \\ 1 & 2 & 3 \end{bmatrix}.$$

Similarly, vectors can be defined as

B = vector([0,3, −1])

To solve the system $AX = B$, we would type

A.solve_right(B)

After clicking on "evaluate" Sage will yield the result

(5,6,−2)

The same result can be achieved by typing

A\ B

and clicking "evaluate."

Sage will also compute many of the combinatorial symbols we have discussed. For instance, typing

```
Permutations(3).list()
```

and clicking "evaluate" will return all permutations on three objects:

[1,2,3], [1,3,2], [2,1,3], [2,3,1], [3,1,2], [3,2,1]

Often, we are not interested in the actual permutations, simply the number of permutations. This can be accomplished with

```
Permutations(3).cardinality()
```

Clicking on "evaluate" will return

6

However, a simpler way to obtain $n!$ is using factorial(n).

We can also obtain all permutations of n of length k. For, instance typing

```
Permutations(5,2).list()
```

and clicking on "evaluate" will return

[1,2], [1,3], [1,4], [1,5], [2,1], [2,3], [2,4], [2,5], [3,1], [3,2], [3,4], [3,5], [4,1], [4,2], [4,3], [4,5], [5,1], [5,2], [5,3], [5,4]

Again, if we simply want the number of such permutations, we instead type

```
Permutations(5,2).cardinality()
```

Clicking on "evaluate" will return

20

Combinations can be achieved by typing

```
Combinations(5,2).list()
```

Clicking "evaluate" will list all 2-subsets of 5:

[1,2], [1,3], [1,4], [1,5], [2,3], [2,4], [2,5], [3,4], [3,5], [4,5]

As usual, if we want to compute $\binom{5}{2}$, we instead type

```
Combinations(5,2).cardinality()
```

Clicking on "evaluate" will yield the desired result

10

Other sequences are built into Sage as well. For instance, the partitions of a number can be found with the command

```
Partitions(5).list()
```

Clicking "evaluate" will list all partitions of 5, namely

[1,1,1,1,1], [1,1,1,2], [1,1,3], [1,2,2], [1,4], [2,3], [5]

Again, to find the number of partitions, we instead use the command

```
Partitions(5).cardinality()
```

Clicking on "evaluate" will yield

7

Sage will also produce partitions of n into k parts. For instance,

```
Partitions(5, length=2).list()
```

and clicking "evaluate" will yield

[1,4], [2,3]

Similarly, if we change our command to

```
Partitions(5, length = 2).cardinality()
```

then this will return the number of partitions of 5 into 2 parts.

Continuing on with applications to the problem of distributions, the Stirling numbers are also built into Sage. For instance,

```
stirling_number1(10,5)
```

and clicking "evaluate" will yield $s(10,5)$, 269325. Stirling numbers of the second kind can be obtained with a similar command.

```
stirling_number2(10,5)
```

and clicking "evaluate" will yield $S(10,5)$, 42525.

Similarly, Sage will evaluate the Fibonacci numbers under the indexing $F_1 = F_2 = 1$ with the command

```
fibonacci(5)
```

Clicking "evaluate" will yield the fifth Fibonacci number (under this indexing), namely 5.

Sage will also give derangements of a set with a little work. We first use the command

```
mset = [1,2,3,4]
```

to define the set we wish to derange. The command

```
derangements(mset)
```

followed by "evaluate" will yield all derangements on [4].

To obtain the number of such derangements can be found with the command

```
number_of_derangements(mset)
```

Sage will also allow us to examine groups as well. The Euler ϕ-function is implemented with the euler_phi(n) command. Several common groups have been coded into Sage. For instance, if you know that G is the eighth symmetric group, this can be implemented with

```
G = SymmetricGroup(8)
```

Similarly, if G is the eighth dihedral group, then use the command

```
G = DihedralGroup(8)
```

Finally, if G is the eighth cyclic group, then use the command

```
G = CylcicPermutationGroup(8)
```

Again, to list the elements of the group, we follow these commands with .list(). By default, Sage lists the elements as products of disjoint cycles. So, if we enter

```
DihedralGroup(5).list()
```

and click "evaluate," Sage returns

```
[(), (2,5)(3,4), (1,2)(3,5), (1,2,3,4,5), (1,3)(4,5), (1,3,5,2,4),
(1,4)(2,3), (1,4,2,5,3), (1,5,4,3,2), (1,5)(2,4)]
```

Similarly, following these commands with .cardinality() will return the number of elements in the group. So, if we enter

```
SymmetricGroup(5).cardinality()
```

and click "evaluate," then Sage will return the number of elements in S_5.

Suppose that we know the generators of a group and wish to know the number of elements in the group. For this example, we will assume that $G = <\rho, \sigma>$, where $\rho = (1,2,3,4,5,6,7)$ and $\sigma = (1)(2,3,5)(4,7,6)$. Both of these permutations are in S_7, the seventh symmetric group. To find the number of elements in G, we input the following into Sage:

```
S7 = SymmetricGroup(7)
p = S7("(1,2,3,4,5,6,7)")
```

```
s = S7("(2,3,5)(4,7,6)")
G = S7.subgroup([p,s])
G.order()
```

Clicking "evaluate" will reveal that G has 21 elements. If instead, we wanted a list of all of the permutations in G, we would use the code:

```
S7 = SymmetricGroup(7)
p = S7("(1,2,3,4,5,6,7)")
s = S7("(2,3,5)(4,7,6)")
G = S7.subgroup([p,s])
G.list()
```

Clicking "evaluate" lists the elements of G as:

```
[(), (2,3,5)(4,7,6), (2,5,3)(4,6,7), (1,2,3,4,5,6,7), (1,2,4)(3,6,5),
(1,2,6)(4,7,5), (1,3,5,7,2,4,6), (1,3,7)(2,5,4), (1,3,4)(2,7,6),
(1,4,2)(3,5,6), (1,4,7,3,6,2,5), (1,4,3)(2,6,7), (1,5,7)(3,6,4),
(1,5,2,6,3,7,4), (1,5,6)(2,7,3), (1,6,2)(4,5,7), (1,6,5)(2,3,7),
(1,6,4,2,7,5,3), (1,7,6,5,4,3,2), (1,7,5)(3,4,6), (1,7,3)(2,4,5)]
```

Notice that Sage does not require us to list the fixed points when we submit our generators. Similarly, it does not list the fixed points for the permutations in the group.

Sage will also give the cycle index polynomial associated with a group G. This can implemented by following the group G with the .cycle_index() command. For instance, if we input

```
CyclicPermutationGroup(12).cycle_index()
```

and click "evaluate," then Sage will return

```
1/12*p[1, 1, 1, 1, 1, 1, 1, 1, 1, 1, 1, 1] + 1/12*p[2, 2, 2, 2, 2, 2] +
1/6*p[3, 3, 3, 3] + 1/6*p[4, 4, 4] + 1/6*p[6, 6] + 1/3*p[12]
```

In Sect. 8.5, we used this polynomial to construct a generating function. So, the cycles of length 2 would be represented as x_2. We would then replace x_2 with m (in the case of Burnside's Lemma), or $x^2 + y^2$ (in the case of Pólya's Enumeration Theorem). In either case, Sage will allow us to compute these values efficiently.

In the case of Burnside's Lemma, suppose that the group acting on our set is the twelfth Dihedral group. If we input

```
D = DihedralGroup(12).cycle_index()
sum([(i[1]*prod([(x) for j in i[0]])) for i in D])
```

and click "evaluate," will return

```
1/24*x^12 + 1/4*x^7 + 7/24*x^6 + 1/12*x^4 + 1/12*x^3 +
1/12*x^2 + 1/6*x
```

We can replace the x in our code with the specific number of colors we wish to use. This will yield the number of distinguishable colorings. A similar technique is used for Pólya's Enumeration Theorem. In this example, we will use two colors, which will be represented by x and y. Again, the group acting on our set will be the twelfth Dihedral group. The pattern inventory is generated by the code

```
y = var('y')
D = DihedralGroup(12).cycle_index()
sum([(i[1]*prod([(x^j + y^j) for j in i[0]])) for i in D])
```

Clicking "evaluate" will yield

```
1/24*(x + y)^12 + 1/12*(x^3 + y^3)^4 + 1/12*(x^4 + y^4)^3 +
1/12*(x^6 + y^6)^2 + 1/6*x^12 + 1/6*y^12 + 1/4*(x + y)^2*(x^2 +
y^2)^5 + 7/24*(x^2 + y^2)^6
```

Usually, we will want a specific coefficient in this polynomial. For instance, if we want the coefficient of x^8y^4, then this is accomplished with the code

```
y = var('y')
D = DihedralGroup(12).cycle_index()
f(x,y) = sum([(i[1]*prod([(x^j + y^j) for j in i[0]])) for i in D])
taylor(f(x,y), (x,0), (y,0), 12).coefficient(x^8).coefficient(y^4)
```

Clicking evaluate will yield the answer of 29.

We end this appendix with a brief introduction to programming. In particular, a loop is necessary t o evaluate the sums. For instance, to evaluate the sum of the first nine positive integers, we would use the code

```
s = 0
for i in range(10):
    s = s + i;
print s
```

We note that the indention in the code is necessary for Sage to perform the loop. Further note that the range is one more than what we actually wanted to sum up to. This is because this command will return

```
[0,1,2,3,4,5,6,7,8,9]
```

References

1. Julian, R., Abel, R., Greig M.: BIBDs with small block size. In: Colbourn, C.J., Dinitz, J.H. (eds.) The CRC Handbook of Combinatorial Designs, pp. 41–47. CRC Press, Inc., Boca Raton (1996)
2. Allis, V.: Searching for solutions in games and artificial intelligence. PhD thesis, University of Limburg, Maastricht, The Netherlands, 1994
3. Appel, K., Haken, W.: The solution of the four-color-map problem. Sci. Am. **237**(4), 108–121, 152 (1977)
4. Rouse Ball, W.W., Coxeter, H.S.M.: Mathematical Recreations and Essays, 13th edn. Dover Publications Inc., New York (1987)
5. Beauregard, R.A., Fraleigh, J.B.: A First Course in Linear Algebra. Houghton Mifflin Co., Boston (1973) (With optional introduction to groups, rings, and fields)
6. Beck, J.: Combinatorial Games, Volume 114 of Encyclopedia of Mathematics and its Applications. Cambridge University Press, Cambridge (2008) (Tic-tac-toe theory)
7. Beeler, R.A.: A note on the number of ways to compute a determinant using cofactor expansion. Bull. Inst. Combin. Appl. **63**, 36–38 (2011)
8. Benjamin, A.T., Quinn, J.J.: Proofs that Really Count, Volume 27 of the Dolciani Mathematical Expositions. Mathematical Association of America, Washington, DC (2003) (The art of combinatorial proof)
9. Biggs, N.L., Lloyd, E.K., Wilson, R.J.: Graph Theory. 1736–1936, 2nd edn. The Clarendon Press/Oxford University Press, New York (1986)
10. Bogart, K.P., Doyle, P.G.: Nonsexist solution of the ménage problem. Am. Math. Monthly **93**(7), 514–519 (1986)
11. Bottomley, H.: How many tic-tac-toe (noughts and crosses) games are possible? http://www.se16.info/hgb/tictactoe.htm (2001). Accessed 3 Feb 2015
12. Bruck, R.H., Ryser, H.J.: The nonexistence of certain finite projective planes. Can. J. Math. **1**, 88–93 (1949)
13. Cayley, A.: A theorem of trees. Q. J. Math. **23**, 376–378 (1889)
14. Chartrand, G., Polimeni, A.D., Zhang, P.: Mathematical Proofs: A Transition to Advanced Mathematics, 3rd edn. Pearson Education, Boston (2013)
15. Chowla, S., Ryser, H.J.: Combinatorial problems. Can. J. Math. **2**, 93–99 (1950)
16. Colbourn, C.J., Dinitz, J.H.: Latin squares. In: Colbourn, C.J., Dinitz, J.H. (eds.) The CRC Handbook of Combinatorial Designs, pp. 97–110. CRC Press, Inc., Boca Raton (1996)
17. Colbourn, C.J., Mathon, R.: Steiner systems. In: Colbourn, C.J., Dinitz, J.H. (eds.) The CRC Handbook of Combinatorial Designs, pp. 66–75. CRC Press, Inc., Boca Raton (1996)
18. DeGroot, M.H.: Probability and Statistics. Addison-Wesley Publishing Co., Reading (1975) (Addison-Wesley Series in Behavioral Science: Quantitative Methods)
19. Dirac, G.A.: Some theorems on abstract graphs. Proc. London Math. Soc. **2**(3), 69–81 (1952)

R. A. Beeler, *How to Count*, DOI 10.1007/978-3-319-13844-2

20. Euler, L.: Solutio problematis ad geometriam situs pertinentis. Comment. Academiae Sci. I. Petropolitanae **8**, 128–140 (1736)
21. Finch, C.: Sage Beginner's Guide. Packt Publishing, Birmingham, UK (2011)
22. Fraleigh, J.B.: A First Course in Abstract Algebra. Addison-Wesley Publishing Co., Reading (1967)
23. Graham, R.L., Rothschild, B.L., Spencer, J.H.: Ramsey theory, 2nd edn. Wiley-Interscience Series in Discrete Mathematics and Optimization. Wiley, New York (1990) (A Wiley-Interscience Publication)
24. Grout, J.: Small graph database. http://sage.cs.drake.edu/grout/graphs/. Accessed 3 Oct 2013
25. Harary, F.: Graph Theory. Addison-Wesley Publishing Co., Reading (1969)
26. Heawood, P.J.: Map-colour theorem. Q. J. Math. **24**, 332–338 (1890) (Oxford)
27. Herstein, I.N.: Topics in Algebra, 2nd edn. Xerox College Publishing, Lexington (1975)
28. Hierholzer, C.: Ueber die Möglichkeit, einen Linienzug ohne Wiederholung und ohne Unterbrechung zu umfahren. Mathematische Annalen **6**(1), 30–32 (1873)
29. Hughes, D.R., Piper, F.C.: Design Theory, 2nd edn. Cambridge University Press, Cambridge (1988)
30. Hwang, F.K., Lin, S.: A direct method to construct triple systems. J. Comb. Theory Ser. A **17**, 84–94 (1974)
31. Jones, G.A., Jones, J.M.: Elementary Number Theory. Springer Undergraduate Mathematics Series. Springer-Verlag London Ltd., London (1998)
32. Kirkman, T.P.: On a problem in combinatorics. Camb. Dublin Math. **2**, 191–204 (1847)
33. Kozaczuk, W., Enigma, S.J.: How the Poles Broke the Nazi Code (Polish Histories). Hippocrene Books, New York (2004)
34. Kuratowski, K.: Sur le problème des courbes gauches en topologie. Fund. Math. **15**, 271–283 (1930)
35. Richeson, D.S.: Euler's Gem. Princeton University Press, Princeton (2012) (The polyhedron formula and the birth of topology, First paperback printing)
36. Rosenhouse, J., Taalman, L.: Taking Sudoku Seriously. Oxford University Press, Oxford (2011) (The math behind the world's most popular pencil puzzle)
37. Shaefer, S.: Mathematical recreations. http://www.mathrec.org/old/2002jan/solutions.html (2002). Accessed 3 Feb 2015
38. Shannon, C.E.: Programming a computer for playing chess. Philos. Mag. **41**(7), 256–275 (1950)
39. Singh, S.: The Code Book: The Science of Secrecy from Ancient Egypt to Quantum Cryptography. Anchor, New York (2000)
40. Stanley, R.P.: Enumerative Combinatorics. Volume 1, Volume 49 of Cambridge Studies in Advanced Mathematics, 2nd edn. Cambridge University Press, Cambridge (2012)
41. Steiner, J.: Kombinatorische aufgabe. J. Reine Angew. Math. **45**, 181–182 (1853)
42. Stinson, D.R.: Cryptography. Discrete Mathematics and its Applications (Boca Raton), 3rd edn. Chapman & Hall/CRC, Boca Raton (2006) (Theory and practice)
43. Wallis, W.D.: Combinatorial Designs, Volume 118 of Monographs and Textbooks in Pure and Applied Mathematics. Marcel Dekker Inc., New York (1988)
44. Wells, D.: Games and Mathematics. Cambridge University Press, Cambridge (2012) (Subtle connections)
45. Wilf, H.S.: Generatingfunctionology, 3rd edn. A K Peters Ltd., Wellesley (2006)
46. Wilson, R.: Four Colors Suffice. Princeton University Press, Princeton (2002) (How the map problem was solved)
47. Winkel, B.J., Deavors, C., Kahn, D., Kruh, L. (eds.): The German Enigma Cipher Machine: Beginnings, Success, and Ultimate Failure. Artech House Print on Demand, Boston (2005)

Index

R. A. Beeler, *How to Count*, DOI 10.1007/978-3-319-13844-2

Printed in the United States
By Bookmasters